Quantum Mechanics Versus Local Realism

The Einstein–Podolsky–Rosen Paradox

PHYSICS OF ATOMS AND MOLECULES

Recent volumes in the series:

ATOMIC INNER-SHELL PHYSICS
Edited by Bernd Crasemann

ATOMS IN ASTROPHYSICS
Edited by P. G. Burke, W. B. Eissner, D. G. Hummer, and I. C. Percival

AUTOIONIZATION: Recent Developments and Applications
Edited by Aaron Temkin

COLLISIONS OF ELECTRONS WITH ATOMS AND MOLECULES
G. F. Drukarev

DENSITY MATRIX THEORY AND APPLICATIONS
Karl Blum

ELECTRON–ATOM AND ELECTRON–MOLECULE COLLISIONS
Edited by Juergen Hinze

ELECTRON–MOLECULE COLLISIONS
Edited by Isao Shimamura and Kazuo Takayanagi

INNER-SHELL AND X-RAY PHYSICS OF ATOMS AND SOLIDS
Edited by Derek J. Fabian, Hans Kleinpoppen, and Lewis M. Watson

INTRODUCTION TO THE THEORY OF LASER-ATOM INTERACTIONS
Marvin H. Mittleman

ISOTOPE SHIFTS IN ATOMIC SPECTRA
W. H. King

PROGRESS IN ATOMIC SPECTROSCOPY, Parts A, B, C, and D
Edited by W. Hanle, H. Kleinpoppen, and H. J. Beyer

QUANTUM MECHANICS VERSUS LOCAL REALISM: The Einstein–Podolsky–Rosen Paradox
Edited by Franco Selleri

RECENT STUDIES IN ATOMIC AND MOLECULAR PROCESSES
Edited by Arthur E. Kingston

THEORY OF MULTIPHOTON PROCESSES
Farhad H. M. Faisal

VARIATIONAL METHODS IN ELECTRON–ATOM SCATTERING THEORY
R. K. Nesbet

ZERO-RANGE POTENTIALS AND THEIR APPLICATIONS IN ATOMIC PHYSICS
Yu. N. Demkov and V. N. Ostrovskii

Quantum Mechanics Versus Local Realism

The Einstein–Podolsky–Rosen Paradox

Edited by

FRANCO SELLERI

University of Bari
Bari, Italy

PLENUM PRESS • NEW YORK AND LONDON

Library of Congress Cataloging in Publication Data

Quantum mechanics versus local realism.

(Physics of atoms and molecules)
Includes bibliographies and index.
1. Quantum theory. I. Selleri, Franco. II. Title: Einstein — Podolsky — Rosen paradox. III. Series.
QC174.12.Q36   1988   530.1'2   88-15124
ISBN 0-306-42739-7

© 1988 Plenum Press, New York
A Division of Plenum Publishing Corporation
233 Spring Street, New York, N.Y. 10013

Printed in the United States of America

Contributors

A. O. *Barut*, Department of Physics, University of Colorado, Boulder, Colorado 80309, United States

D. *Bohm*, Physics Department, Birkbeck College, University of London, London WC1E 7HX, England, United Kingdom

V. *Buonomano*, Institute of Mathematics, State University of Campinas, Campinas, São Paulo, Brazil

C. *Dewdney*, Department of Applied Physics and Physical Electronics, Portsmouth Polytechnic, Portsmouth PO1 2DZ, England, United Kingdom

A. J. *Duncan*, Atomic Physics Laboratory, University of Stirling, Stirling FK9 4LA, Scotland, United Kingdom

F. *Falciglia*, Institute of Physics, Viale Andrea Doria, 95125 Catania, Italy

M. *Ferrero*, Department of Physics, University of Oviedo, Oviedo, Spain

A. *Garuccio*, Department of Physics, University of Bari, 70216 Bari, Italy

B. J. *Hiley*, Physics Department, Birkbeck College, University of London, London WC1E 7HX, England, United Kingdom

P. R. *Holland*, Henri Poincaré Institute, 75231 Paris Cedex 05, France

D. *Home*, Department of Physics, Bose Institute, Calcutta 700009, India

H. *Kleinpoppen*, Atomic Physics Laboratory, University of Stirling, Stirling FK9 4LA, Scotland, United Kingdom

A. *Kyprianidis*, Laboratory of Theoretical Physics, Henri Poincaré Institute, 75231 Paris Cedex 05, France

T. W. *Marshall*, Department of Theoretical Physics, University of Cantabria, Santander 39005, Spain

W. *Mückenheim*, Südring 10, D-3400 Göttingen, Federal Republic of Germany

L. *Pappalardo*, Department of Physics, University of Catania, 95129 Catania, Italy

S. *Pascazio*, Department of Mathematics, Free University of Brussels, 1050 Brussels, Belgium

J. Rayski, Institute of Physics, Jagellonian University, Cracow, Poland

E. Santos, Department of Theoretical Physics, University of Cantabria, Santander 39005, Spain

F. Selleri, Department of Physics, University of Bari, 70126 Bari, Italy

H. P. Stapp, Lawrence Berkeley Laboratory, Berkeley, California 94720, United States

N. A. Törnqvist, Department of High Energy Physics, University of Helsinki, SF-00170 Helsinki 17, Finland

J. P. Vigier, Laboratory of Theoretical Physics, Henri Poincaré Institute, 75231 Paris Cedex 05, France

Preface

If you have two small objects, one here on Earth and the other on the planet Pluto, what would you say of the following statement: No modification of the properties of the object on the earth can take place as a consequence of an interaction of the distant object with a third body also located on Pluto?

The opinion that the previous statement is correct is very natural, but modern quantum theory implies that it must be wrong in certain cases.

Consider in fact two arbitrary objects separated by such a large distance that they are unable to exert any important mutual influence. It is possible to show rigorously that a measurable physical quantity exists, with a value more than 40% different from the value theoretically predicted by quantum mechanics.

Necessarily then, either space is largely an illusion of our senses and it does not exist objectively, or information can be sent from the future to the past, or ... something important has to be changed in modern physics.

This is the essence of the Einstein-Podolsky-Rosen (EPR) paradox. A paradox is an argument that derives absurd conclusions by valid deduction from acceptable premises. In the case of the EPR paradox the absurd conclusion is that Bell's observable Δ should have two different values

$$\Delta = 2\sqrt{2} \qquad \text{and} \qquad \Delta \leq 2$$

The "acceptable premises" are the following:

1. All the empirical predictions of the existing quantum theory are correct.
2. Local realism has an unlimited validity, where by *local realism* one means a set of three reasonable ideas (reality criterion, separability, time arrow) that are discussed in depth in this book.
3. Probabilities are well behaved, that is, they are positive, do not exceed unity, and satisfy the ergodic hypothesis.

The incompatibility among (1), (2), and (3) shows that the EPR paradox is not merely a matter of interpretation of the quantum formalism, but holds for all the conceivable interpretations, since it is a consequence of the empirical predictions of the theory only. Thus the EPR paradox leads to the fantastic conclusion that some empirical predictions of the existing quantum theory are incompatible with deep-rooted conceptions of modern science. This is the reason why the solution of the paradox is expected to have a profound effect on future physics. It is then understandable that Stapp could write that "Bell's theorem is the most profound discovery of science."

In preparing their contributions to this volume the authors have kept in mind that it is directed to a broad spectrum of readers, including physicists, mathematicians, chemists, and philosophers of science. Therefore the book is not primarily directed to people already active on EPR (although they are welcome to buy a copy!). It has only been assumed that our readers are generally interested in EPR matters but know nothing about them and wish to learn from this book, starting from a mere knowledge of elementary quantum mechanics.

Every chapter of the book considers as known and "given" only a few very general facts about the EPR paradox. These are contained in the introductory chapter, which consists of a very simple and general historical review of the most important ideas, those that are "obvious" to researchers active on the problem. There is no other book on the market entirely devoted to the EPR paradox. In order to fill the gap this book provides an almost complete review of all the lines of research which are today trying to solve the paradox in different ways. The following proposed solutions are presented:

1. Existence of superluminal connections between atomic objects separated by a large distance.
2. Retroactions in time, that is, the idea that the past can be actively influenced by choices made in the future.
3. Variable detection probability, that is, the idea that some quantum probabilities are different for different individual quantum systems, so that new physical features arise only for two (or more) correlated systems.
4. Breakdown of the "ergodic hypothesis," that is, the idea that ensemble averages can be different from time averages.
5. Negative probabilities, that is, the idea that the usual numerical bounds for probabilities can sometimes break down. Such a possibility has been advocated by Dirac and by Feynman.

The book shows that it will be possible to decide *experimentally* on very fundamental conceptual matters, such as local realism and the existence of superluminal connections and their properties. New experiments are proposed in some chapters, for instance in the domain of particle physics. It is also shown that, contrary to a rather widespread opinion, the question of locality in atomic-cascade experiments is far from settled, and that it will require an entirely new generation of experiments.

Every conceivable solution of the EPR paradox is incredibly revolutionary: There is a definite possibility that its study will lead to a new start in fundamental physics. If that happens we do not know what direction the field will take. It is, however, tempting to say that the solution of the paradox will be physical, that is, along the natural ways of science, and that the seemingly strange proposals which are popular today reflect, more than anything else, the depth of the problem which is being faced and the great expectations which everyone has for its future solution.

Franco Selleri

Bari, Italy

Contents

Chapter 3
All the Inequalities of Einstein Locality
Augusto Garuccio

Chapter 4
Einstein–Podolsky–Rosen Experiments Using the Decays of η_c or J/ψ into $\Lambda\bar{\Lambda} \to \pi^- p\pi^+ \bar{p}$
Nils A. Törnqvist

Chapter 8
Rapisarda's Experiment: Testing Quantum Mechanics versus Local Hidden-Variable Theories with Dichotomic Analyzers   219
Lorenzo Pappalardo and Filippo Falciglia

Chapter 9
Nonlocality and the Einstein–Podolsky–Rosen Experiment as Understood through the Quantum-Potential Approach   235
D. Bohm and B. J. Hiley

Chapter 10
Interpretation of the Einstein–Podolsky–Rosen Effect in Terms of a Generalized Causality
Jerzy Rayski

257

Chapter 11
Quantum Action-at-a-Distance: The Mystery of Einstein–Podolsky–Rosen Correlations
A. Kyprianidis and J. P. Vigier

273

Chapter 12
Particle Trajectories and Quantum Correlations
C. Dewdney and P. R. Holland

301

Chapter 13
*Bell's Inequality and the Nonergodic Interpretation of Quantum
Mechanics*   *327*
Vincent Buonomano

Chapter 14
*An Extended-Probability Response to the Einstein–Podolsky–Rosen
Argument*   *345*
W. Mückenheim

Chapter 19
Symmetric and Asymmetric Models for Atomic Cascade Experiments   447
Miguel Ferrero, Trevor Marshall, and Emilio Santos

History of the Einstein-Podolsky–Rosen Paradox

1. Early Formulations

Attention has recently (A. Einstein, B. Podolsky, and N. Rosen, *Phys. Rev.* 47 (1935), 777) been called to the obvious but very disconcerting fact that even though we restrict the disentangling measurement to *one* system, the representative obtained for the *other* system is by no means independent of the particular choice of observations which we select for that purpose and which by the way are *entirely* arbitrary. It is rather discomforting that the theory should allow a system to be steered or piloted into one or the other type of state at the experimenter's mercy in spite of his having no access to it.

(E. SCHRÖDINGER, 1935)

1.1. The Einstein-Podolsky-Rosen Paper (1935)

Position, in quantum theory, is described by a linear Hermitian operator Q whose action on the wave function is equivalent to a multiplication by the position parameter x. Therefore the eigenvalue equation

$$Q u(q; x) = q u(q; x)$$

is solved by an arbitrary real value of q and by the corresponding eigenfunction (normalized on the unit length in wave number space)

$$u(q; x) = \delta(x - q) \qquad (1)$$

1

which is Dirac's δ-function. The wave function (1) predicts a fixed position *q*. As one can write

$$u(q; x) = \frac{1}{h} \int dp' \exp[ip'(x - q)/\hbar] \qquad (2)$$

it follows that all possible values of momentum have equal probability.

With consideration next of the momentum operator

$$P = \frac{\hbar}{i} \frac{\partial}{\partial x}$$

the eigenvalue equation

$$P v(p; x) = p \, v(p; x)$$

is solved by arbitrary real values of *p* and by the corresponding eigenfunctions (i.e., plane waves, normalized on the unit length):

$$v(p; x) = \exp(ipx/\hbar) \qquad (3)$$

The position density $|v(p; x)|^2$ of the previous wave function is a constant, meaning that all conceivable positions can be found with equal probability.

In the state $u(q; x)$ the position is known exactly, but nothing can be said about momentum. The opposite holds for $v(p; x)$, which leads to an exact prediction of momentum, but to a completely undefined position. All this is, of course, consistent with the nonvanishing commutator of *Q* and *P*:

$$[Q, P] = \hbar/i$$

When, in physics, a situation is met in which one can predict with certainty the value of a measurable quantity, it is very common to assume that there is something real in the considered object, which is reflected in the exactly predictable value. This "something real" is, however, attributed to the object before, or even in the absence of, an act of measurement. The latter attribution marks the difference between a realistic attitude and the strictly positivistic point of view which considers as real only acts of observation.

Einstein, Podolsky, and Rosen gave a precise form to this idea when they wrote their famous reality criterion: *If, without in any way disturbing a system, we can predict with certainty (i.e., with probability equal to unity) the value of a physical quantity, then there exists an element of physical reality corresponding to this physical quantity.*

The reality criterion can be applied to the wave function (1) and it can be concluded that an element of reality belongs to the physical object described by the wave function (1), which corresponds to the predicted value q of position. A similar application to the wave function (3) leads instead to the attribution of an element of reality corresponding to the fixed value p of momentum. The notations $u(q; x)$ and $v(p; x)$ then show explicitly the objective physical properties q and p of the respective wave functions.

All these considerations can be made only at one particular time t_0: the wave function (1), for instance, would be an imploding (exploding) wave function for $t < t_0$ ($t > t_0$) if the Schrödinger time evolution were considered.

When a single quantum object is considered, it is not possible to attribute to it, *simultaneously*, the two elements of reality corresponding to P and Q, *on the basis of the Einstein-Podolsky-Rosen (EPR) reality criterion*. Therefore, it is consistent with the quantum formalism to assume that when P is measured and obtains a definite value, an eventual previous element of reality corresponding to Q is destroyed. Furthermore it is natural to assume that this destruction is brought about by the action quanta exchanged between the measuring apparatus and the observed atomic object. In this way the attribution of elements of reality becomes a rather innocuous and probably useless exercise.

Things change dramatically, however, when two correlated quantum objects (α and β) are considered. As is well known, the wave function can be assigned almost arbitrarily *at a single time*, the ensuing evolution being given by the time-dependent Schrödinger equation. Therefore, possible fixed-time wave functions for the system $\alpha + \beta$ are

$$\phi(q_0; x_1, x_2) = \int dq' \, c(q')u_\alpha(q'; x_1)u_\beta(q_0 + q'; x_2) \tag{4}$$

$$\tilde{\phi}(p_0; x_1, x_2) = \int dp' \, \tilde{c}(p')v_\alpha(p'; x_1)v_\beta(p_0 - p'; x_2) \tag{5}$$

where the notation for fixed-position and fixed-momentum wave functions is the same as before, the only change being the specification of the quantum object (α or β) to which they refer.

The meaning of $\phi(q_0; x_1, x_2)$ is the usual one; for instance, a position measurement on α will give the result q' with probability $|c(q')|^2$. However, if q' has been found for α it can *then* be predicted that a position measurement for β will certainly give the result $q_0 + q'$. In other words, correlated position measurements made on α and on β will lead to results whose difference certainly equals q_0. It can then be concluded that to q_0 there corresponds an element of reality of $(\alpha + \beta)$.

A similar reasoning applied to $\tilde{\phi}$ leads to the conclusion that to the *sum* p_0 of the momenta there corresponds an element of reality of $(\alpha + \beta)$.

The simultaneous attribution of q_0 and p_0 to a pair of quantum objects is no longer excluded *a priori*—as q and p were excluded in the case of a single object—since the difference of positions and the sum of momenta are represented by commuting operators:

$$[Q_\alpha - Q_\beta, P_\alpha + P_\beta] = 0 \qquad (6)$$

as one can easily check. Einstein, Podolsky, and Rosen were able to find a wave function which allows the simultaneous attribution of the two elements of reality. It is given by

$$\psi(q_0, p_0; x_1, x_2) = \frac{1}{h} \int dp' \exp[i(x_1 - x_2 + q_0)p'/\hbar] \qquad (7)$$

and it can immediately be written in the form (5) with $p_0 = 0$ and with $\tilde{c}(p') = h^{-1}$ (apart from a constant phase factor). It can also be written in the form (4) [with $c(q') = 1$] since the integral in equation (7) gives

$$\psi(q_0, 0; x_1, x_2) = \delta(x_1 - x_2 + q_0)$$

$$= \int dq' \, \delta(x_1 - q')\delta(q' - x_2 + q_0) \qquad (8)$$

These results were used by Einstein, Podolsky, and Rosen for their proof that the quantum-mechanical description of physical reality cannot be complete. A theory is considered complete when it satisfies the following definition: *every element of physical reality attributable to a certain physical system in a given state must have a counterpart in the mathematical description provided by the theory for that physical situation.* For example, the wave function (7) would provide a complete description of the pair (α, β) if no further elements of reality beyond q_0 and p_0 could be attributed to the pair.

There is, however, a reasoning proving that *individual positions and momenta* of α and β do possess a physical reality, thus leading to the conclusion that the quantum-mechanical description provided by equation (7) is not complete.

The argument goes as follows: Suppose we are given a very large set E of similar pairs (α, β) all described by the wave function (7). It can then be predicted that measurements of the positions of α and β, performed on individual pairs, will give results that *always* satisfy the relation $x_2 - x_1 = q_0$. Similarly, measurements of the momenta of α and β performed on (other) individual decay processes will give results p_1 and p_2 that *always* satisfy the relation $p_1 + p_2 = p_0 = 0$.

We consider now a subset E_1 not previously subjected to measurements and perform position measurements on every α of E_1, where $x_1', x_1'', \ldots$ denote the values obtained. It can then be predicted *with certainty* that simultaneous measurements of the position of every β will give $x_1' - q_0$ for the first pair, $x_1'' - q_0$ for the second pair, and so on. Consequently, one can invoke the EPR criterion of physical reality and conclude that to the position of β there corresponds an element of physical reality. It is natural to conclude that this element of physical reality exists regardless of whether or not a measurement on α has been made because, if this were not true, the only alternative (which Einstein, Podolsky, and Rosen wished to exclude) would be to say that the measurement made on α has created at a distance, and instantaneously, the element of reality of β. Therefore, to the position of β there corresponds an element of reality *for all the pairs of the full ensemble E.*

A parallel argument can be made for momenta by considering a subset E_2 of E and performing a momentum measurement on every α of E_2, where $p_1', p_1'', \ldots$ denote the obtained results. Since it can be predicted with certainty that subsequent measurements of the momentum of β will give $-p_1'$ for the first pair, $-p_1''$ for the second pair, and so on, it can also be concluded that to the momentum of β there corresponds an element of reality for every β of E_2. Unless this element of reality is created at a distance and instantaneously by the measurements made on α, one can then extend the previous conclusion to the whole of E.

Obviously the choice of the system (α or β) on which measurements are performed is arbitrary. A symmetrical reasoning thus leads to the conclusion that to both the position and the momentum of particle α there correspond simultaneous elements of reality in the whole ensemble E.

Individual positions and momenta are therefore seen to be real *before measurements*, in an indirect sense, for all objects (α and β) of E, the sense being that there exists something in the physical reality of α and β that leads necessarily to preassigned results if and when a measurement of one or the other of the two observables is made.

Since the wave function (7) describes these quantities *a priori* as indeterminate, one must necessarily conclude that the description of the physical reality provided by the wave function (7) is not complete, on the basis of the given definition of completeness. This was, in 1935, the essence of the EPR paradox, which was then only a paradox of incompleteness of the existing theory.

1.2. Bohr's Answer (1935)

Bohr stressed that the EPR paradox disappeared if one worked consistently within the notion of *complementarity*, which was for him "a new feature of natural philosophy." He showed that complementarity implied (1) a final renunciation of the classical ideal of causality, and (2) a radical revision of our attitude as regards physical reality.

Basically, one can therefore say that Bohr did not question the correctness of the EPR reasoning once all its implicit and explicit premises are
accepted. But it is exactly these premises that, in his opinion, are not valid
in the atomic domain. In order to show this, Bohr formulated causality as
follows: A process is causal if it takes place according to well-defined and
identifiable rules, the most important one being the law of conservation of
energy and momentum. The physicist who studies the phenomena of the
atomic domain will naturally *try* to use his macroscopic preconceptions
and will *try* to describe atomic processes as taking place both in space and
time and according to energy and momentum conservation. However, he
will discover that it is not possible to do so because quantum observables
described by noncommuting operators cannot be measured simultaneously.
The measurement of one of them, in general, destroys previous knowledge
of other ones.

The roots of complementarity can best be exposed by discussing space
localization (position measurement) and causality implementation (momentum measurement). Space localization can be obtained by measuring position with infinite precision ($\Delta x = 0$). After such a measurement the wave
function becomes the δ-function $\delta(x - q)$, if q is the obtained result. But
a δ-function can be written as the superposition of all possible plane waves
with constant weight [see equation (2)] and this means that absolutely
nothing is known about momentum. All eventual knowledge about momentum prior to the position measurement is, in this way, lost. No evidence
can therefore exist about momentum conservation, if no knowledge about
momentum is available. A concrete localization in space of the phenomenon
thus implies a necessary abandonment of the causal description.

Symmetrically, in a different experiment, one could decide to implement
the causal description by measuring momentum with infinite precision: the
wave function would therefore *become*, as a result of the measurement, a
plane wave. But this would immediately imply that nothing could be known
about position, with a complete loss of the description of the quantum
phenomenon in space. A concrete implementation of the causal description
would thus force the physicist to abandon the description in space.

The two possibilities (space–time and causality) are thus seen to be
mutually incompatible. Bohr concludes that in the atomic world it is, in
principle, impossible to give a picture of quantum processes as developing
causally in space and time, and that this element of irrationality is introduced
into quantum physics by the finite value of Planck's constant. For these
reasons it becomes, in his opinion, necessary to limit the interest of the
physicist to the *exclusive* consideration of the acts of observation.

Obviously, then, no paradox exists when one considers two correlated
systems described by the wave function (7). Let us consider, in fact, two
apparatuses Q_1 and P_1 (Q_2 and P_2) capable of performing position and

momentum measurements, respectively, on the system α (β). If one choses to use Q_1 and Q_2, the wave function (7) predicts that the results x_1 and x_2 will be precisely correlated: $x_1 - x_2 = q_0$. If instead one chooses to use P_1 and P_2, the wave function (7) predicts a precise correlation of the results p_1 and p_2: $p_1 + p_2 = 0$. The two apparatuses Q_1 and P_1 are mutually incompatible: One can choose to employ either Q_1 or P_1, but never the two of them simultaneously; the same holds for Q_2 and P_2. From this point of view the EPR assumption about the elements of reality becomes useless: it now merely concludes that an *element of reality is associated with a concretely performed act of measurement*, since there is no other reality to speak of. In particular, the EPR conclusion that position and momentum correspond to two simultaneously existing elements of reality appears totally unjustified (Bohr says that it contains "an essential ambiguity"), because one can never perform simultaneous measurements of position and momentum.

Einstein, Podolsky, and Rosen anticipated the possibility of such a refutation, but they considered it as unacceptable. In the conclusive part of their paper one can read:

> One could object to this conclusion on the grounds that our criterion of reality is not sufficiently restrictive. Indeed, one would not arrive at our conclusion if one insisted that one or more physical quantities can be regarded as simultaneous elements of reality *only when they can be simultaneously measured or predicted*. On this point of view, since either one or the other, but not both simultaneously, of the quantities P and Q can be predicted, they are not simultaneously real. This makes the reality of P and Q depend upon the process of measurement carried out on the first system, which does not disturb the second system in any way. No reasonable definition of reality could be expected to do this.

The previous considerations apply to any two noncommuting operators. Consider, for instance, a spin-$\frac{1}{2}$ particle and its spin-component operators S_x, S_y, and S_z. It is well known that any two of them do not commute. This means that the corresponding spin observables cannot be assumed to be simultaneously measurable. Consider an electron in the spin state

$$u(+) = \begin{pmatrix} 1 \\ 0 \end{pmatrix}$$

which is the eigenstate of S_z with eigenvalue $+\hbar/2$. If the observable associated with S_x is measured, there can be only two results:

$$S_x = \pm\hbar/2$$

The spin state after the measurement becomes an eigenstate of S_x, that is,

$$\text{either} \quad \frac{1}{\sqrt{2}} \begin{pmatrix} 1 \\ 1 \end{pmatrix} \quad \text{or} \quad \frac{1}{\sqrt{2}} \begin{pmatrix} 1 \\ -1 \end{pmatrix}$$

In either case the S_y component is totally unknown, as one can easily check. Bohr would say that the implementation of the reality of S_x has made S_y completely undetermined. The opposite reasoning can obviously be made: S_y can become known, but then it is S_x which becomes necessarily completely unknown. One can thus say, with Bohr, that S_x and S_y are complementary aspects of reality: either S_x is real, or S_y is real, but never both of them at the same time.

1.3. Schrödinger's Extension (1935)

Schrödinger (1935) considered a wave function [like equation (7)] satisfying the two eigenvalue equations

$$Q\,\Psi(x_1, x_2) = q_0\Psi(x_1, x_2)$$

$$P\,\Psi(x_1, x_2) = p_0\Psi(x_1, x_2)$$

(9)

where $Q = Q_\alpha - Q_\beta$ and $P = P_\alpha + P_\beta$ [the notation is the same as in Section 1.1: see equation (6)] and showed that to every Hermitian operator $F(Q_\alpha, P_\alpha)$ of the first particle of an EPR pair there corresponds another Hermitian operator $G(Q_\beta, P_\beta)$ of the second particle, such that

$$[F(Q_\alpha, P_\alpha) - G(Q_\beta, P_\beta)]\Psi(x_1, x_2) = 0 \qquad (10)$$

which can be read as follows: $\Psi(x_1, x_2)$ is an eigenfunction of $F - G$ with eigenvalue zero. Therefore measurements of F on α and of G on β must give equal results if α and β are described by $\Psi(x_1, x_2)$.

The proof of Schrödinger's theorem is easy if one starts from the operator

$$F_{mn}(Q_\alpha, P_\alpha) = Q_\alpha^m P_\alpha^n + \text{H.c.} \qquad (11)$$

and assumes that it corresponds to

$$G_{mn}(Q_\beta, P_\beta) = (Q_\beta + q_0)^m(p - P_\beta)^n + \text{H.c.}$$

which can also be written

$$G_{mn}(Q_\beta, P_\beta) = (Q_\alpha - Q + q_0)^m(p_0 + P_\alpha - P)^n + \text{H.c.}$$

by definition of Q and P. Therefore, when G_{mn} is applied to Ψ the factor $(p_0 + P_\alpha - P)^n$ becomes P_α^n because of equation (9). Since P_α obviously commutes with $Q_\beta + q_0 = Q_\alpha - Q + q_0$, one can commute P_α^n to the extreme left of G_{mn}. On the right, there remains a factor $(Q_\alpha - Q + q_0)^m$ which, applied to $\Psi(x_1, x_2)$, gives Q_α^m. One thus obtains

$$G_{mn}\Psi(x_1, x_2) = [P_\alpha^n Q_\alpha^m + \text{H.c.}]\Psi(x_1, x_2)$$

The right-hand side coincides with $F_{mn}\Psi(x_1, x_2)$ and thus relation (10) holds for the operator (11).

The previous result can obviously be generalized to functions of the type

$$F(Q_\alpha, P_\alpha) = \sum_{mn} c_{mn} Q_\alpha^m P_\alpha^n + \text{H.c.} \tag{12}$$

where the c_{mn} are numerical coefficients. There is, thus, a wide class of operators, containing infinitely many terms, which satisfy Schrödinger's theorem: in practice, every analytic function $F(Q_\alpha, P_\alpha)$ can be developed as in equation (12) and must therefore satisfy Schrödinger's theorem.

In general two such operators, $F_1(Q_\alpha, P_\alpha)$ and $F_2(Q_\alpha, P_\alpha)$, [and their corresponding operators for β: $G_1(Q_\beta, P_\beta)$ and $G_2(Q_\beta, P_\beta)$] do not commute with one another. Still, since by Schrödinger's theorem the measurements of an F operator and of its corresponding G operator must in all cases give equal results, a measurement of F on α steers β into an eigenstate of G. Schrödinger concludes with his deep understanding of the EPR paradox:

> It is rather discomforting that the theory should allow a system to be steered or piloted into one or the other type of state at the experimenter's mercy in spite of his having no access to it.

1.4. Furry's Hypothesis (1936)

If there are two quantum objects: α in the state $|\Psi\rangle$ and β in the state $|\Phi\rangle$, then the global system $\varepsilon = (\alpha, \beta)$ is described by the state vector $|\Psi\rangle|\Phi\rangle$.

The opposite problem is also of interest: Given a general state vector for ε, can one always write it as a direct product of two states separately describing the objects composing ε ?

That the answer is negative can be seen in the following way: Let $\{|\Psi_i\rangle\}$ and $\{|\Phi_j\rangle\}$ be two orthonormal and complete sets of states for α and β, respectively. If $|\eta\rangle$ is the state vector for $\varepsilon = (\alpha, \beta)$, one can carry out a (double) development of it over the sets $\{|\Psi_i\rangle\}$ and $\{|\Phi_j\rangle\}$ and write

$$|\eta\rangle = \sum_{ij} c_{ij} |\Psi_i\rangle |\Phi_j\rangle \tag{13}$$

where the c_{ij}'s constitute a set of generally complex coefficients.

It is important to observe that in the absence of superselection rules every vector $|\eta\rangle$ of the previous form represents a possible state for ε. This is a consequence of the superposition principle. In fact, every possible vector $|\Psi_i\rangle|\Phi_j\rangle$ is a conceivable state vector for ε, given the fact that it describes α in the state $|\Psi_i\rangle$ and β in the state $|\Phi_j\rangle$. Therefore every linear combination of these vectors, such as the state vector (13), is also a possible state vector of ε.

Hence the coefficients c_{ij} are totally unrestricted, except for the normalization condition

$$\sum_{ij} |c_{ij}|^2 = 1 \qquad (14)$$

In particular, there is no way, if the superposition principle is to be of general validity, to restrict the c_{ij}'s to numbers of the type

$$c_{ij} = x_i y_j \qquad (15)$$

Therefore, there is no way to guarantee that not only ε but also α and β be in a well-defined state. The factorization condition (15) is in fact *necessary and sufficient for having α and β in a quantum state*, as can easily be seen.

In conclusion: the superposition principle forces us to consider vectors $|\eta\rangle$, of the type in equation (13), for which condition (15) does not hold. This implies that some states for ε are such that neither α nor β are in a quantum state.

States for $\varepsilon = (\alpha, \beta)$ with the latter property are said to be of the second type, while those for which condition (15) holds are said to be of the first type. One also says that state vectors of the first (second) type are the factorizable (nonfactorizable) state vectors. Since the state vector is the only link between the quantum-mechanical formalism and the microphysical reality, one thus sees that *quantum theory does not ascribe any separate reality to the objects α and β whose complex (α, β) is described by a state vector of the second type.* An embryo of the EPR paradox is already visible in this strange fact.

A general discussion of the EPR paradox was given by Furry shortly after the publication of the EPR paper (Furry, 1936). His starting point was a theorem proved by von Neumann, according to which the state vector (13) can always be written in the form

$$|\eta\rangle = \sum_i c_i |\Psi_i\rangle |\Phi_i\rangle \qquad (16)$$

if the complete orthonormal sets $\{|\Psi_i\rangle\}$ and $\{|\Phi_j\rangle\}$ [in general different from those entering into the state vector (13)] are suitably chosen. If $|\eta\rangle$ is of the first (second) type only one (more than one) of the coefficients c_i will be different from zero. Let these two new sets of state vectors constitute sets of eigenstates of two linear Hermitian operators, A and B, respectively, so that

$$A |\Psi_i\rangle = a_i |\Psi_i\rangle, \qquad B |\Phi_i\rangle = b_i |\Phi_i\rangle \qquad (17)$$

these relations being valid for all values of the index i. One can thus say that A (B) represents an observable of the object α (β) and that the possible values of such an observable are the eigenvalues a_i (b_i).

The state vector (16) predicts a strict correlation of the measured values of the two observables. In fact, since the structure (16) is preserved during time evolution, if a first observer measures A at time t_1 on the α component of a certain pair and finds $A = a_k$, then a second observer will necessarily find $B = b_k$ (where the index k is the same) for a measurement made at time $t \geq t_1$ on the β component of the same pair.

If this prediction has been checked many times and always found to be correct, one can conclude that from the time t_1 onward the outcome of the B measurement is certain (and equal to b_k) for the object β. But the two objects can be assumed to be very far from one another, so that no change of the state of β can have taken place as a consequence of the measurement performed on α. As a consequence, the result of an eventual measurement of B on β must have been fixed and equal to b_k even before the measurement on α was performed. Quantum theory has a very precise way to describe a physical situation (or state) of an atomic object for which the result of a measurement is known *a priori*: It attributes as state vector the eigenvector corresponding to the known value of the considered physical quantity.

If one recalls equation (17), it is obvious that the state of β must have been $|\Phi_k\rangle$ even before time t_1. Given the predicted correlation of values for the observables A and B, one must conclude that the state vector for (α, β) even before time t_1 was

$$|\Psi_k\rangle|\Phi_k\rangle$$

Repeating the previous argument for a statistical ensemble of pairs (α, β) one concludes that the state vectors actually were

$$|\eta_1\rangle = |\Psi_1\rangle|\Phi_1\rangle \qquad \text{in } |c_1|^2\% \text{ of the cases}$$

$$|\eta_2\rangle = |\Psi_2\rangle|\Phi_2\rangle \qquad \text{in } |c_2|^2\% \text{ of the cases}$$

$$\vdots \qquad \qquad \vdots \qquad \qquad \qquad \vdots \qquad \qquad \quad (18)$$

$$|\eta_k\rangle = |\Psi_k\rangle|\Phi_k\rangle \qquad \text{in } |c_k|^2\% \text{ of the cases}$$

$$\vdots \qquad \qquad \vdots \qquad \qquad \qquad \vdots$$

We started from the assumption that all the pairs of the considered statistical ensemble had the state vector $|\eta\rangle$ and concluded, instead, that the different state vectors $|\eta_1\rangle, |\eta_2\rangle, \ldots, |\eta_k\rangle, \ldots$ applied, with the stated frequencies, to different pairs.

Now, the states (16) and (18) are obviously different *mathematical* descriptions of the ensemble. Could it be that they are nevertheless equivalent descriptions for all practical purposes? Furry could show that the answer is negative.

An elegant way to express the incompatibility, at the empirical level between the state vectors of the second type and the mixtures was found by Fortunato (1976). Consider the projection operator

$$P_\eta = |\eta\rangle\langle\eta| \tag{19}$$

which is Hermitian and can be assumed to correspond to an observable. Its expectation value over the state (16) is obviously unity:

$$\langle\eta|P_\eta|\eta\rangle = 1 \tag{20}$$

The expectation value of the same operator over the mixture (18) is instead

$$\langle P_\eta\rangle = |c_1|^2\langle\Psi_1\Phi_1|P_\eta|\Psi_1\Phi_1\rangle + |c_2|^2\langle\Psi_2\Phi_2|P_\eta|\Psi_2\Phi_2\rangle + \cdots$$

$$+ |c_k|^2\langle\Psi_k\Phi_k|P_\eta|\Psi_k\Phi_k\rangle + \cdots$$

Since it is a simple matter to show that

$$\langle\Psi_k\Phi_k|P_\eta|\Psi_k\Phi_k\rangle = |c_k|^2$$

it follows that

$$\langle P_\eta\rangle = |c_1|^4 + |c_2|^4 + \cdots + |c_k| + \cdots \tag{21}$$

But the sum of the fourth powers of the moduli of numbers such that the sum of their *squared* moduli is one will certainly be less than one if there are at least two numbers different from zero. The latter condition is, however, precisely that of having a state vector of the second type. Therefore the observable corresponding to P_η has an expectation value equal to one (less than one) over the state vector of the second type $|\eta\rangle$ [over the mixture of state vectors of the first type (18)].

1.5. Bohm's Formulation (1951)

Let a physical system (atom, molecule, etc.) M be given decaying into two spin-$\frac{1}{2}$ "particles" α and β. Let $u_\alpha(+)$ and $u_\alpha(-)$ be eigenvectors corresponding to the eigenvalues $+1$ and -1, respectively, of the Pauli matrix $\sigma_3(\alpha)$ representing the third component of the spin angular momentum for α; and let $u_\beta(+)$ and $u_\beta(-)$ be the corresponding eigenvectors of the Pauli matrix $\sigma_3(\beta)$ for β.

The only factorizable spin states for (α, β) that one can construct with these four spinors are

$$u_\alpha(+)u_\beta(+), \qquad u_\alpha(+)u_\beta(-), \qquad u_\alpha(-)u_\beta(+), \qquad u_\alpha(-)u_\beta(-) \quad (22)$$

where the first one applies when the spin vectors of both particles α and β point along the positive z direction, and so on.

There are some concrete physical situations in which the spin state vector for (α, β) must be the "singlet" state vector η_0 given by

$$\eta_0 = (1/\sqrt{2})[u_\alpha(+)u_\beta(-) - u_\alpha(-)u_\beta(+)] \quad (23)$$

Four important properties of η_0 will be used in the following proof of Bohm of the EPR paradox

(P1) *It is not a factorizable state.*
(P2) *It predicts the result zero for a measurement of the total squared spin of particles α and β.*
(P3) *It is rotationally invariant.*
(P4) *It predicts opposite results for measurements of the components along $\hat{n}$ of the spins of particles α and β, $\hat{n}$ being an arbitrary unit vector.*

Property (P1) is not difficult to prove, since the most general spin state for α is

$$u_\alpha = au_\alpha(+) + bu_\alpha(-) \qquad (|a|^2 + |b|^2 = 1) \quad (24)$$

where a and b are constants. Similarly, the most general spin state for β is given by

$$u_\beta = cu_\beta(+) + du_\beta(-) \qquad (|c|^2 + |d|^2 = 1) \quad (25)$$

where c and d are some other constants. Obviously, the most general *factorizable* spin state for the combined system is

$$u_\alpha u_\beta = acu_\alpha(+)u_\beta(+) + adu_\alpha(+)u_\beta(-)$$

$$+ bcu_\alpha(-)u_\beta(+) + bdu_\alpha(-)u_\beta(-) \quad (26)$$

Now, since $u_\alpha(+)u_\beta(+)$ does not enter into η_0, $u_\alpha u_\beta$ can equal η_0 only if $ac = 0$. Thus $a = 0$, which implies that $u_\alpha(+)u_\beta(-)$ also disappears from $u_\alpha u_\beta$, and/or $c = 0$, which implies that $u_\alpha(-)u_\beta(+)$ disappears. It is therefore impossible, by any choice of u_α and u_β, to satisfy $\eta_0 = u_\alpha u_\beta$.

As for property (P2), it can be verified by introducing the total squared spin operator, defined by

$$\Sigma^2 = [\sigma_1(\alpha) + \sigma_1(\beta)]^2 + [\sigma_2(\alpha) + \sigma_2(\beta)]^2 + [\sigma_3(\alpha) + \sigma_3(\beta)]^2$$

$$= 6I_\alpha I_\beta + 2\boldsymbol{\sigma}(\alpha) \cdot \boldsymbol{\sigma}(\beta) \tag{27}$$

where I_α and I_β are the unit operators in the spin spaces of α and β, respectively. One can easily check that

$$\Sigma^2 \eta_0 = 0 \tag{28}$$

from which it follows that a measurement of the observable corresponding to Σ^2, on a pair described by the state η_0, will certainly give the result zero.

The third fundamental property of η_0 (i.e., rotational invariance) can be proved by introducing the new vectors $u_\alpha(n\pm)$ and $u_\beta(n\pm)$, which denote eigenvectors of $\boldsymbol{\sigma}(\alpha) \cdot \hat{\mathbf{n}}$ and $\boldsymbol{\sigma}(\beta) \cdot \hat{\mathbf{n}}$, respectively ($\hat{\mathbf{n}}$ being an arbitrary unit vector) and showing that η_0 transforms into

$$\eta_0 = (1/\sqrt{2})[u_\alpha(n+)u_\beta(n-) - u_\alpha(n-)u_\beta(n+)] \tag{29}$$

which has the same structure as (23) with different states.

As for property (P4) it can be checked that η_0 is an eigenstate of the $\hat{\mathbf{n}}$ component of the total spin operator with eigenvalue zero, that is,

$$[\boldsymbol{\sigma}(\alpha) \cdot \hat{\mathbf{n}} + \boldsymbol{\sigma}(\beta) \cdot \hat{\mathbf{n}}]\eta_0 = 0 \tag{30}$$

From the physical interpretation of eigenvalue relations, it then follows that measurements of the $\hat{\mathbf{n}}$ components of the spins of α and β must always give opposite results.

Another state important for the discussion of the EPR paradox is the "triplet" state, given by

$$\eta_1 = (1/\sqrt{2})[u_\alpha(+)u_\beta(-) + u_\alpha(-)u_\beta(+)] \tag{31}$$

One can show that η_1 shares with η_0 the properties (P1) and (P4), but not (P3); it is not rotationally invariant. Moreover, in place of property (P2), η_1 has the following property: *Any measurement of the total squared spin of the two particles described by η_1 will give the result $2\hbar^2$ (spin 1).*

One has

$$u_\alpha(+)u_\beta(-) = (1/\sqrt{2})(\eta_0 + \eta_1)$$

$$u_\alpha(-)u_\beta(+) = (1/\sqrt{2})(\eta_0 - \eta_1) \tag{32}$$

as can be proved simply by adding and subtracting the states (23) and (31). Furthermore, on invoking the quantum-mechanical interpretation of super-positions, one sees that:

> Measurements of the total squared spin on a set of (α, β) pairs described as a mixture of the factorizable state vectors (32) will produce with equal probability the results 0 and $2\hbar^2$.

This large observable difference between an ensemble which is an arbitrary mixture of the states (32) and an ensemble whose elements are all described by η_0 is the basis of the Bohm formulation of the paradox, given below. The Bohm formulation has several important advantages over the original one. First, it deals with dichotomic observables and therefore allows sharper definition of the results. Second, it allows the introduction of time, which enters only in the space-dependent part of the wave function, while the spin part is in most cases time-independent. Therefore the singlet state is stable, so to say, while the original wave function, equations (7) and (8), introduced by Einstein, Podolsky, and Rosen holds only at one particular time, and blows up immediately after. The third advantage of the Bohm formulation is that it allows one to deal with clearly separated objects (in space), while the wave function (7), based on plane waves, described the two correlated objects as present, with constant probability, in all points of space.

In order to establish the EPR paradox in conceptually clear conditions, we consider only (α, β) pairs with the following wave function:

$$\Psi(x_1, x_2) = \eta_0 \Psi_\alpha(x_1) \Psi_\beta(x_2) \tag{33}$$

where η_0 is the singlet state (23) and $\Psi_\alpha(x_1)$ and $\Psi_\beta(x_2)$ are the space parts of the wave functions for α and β, respectively. Suppose, furthermore, that $\Psi_\alpha(x_1)$ is a Gaussian function with modulus appreciably different from zero only in a region R_1 of width Δ_1, centered around the point x_{10}. Similarly, let $\Psi_\beta(x_2)$ be a Gaussian function localized in the region R_2 of width Δ_2 centered around x_{20}. We shall consider a sufficient condition for separability of the systems α and β to be the validity of the condition

$$|x_{20} - x_{10}| \gg \Delta_1, \Delta_2 \tag{34}$$

If now particles α and β are supposed to move to the left and to the right, respectively, so that the distance between the centers of the two wave packets increases linearly with time, it is not difficult to show that Schrödinger's equation allows for situations in which the condition for separability does not deteriorate with time. One can thus say that α and β are located within two small regions R_1 and R_2, respectively, very far from one another, so that all interactions (gravitational, electromagnetic, strong, and weak) are known to be vanishingly small. In such conditions it is natural to conclude that a measurement performed on α does not give rise to any modification of the physical properties of β, and *vice versa*. The presence of η_0 in the wave function (33) leads instead to paradoxical conclusions.

The EPR reasoning is as follows: Consider a large set E of (α, β) pairs in the state (33). Measure $\sigma_3(\alpha)$ at time t_0 on all α's of a subset E_1 of E. If $+1$ (-1) is found, a future $(t > t_0)$ measurement of $\sigma_3(\beta)$ will certainly give -1 $(+1)$. Using the EPR reality criterion, one can assign to the β's of E_1 an element of reality λ_1 (λ_2) fixing *a priori* the result -1 $(+1)$ of the $\sigma_3(\beta)$ measurement.

But the quantum mechanics treats an object β with a predetermined value of $\sigma_3(\beta)$, by assigning it the state $u_\beta(-)$ $[u_\beta(+)]$: This is the completeness assumption. The strict correlation (P4), applied to the z-axis, implies then, even for $t < t_0$, that the ensemble E_1 had to be described in spin space by the mixture (32). Excluding the possibility that λ_1 (λ_2) is created at a distance by the measurement of $\sigma_3(\alpha)$, it must be concluded that λ_1 (λ_2) actually belongs to all β's of E. Applying completeness again, one concludes, as before, that the mixture (32) applies to all pairs of E.

But this contradicts the description (33) at the empirical level, as was shown above. One thus reaches an absurd conclusion (the EPR paradox).

1.6. The Bohm-Aharonov Conclusion (1957)

As early as 1936 it was clear that the EPR paradox had evidenced the existence of a striking

> ... disagreement between the results of quantum-mechanical calculations and those to be expected on the assumption that a system once freed from dynamical interference can be regarded as possessing independently real properties (Furry, 1936).

This conclusion is, for some people, very difficult to accept and can lead to the idea that something must be wrong with the existing quantum theory. Even Einstein entertained such a point of view:

> ... Einstein has (in private communication) actually proposed such an idea; namely, that the current formulation of the many-body problem in quantum mechanics may break down when particles are far enough apart (Bohm and Aharonov, 1957).

The first organic examination of an eventual breakdown of quantum theory (with immediately negative conclusions) was made by Bohm and Aharonov in 1957. They considered the annihilation of a positron–electron pair into two energetic photons ("gamma rays") and showed that the quantum state produced is

$$|0^-\rangle = (|x_\alpha\rangle|y_\beta\rangle - |y_\alpha\rangle|x_\beta\rangle)/\sqrt{2} \tag{35}$$

that is, the zero angular momentum negative-parity state, where x and y denote the direction of linear polarization of photon α and photon β. Also the latter state, like the singlet state of the two spin-$\frac{1}{2}$ particles considered before, is rotationally invariant. In practice this means that each photon is always found in a state of linear polarization, orthogonal to that of the other, no matter what the choice of axes with respect to which the state of polarization is expressed. Bohm and Aharanov calculated the ratio $R = \Gamma_1/\Gamma_2$, where Γ_1 is the rate of double scattering of the two photons through a fixed angle θ, when the planes π_1 and π_2 formed by the lines of motion of the first and the second photon (after scattering) with their common original direction of motion are *perpendicular*; and Γ_2 is the same rate when the planes π_1 and π_2 are *parallel*. The value of R predicted by the $|0^-\rangle$ state is

$$R = \frac{(\gamma - 2\sin^2\theta)^2 + \gamma^2}{2\gamma(\gamma - 2\sin^2\theta)} \tag{36}$$

where

$$\gamma = (k_0/k) + (k/k_0) \tag{37}$$

Here k_0 is the wave number of the incident photon and k is that of the final photon. Bohm and Aharonov considered an angle of 82° for which the ratio k_0/k can easily be calculated from Compton scattering kinematics, and obtained $R = 2.85$. This figure could not be compared directly with the experiment of Wu (1950), because there photons were detected with an angular spread around the ideal value of 82°. For such a concrete situation the prediction $R = 2.00$ applied instead, obtained with a suitable angular average of equation (36), which agrees very well with the experimental result $R = 2.04 \pm 0.08$. Bohm and Aharonov could also show that *the hypothesis of a breakdown of the $|0^-\rangle$ state vector with the increasing distance between the two photons, and of its substitution with mixtures of factorizable vectors led necessarily to considerably smaller values of R, always satisfying $R \leq 1.5$.*

These results show that Wu's experiment is adequately explained by existing quantum theory, which implies distant correlations of the type leading to the EPR paradox, but not by any hypothesis implying a simple-minded breakdown of the quantum theory that could avoid the paradox.

It would however not be correct to conclude that this experimental evidence constitutes an argument against local realism, since there are well-known local models capable of reproducing the quantum-mechanical predictions for the experiments on double rescattering following electron-positron annihilation into two gamma rays. One such model was proposed by Kasday (1971). Suppose there are two "hidden" vectors $\hat{\lambda}_1$ and $\hat{\lambda}_2$ associated with photon α and photon β, respectively, and let the photons ultimately scatter in the directions of these vectors. Then simply give $\hat{\lambda}_1$ and $\hat{\lambda}_2$ the same probability distribution as that of the momenta $\hat{k}_1$ and $\hat{k}_2$ of the scattered photons:

$$\rho(\hat{\lambda}_1, \hat{\lambda}_2) = H(\hat{\lambda}_1, \hat{\lambda}_2)$$

where $H(\hat{k}_1, \hat{k}_2)$ is the probability distribution of the momenta $\hat{k}_1$ and $\hat{k}_2$ of the scattered photons, as predicted by quantum theory. The assumption is clearly that the photons have "decided in advance," at the time of annihilation, in which direction they would ultimately scatter. The model is local: changing the position of "detector 1" does not affect the parameter $\hat{\lambda}_2$, for example, and therefore it does not change the response of "detector 2." Furthermore the model reproduces the results of *all* measurements that can be made on the scattered photons.

Given the conclusive evidence found by Bohm and Aharonov against a simple-minded breakdown of quantum theory, it is surprising that several authors rediscovered their idea long after it had been discarded by its proponents.

Thus Jauch (1971) developed an ambitious approach to quantum theory based on an "algebra of propositions," where he defined as "mixtures of the 2nd kind" the quantum-mechanical description of EPR pairs based on nonfactorizable state vectors, that is, the states which lead to the EPR paradox, and concluded that "Mixtures of the 2nd kind do not exist." (Jauch, 1971.)

Similarly de Broglie (1974) objected to Bell's proof of the incompatability between local realism and quantum-mechanical predictions that the latter had been deduced from a formalism implying that the two particles are transported by one and the same "wave train." However, according to de Broglie, a unique wave train is possible, only if the two particles are *near* to one another. Therefore the usual quantum formulas should break down at distances larger than the coherence length of the two wave packets.

The Bohm–Aharonov hypothesis has recently been rediscovered by Piccioni and Mehlhop (1987), who also came to the conclusion that the singlet state does not exist because it leads to unacceptable action at a distance.

2. *Inequalities from Einstein Locality*

> I hope that the rigor and beauty of the argument of EPR is apparent. If one does not recognize how good an argument it is—proceeding rigorously from premises which are thoroughly reasonable—then one does not experience an adequate intellectual shock when one finds out that the experimental evidence contradicts their conclusion. The shock should be as great as the one experienced by Frege when he read Russell's theoretical paradox and said, "Alas, arithmetic totters!"
>
> (A. SHIMONY, 1978)

2.1. *Bell's Inequality (1965)*

Consider again an ensemble formed by a very large number N of decays $\varepsilon \to \alpha + \beta$ and suppose that the observer O_α measures on α the dichotomic observable $A(a)$, while in a distance region of space a second observer O_β measures on β another dichotomic observable $B(b)$.

The observables $A(a)$ and $B(b)$ have been taken to depend on the arguments a and b, respectively, which are assumed to be experimental parameters, fixed in the structure of the apparatuses in any given experiment, but possibly variable over different experiments. Examples of such dichotomic observables are those represented by the spin matrices $\boldsymbol{\sigma}(\alpha) \cdot \hat{\mathbf{a}}$ and $\boldsymbol{\sigma}(\beta) \cdot \hat{\mathbf{b}}$, where the experimental parameters are the unit vectors $\hat{\mathbf{a}}$ and $\hat{\mathbf{b}}$. They could be fixed experimentally, for example, by the directions of the inhomogeneous magnetic fields of two Stern-Gerlach apparatuses.

In practice, any physical quantity can be used to define a dichotomic observable: for instance one could say that $A(a) = \pm 1$ if the energy of an atom is above or below a certain level a.

In any event, when measurements of such observables are made on all the N pairs of the given ensemble, O_α will obtain a set of results $\{A_1, A_2, \ldots, A_N\}$, while O_β will collect a similar set $\{B_1, B_2, \ldots, B_N\}$, all relative to fixed values of the parameters a and b. The results of the two sets are correlated in the sense that A_1 and B_1 pertain to the particles α and β, respectively, arising from the first decay; A_2 and B_2 are similarly associated with the second decay; and so on. By definition, these results in every case equal ± 1.

The *correlation function* $P(a, b)$ of the results A_i and B_i is defined as the average product of the results obtained by O_α and O_β:

$$P(a, b) = \frac{1}{N} \sum_{i=1}^{N} A_i B_i \tag{38}$$

Since every product $A_i B_i$ is ± 1, it follows that

$$-1 \le P(a, b) \le +1 \tag{39}$$

The quantum-mechanical correlation function, in the case of the singlet state, is given by

$$P(\hat{\mathbf{a}}, \hat{\mathbf{b}}) = \eta_0^+[\boldsymbol{\sigma}(\alpha) \cdot \hat{\mathbf{a}} \otimes \boldsymbol{\sigma}(\beta) \cdot \hat{\mathbf{b}}]\eta_0 = -\hat{\mathbf{a}} \cdot \hat{\mathbf{b}} \tag{40}$$

This result is simple and elegant, but incompatible with local realism as we will soon see.

Define now the quantity

$$\Delta = |P(\hat{\mathbf{a}}, \hat{\mathbf{b}}) - P(\hat{\mathbf{a}}, \hat{\mathbf{b}}')| + |P(\hat{\mathbf{a}}', \hat{\mathbf{b}}) + P(\hat{\mathbf{a}}', \hat{\mathbf{b}}')| \tag{41}$$

Consider two orthogonal unit vectors $\hat{\mathbf{a}}$ and $\hat{\mathbf{a}}'$, associated with particle α, and two orthogonal unit vectors $\hat{\mathbf{b}}$ and $\hat{\mathbf{b}}'$, associated with particle β, and suppose that their orientation is such that they can be found by clockwise rotations of $\pi/4$ in the order $\hat{\mathbf{a}}$, $\hat{\mathbf{b}}$, $\hat{\mathbf{a}}'$, $\hat{\mathbf{b}}'$. One can then easily see that the substitution of equation (40) into equation (41) leads to

$$\Delta = |\hat{\mathbf{a}} \cdot \hat{\mathbf{b}} - \hat{\mathbf{a}} \cdot \hat{\mathbf{b}}'| + |\hat{\mathbf{a}}' \cdot \hat{\mathbf{b}} + \hat{\mathbf{a}}' \cdot \hat{\mathbf{b}}'| = 2\sqrt{2}$$

It can, moreover, be shown that $2\sqrt{2}$ is the maximum value of Δ for all conceivable orientations of the vectors $\hat{\mathbf{a}}$, $\hat{\mathbf{a}}'$, $\hat{\mathbf{b}}$, $\hat{\mathbf{b}}'$. This result is of great interest because, as we will see next, local realism allows Δ to have a maximum value of 2. The inequality $\Delta \le 2$ is *Bell's inequality*, and it has been called "the most profound discovery of science." (Stapp, 1977.)

In a theory developed according to the EPR reality criterion there are elements of reality λ which fix all observables. In general, they can be expected to vary, with density $\rho(\lambda)$, over the set Λ. Of course

$$\int_\Lambda d\lambda\, \rho(\lambda) = 1 \tag{42}$$

The role of the new variable λ is naturally that of fixing a priori the values of the dichotomic observables; for example,

$$\boldsymbol{\sigma}(\alpha) \cdot \hat{\mathbf{a}} \to A(a, \lambda)$$
$$\boldsymbol{\sigma}(\beta) \cdot \hat{\mathbf{b}} \to B(b, \lambda) \tag{43}$$

where the discontinuous functions $A(a, \lambda)$ and $B(b, \lambda)$ can assume only the values ± 1. The correlation function as defined in equation (38) (average product of the two observables) can obviously be written

$$P(a, b) = \int d\lambda \, \rho(\lambda) A(a, \lambda) B(b, \lambda) \tag{44}$$

This is a local expression, in the sense that A does not depend on b, nor does B depend on a.

It is a simple matter to show that

$$|P(a, b) - P(a, b')| \leqslant \int d\lambda \, \rho(\lambda) |B(b, \lambda) - B(b', \lambda)| \tag{45}$$

since $|A(a, \lambda)| = 1$, and that

$$|P(a', b') + P(a', b')| \leqslant \int d\lambda \, \rho(\lambda) |B(b, \lambda) + B(b', \lambda)| \tag{46}$$

By adding equations (45) and (46) and using

$$|B(b, \lambda) - B(b', \lambda)| + |B(b, \lambda) + B(b', \lambda)| = 2$$

which is a consequence of $|B(b, \lambda)| = |B(b', \lambda)| = 1$, one obtains from equation (42) Bell's inequality

$$\Delta \equiv |P(a, b) - P(a, b')| + |P(a', b) + P(a', b')| \leqslant 2 \tag{47}$$

The practical meaning of the inequality (47) will be discussed at length in the next and in subsequent sections.

2.2. The Strong Inequalities (1969)

A practical way of testing experimentally the validity in nature of Bell's inequality could be the following: a source is built in such a way that the decays $\varepsilon \to \alpha + \beta$ lead to the emission of the pair only when the object α (β) flies to the right (to the left) where a two-channel analyzing apparatus can transmit it or reflect it at 90° depending on its physical properties. The dichotomic choice forced in this way upon the atomic objects can then be used for defining Bell's dichotomic observables, by saying that $A(a) = \pm 1$ [$B(b) = \pm 1$], depending on the channel, first or second, chosen by the object α (β).

In 1969 Clauser, Horne, Shimony, and Holt (CHSH) suggested the use of pairs of optical photons emitted by atomic cascades. For such photons they assumed that the *binary choice* was the one *between transmission and absorption in a polarizer*. For every choice of the polarizers' orientations a and b, they introduced four probabilities $\omega(a_{\pm}, b_{\pm})$, where, for instance, $\omega(a_+, b_-)$ is the probability that observer O_α finds $A(a) = +1$ (photon α transmitted through polarizer with axis a) and that O_β finds $B(b) = -1$

(photon β absorbed by polarizer with axis b). The correlation function can then be written

$$P(a, b) = \omega(a_+, b_+) - \omega(a_+, b_-) - \omega(a_-, b_+) + \omega(a_-, b_-) \qquad (48)$$

since the product of the results obtained by O_α and O_β is $+1$ (-1) in the cases of $\omega(a_+, b_+)$ and $\omega(a_-, b_-)$ [in the cases of $\omega(a_+, b_-)$ and $\omega(a_-, b_+)$]. Of course

$$\omega(a_+, b_+) + \omega(a_+, b_-) + \omega(a_-, b_+) + \omega(a_-, b_-) = 1 \qquad (49)$$

Considering further the case in which the second polarizer has been removed (the symbol ∞ is used to denote this), one will obviously get

$$\omega(a_+, b_+) + \omega(a_+, b_-) = \omega(a_+, \infty_+) \qquad (50)$$

If, instead, the first polarizer has been removed, one similarly gets

$$\omega(a_+, b_+) + \omega(a_-, b_+) = \omega(\infty_+, b_+) \qquad (51)$$

Finally, if both polarizers have been removed both photons will certainly be transmitted, so that

$$\omega(\infty_+, \infty_+) = 1 \qquad (52)$$

Using now equations (49) to (51), we find it a simple matter to show that the correlation function can be written

$$P(a, b) = 4\omega(a_+, b_+) - 2\omega(a_+, \infty_+) - 2\omega(\infty_+, b_+) + 1 \qquad (53)$$

In the latter expression only cases of double transmission appear, which are nearer to experimental observation, since it is impossible to detect the absorption of a photon in a polarizer. However at this point one must face a very important problem: can one really measure the right-hand side of equation (53) with an error, say, of a few percent? Obviously, the only way to know that a photon has been transmitted through a polarizer is to detect its presence beyond that instrument, but the problem is that photon detectors have an efficiency of only 10 to 20%. This means that one cannot really measure a double-transmission probability, but only a joint probability for double transmission *and* double detection of the two photons. This is not what enters in equation (53)!

One could attempt to redefine the correlation function by using only the measurable joint probabilities for detection and transmission. This can certainly be done, but the trouble is that the values of $P(a, b)$ turn out to be of the order of 10^{-2} which is far too small to lead to a violation of Bell's inequality (47).

This problem has traditionally been "solved" by means of *ad hoc* assumptions concerning the nature of the transmission/detection process. The additional assumption made by CHSH is the following:

Given that a pair of photons emerges from two regions of space where two polarizers can be located, the probability of their joint detection from two photomultipliers is independent of the presence and of the orientation of the polarizers.

Calling D_0 the double-detection probability dealt with in the previous assumption, and denoting by the letter D the joint probability for transmission *and* detection of both photons, one can translate the previous assumption into the following relations:

$$D(a, b) = D_0\omega(a_+, b_+)$$

$$D(a, \infty) = D_0\omega(a_+, \infty_+)$$

$$D(\infty, b) = D_0\omega(\infty_+, b_+) \tag{54}$$

$$D(\infty, \infty) = D_0\omega(\infty_+, \infty_+)$$

where $D(a, b)$ is the joint probability in the case of polarizers with orientations a and b, $D(a, \infty)$ is the joint probability with the second polarizer removed and the first one oriented along a, and so on.

The rates of double detections depend of course on the number N_0 of photon pairs entering, per second, into the right solid angles defined by the optical apparatuses. Using the letter R for denoting rates, one has

$$R(a, b) = N_0 D(a, b)$$

$$R(a, \infty) = N_0 D(a, \infty)$$

$$R(\infty, b) = N_0 D(\infty, b) \tag{55}$$

$$R_0 = N_0 D_0$$

where $R(\infty, \infty)$ has been called R_0 and the meaning of the new symbols is obvious. If one obtains the ω functions from the relations (54) and (55) and substitutes them in equation (53) one gets

$$P(a, b) = 4\frac{R(a, b)}{R_0} - 2\frac{R(a, \infty)}{R_0} - 2\frac{R(\infty, b)}{R_0} + 1 \tag{56}$$

Only coincidence rates enter into equation (56): by virtue of the CHSH additional assumption the correlation function has therefore become measurable!

Equation (56) allows us to transform Bell's inequality into a directly measurable expression. In fact from the inequality (47) it is easy to deduce that

$$-2 \leq P(a, b) - P(a, b') + P(a', b) + P(a', b') \leq +2 \qquad (57)$$

Substituting into the previous inequalities the expressions of the type (56) for the four correlation functions one obtains

$$-1 \leq \frac{R(a, b)}{R_0} - \frac{R(a, b')}{R_0} + \frac{R(a', b)}{R_0} + \frac{R(a', b')}{R_0} - \frac{R(a', \infty)}{R_0} - \frac{R(\infty, b)}{R_0} \leq 0$$
$$(58)$$

Only coincidence rates enter into the previous inequalities, which can therefore be checked experimentally. Historically, the second one was obtained in the 1969 CHSH paper, while the first one first appeared in print in the 1972 letter reporting on the Freedman-Clauser experiment.

A useful simplification is obtained if two qualitative predictions of quantum theory, which have nothing paradoxical and which can be checked directly in experiments, are accepted:

1. The prediction that $R_1 \equiv R(a', \infty)$ does not depend on a', and that $R_2 \equiv R(\infty, b)$ does not depend on b.
2. The prediction that every R function should depend only on the relative angle between the polarizers' axes. For example,

$$R(a, b) = R(a - b)$$

Adopting these simplifications, one gets

$$-1 \leq \frac{R(a - b)}{R_0} - \frac{R(a - b')}{R_0} + \frac{R(a' - b)}{R_0} + \frac{R(a' - b')}{R_0} - \frac{R_1}{R_0} - \frac{R_2}{R_0} \leq 0 \quad (59)$$

The axes of the polarizers can be chosen in such a way that

$$a - b = a' - b = a' - b' = \phi, \qquad a - b' = 3\phi \qquad (60)$$

Therefore from equation (59) it follows that

$$-1 \leq \frac{3R(\phi)}{R_0} - \frac{R(3\phi)}{R_0} - \frac{R_1 + R_2}{R_0} \leq 0 \qquad (61)$$

Considering the previous inequalities for $\phi = 22\frac{1}{2}°$ and for $\phi = 67\frac{1}{2}°$, for which the maximal quantum-mechanical violations take place, one can easily obtain the so-called Freedman inequality

$$\delta \equiv |R(22\frac{1}{2}°)/R_0 - R(67\frac{1}{2}°)/R_0| - \tfrac{1}{4} \leqslant 0 \tag{62}$$

which does not involve R_1 or R_2.

It is perhaps worthwhile to repeat that all the new results deduced, starting from equation (56) and ending with inequality (62), *have become possible only because the CHSH assumption has been made.* It is therefore not correct to confuse the original Bell's inequality with the *much stronger* inequalities which were now deduced. In future we will therefore adopt the following definitions:

1. *Weak inequality*: An inequality deduced from the sole assumption of local realism and violated by quantum mechanics in the case of nearly perfect instruments.
2. *Strong inequality*: An inequality deduced from local realism and from *ad hoc* additional assumptions, such as the CHSH hypothesis stated above, or other hypotheses to be seen later, and violated by quantum mechanics in the case of real instruments.

In order to see the difference, suppose that the CHSH hypothesis had not been made. One could then have redefined the joint probabilities ω in equation (48) in the following way:

$\omega(a_+, b_+)$ = Probability that both photons are transmitted and detected

$\omega(a_+, b_-)$ = Probability that the first photon is transmitted and detected, and that the second photon either is not transmitted or is transmitted but not detected

and so on.

In place of equation (53) one would then have obtained

$$P(a, b) = 4\omega(a_+, b_+) - 2\omega_1(a_+) - 2\omega_2(b_+) + 1$$

where $\omega_1(a_+)$ $[\omega_2(b_+)]$ is the probability that the first [second] photon be transmitted and detected. These are single-photon probabilities and do not refer to joint events for the two photons.

Bell's inequality (57) could then have been written

$$-1 \leqslant \omega(a_+, b_+) - \omega(a_+, b'_+) + \omega(a'_+, b_+) + \omega(a'_+, b'_+)$$

$$- \omega_1(a_+) - \omega_2(b_+) \leqslant 0$$

The trouble with the latter inequality is that in the case of real experiments, the single-photon probabilities are larger, by about an order of magnitude, than the joint probabilities. This results in the fact that the quantum-mechanical predictions do *not* violate the latter inequalities. They are thus weaker than inequality (58), which is also violated in the case of real instruments.

2.3. Wigner's Proof of Bell's Inequality (1970)

In the present section Wigner's 1970 proof of Bell's inequality will be reviewed in a form which is somewhat simpler and more general than the original one. Of course, the basic ideas are strictly the same.

Wigner made two basic assumptions. The first one was that the results of all conceivable measurements are simultaneously prefixed (even in the case of incompatible observables). This realistic standpoint does not contradict Heisenberg's relations because the latter can be taken simply to mean that a concrete measurement made on a given object modifies the pre-fixed values of other observables of that object, not compatible with the measured one. But, *before the action of the instrument*, it is possible that the results of all conceivable measurements are predetermined.

The second assumption was locality. A measurement made on α (β) does not modify the pre-fixed values of the observables $B(b)$, $B(b')$ $[A(a)$, $A(a')]$ of β (α). If one writes

$$A(a) = s, \qquad A(a') = s'$$
$$B(b) = t, \qquad B(b') = t' \tag{63}$$

where s, s', t, t' all equal ± 1, locality means that these four parameters, preassigned by the realistic assumption, are not modified at a distance by measurements. Therefore, if $A(a)$ is measured on an α particle, for example, and the value s is found, the preassigned values t and t', associated with the correlated β particle, are in no way modified.

We are obviously dealing with a realistic and *deterministic* approach, since the result of every possible measurement is predetermined by some concrete properties of the measured objects ("hidden variables"). This does not mean, however, that an active role of the apparatus is excluded, but only that the interaction between object and apparatus is driven to a pre-fixed outcome ("result of measurement") by the hidden variables of the object.

As a consequence of these assumptions, the set E of N (α, β) pairs splits into 2^4 subsets with well-defined populations in which the outcome of the four possible measurements is predetermined. Let $E(s, s'; t, t')$ be a subset of E with prefixed values of the four observables (63) and $n(s, s'; t, t')$ be its population. Naturally

$$\sum n(s, s'; t, t') = N \qquad (64)$$

where $\sum$ denotes a sum over the 2^4 different sets of values of the dichotomic parameters s, s', t, and t'.

By virtue of the locality assumption, the concrete performance of the measurement of $A(a)$, or of $A(a')$, on the α objects of a subset $E_1 \subset E$ does not in any way modify the pre-fixed values of $B(b)$ and $B(b')$ in that subset. In other words, there is no action at a distance modifying $B(b)$ and/or $B(b')$ arising from the measurements of $A(a)$ or $A(a')$ (and *vice versa*).

The *a priori* probabilities

$$\omega(s, s'; t, t') = \frac{1}{N} n(s, s'; t, t') \qquad (65)$$

can therefore be used for the calculation of correlations of concretely performed experiments. Therefore:

$$P(a, b) = \sum \omega(s, s'; t, t') st$$

$$P(a, b') = \sum \omega(s, s'; t, t') st'$$

$$\qquad (66)$$

$$P(a', b) = \sum \omega(s, s'; t, t') s't$$

$$P(a', b') = \sum \omega(s, s'; t, t') s't'$$

where $\sum$ again denotes a sum over the dichotomic variables s, s', t and t'. It is now a simpler matter to show that

$$|P(a, b) - P(a, b')| \leq \sum \omega(s, s'; t, t') |t - t'| \qquad (67)$$

since $|s| = 1$. Similarly, from $|s'| = 1$ it follows that

$$|P(a', b) + P(a', b')| \leq \sum \omega(s, s'; t, t') |t + t'| \qquad (68)$$

By adding equations (67) and (68) and using the equality

$$|t - t'| + |t + t'| = 2$$

which is a consequence of $|t| = |t'| = 1$, Bell's inequality [equation (47)] follows, since equation (64) is equivalent to

$$\sum \omega(s, s'; t, t') = 1$$

With Wigner's proof, probabilities entered for the first time into the EPR paradox. They were, however, deduced from a deterministic background, much in the same way as was done by Laplace with his formulation of probability calculus.

2.4. Bell's Inequality within Quantum Theory (1973)

A remarkable property of mixtures of factorizable state vectors is that in all cases they satisfy Bell's inequality, as first shown by Capasso, Fortunato, and Selleri (1973). A simple proof is the following: Consider an ensemble E of N quantum pairs (α, β) and suppose that they are described by factorizable state vectors $|\Psi_k\rangle|\Phi_k\rangle$ with frequencies n_k/N ($k = 1, 2, \ldots$). Therefore in the ensemble E

$$|\Psi_1\rangle|\Phi_1\rangle \text{ applies to } n_1 \text{ pairs}$$

$$|\Psi_2\rangle|\Phi_2\rangle \text{ applies to } n_2 \text{ pairs}$$
$$\vdots \qquad\qquad\qquad \vdots \tag{69}$$

$$|\Psi_k\rangle|\Phi_k\rangle \text{ applies to } n_k \text{ pairs}$$
$$\vdots \qquad\qquad\qquad \vdots$$

and one has

$$\sum_k n_k = N \tag{70}$$

Suppose that the dichotomic observables to be measured on α and β are described quantum-mechanically by the operators $\tilde{A}(a)$ and $\tilde{B}(b)$, respectively, so that the operator corresponding to the product of the joint measurements on the two systems is $\tilde{A}(a) \otimes \tilde{B}(b)$. The correlation function predicted by quantum mechanics is precisely the average of the latter observable over the mixture (69), so that

$$P(a, b) = \sum_k p_k \langle\Psi_k|\langle\Phi_k|\tilde{A}(a) \otimes \tilde{B}(b)|\Phi_k\rangle|\Psi_k\rangle \tag{71}$$

where

$$p_k = n_k/N, \qquad \sum_k p_k = 1 \qquad (72)$$

The four correlation functions entering into Bell's inequality can then be written

$$P(a, b) = \sum_k p_k \bar{A}_k \bar{B}_k \qquad P(a, b') = \sum_k p_k \bar{A}_k \bar{B}'_k$$

$$P(a', b) = \sum_k p_k \bar{A}'_k \bar{B}_k \qquad P(a', b') = \sum_k p_k \bar{A}'_k \bar{B}'_k \qquad (73)$$

where

$$\bar{A}_k = \langle \Psi_k | \tilde{A}(a) | \Psi_k \rangle$$

$$\bar{A}'_k = \langle \Psi_k | \tilde{A}(a') | \Psi_k \rangle$$

$$\bar{B}_k = \langle \Phi_k | \tilde{B}(b) | \Phi_k \rangle \qquad (74)$$

$$\bar{B}'_k = \langle \Phi_k | \tilde{B}(b') | \Phi_k \rangle$$

The previous quantities are expectation values of operators having ± 1 as only possible eigenvalues. Therefore

$$|\bar{A}_k| \leq 1, \qquad |\bar{A}'_k| \leq 1, \qquad |\bar{B}_k| \leq 1, \qquad |\bar{B}'_k| \leq 1 \qquad (75)$$

these inequalities being true for all k.

By inserting equation (73) into equation (41) one easily gets

$$\Delta \leq \sum_k p_k \Delta_k \qquad (76)$$

where

$$\Delta_k \equiv |\bar{A}_k \bar{B}_k - \bar{A}_k \bar{B}'_k| + |\bar{A}'_k \bar{B}_k + \bar{A}'_k \bar{B}'_k| \qquad (77)$$

Recalling the inequalities (75), one can immediately deduce that

$$\Delta_k \leq |\bar{B}_k - \bar{B}'_k| + |\bar{B}_k + \bar{B}'_k| \qquad (78)$$

from which it follows that

$$\Delta_k \leq 2 \qquad (79)$$

since any two real numbers x and y such that $|x| \leq 1$ and $|y| \leq 1$ always satisfy $|x - y| + |x + y| \leq 2$. If equation (79) is inserted into equation (76) one finally gets $\Delta \leq 2$, that is, Bell's inequality as given by equation (47).

2.5. Factorizable Probabilities (1974)

In the considerations developed in Section 2.2, probabilities entered in an essential way; for example, double-detection probabilities were introduced in equation (53). This was of course a necessity, since the deterministic scheme, considered originally by Einstein, Podolsky, and Rosen and by Bell, cannot really apply to a concrete physical situation.

A systematic probabilistic approach was proposed by Clauser and Horne (1974). They characterized pairs of correlated objects with a variable λ representing their physical state, and introduced probabilities for describing the interaction of a quantum object with an analyzer (e.g., the interaction of a photon with a polarizer) and the subsequent detection. Thus $p(a, \lambda)$ is the probability that object α crosses the analyzer with parameter a *and* that it is subsequently detected; $q(b, \lambda)$ is the similar probability for object β; and $D(a, b, \lambda)$ is the probability that both objects α and β cross their respective analyzers with parameters a and b, and that they are both detected. Furthermore, Clauser and Horne proposed that the very definition of the locality condition should be written

$$D(a, b, \lambda) = p(a, \lambda)q(b, \lambda) \tag{80}$$

It is not obvious that this definition should exhaust all possible local situations, but this important problem will be discussed later. Supposing that the variable λ can vary in the set Λ with density $\rho(\lambda)$, both independent of a and b, one can write

$$p(a) = \int d\lambda \, \rho(\lambda)p(a, \lambda)$$

$$q(b) = \int d\lambda \, \rho(\lambda)q(b, \lambda) \tag{81}$$

$$D(a, b) = \int d\lambda \, \rho(\lambda)p(a, \lambda)q(b, \lambda) \tag{82}$$

for the ensemble probabilities, expressed as weighted averages of the individual probabilities. The integrals in equations (81) and (82) are taken over Λ.

In order to deduce inequalities from their definition of locality, Clauser and Horne considered the following simple algebraic theorem: Given six real numbers x, x', X, y, y', and Y, such that

$$0 \leqslant x, x' \leqslant X, \qquad 0 \leqslant y, y' \leqslant Y$$

one must always have

$$-XY \leqslant xy - xy' + x'y + x'y' - x'Y - Xy \leqslant 0 \qquad (83)$$

The proof of equation (83) is straightforward, since the intermediate quantity in it is linear in each of the four variables x, x', y and y', so that its extremes must be looked for on the boundary of the variables.

These inequalities can now be applied to the EPR paradox by making the identifications

$$x = p(a, \lambda)$$

$$x' = p(a', \lambda)$$

$$y = q(b, \lambda) \qquad (84)$$

$$y' = q(b', \lambda)$$

Introducing equation (84) into equation (83), multiplying the result by $\rho(\lambda)$, and integrating over λ, one obtains

$$-XY \leqslant D(a, b) - D(a, b') + D(a', b) + D(a', b') - p(a')Y - Xq(b) \leqslant 0 \qquad (85)$$

We may ask what the correct values of X and Y are in the inequalities (85). The straightforward answer is, of course, $X = Y = 1$, since the probabilities on the right-hand sides of equations (84) might reach the value 1 for some value of λ. This leads to inequalities of Bell's type (with no additional assumption),

$$-1 \leqslant D(a, b) - D(a, b') + D(a', b) + D(a', b') - p(a') - q(b) \leqslant 0 \quad (86)$$

which could also have been deduced directly from Bell's inequality (47). Inequalities of the previous type, which we decided to call "weak type" in Section 2.2, are sometimes also called "inhomogeneous inequalities" since they are based both on double- and on single-detection probabilities. "Homogeneous inequalities," which are based on double-detection probabilities only, will be deduced next.

The problem with the inequalities (86) is the usual one: for real detectors they are not violated by the quantum-mechanical predictions. For this reason Clauser and Horne proposed the following additional hypothesis, formulated for the case in which α and β are photons:

> For every photon in the state λ, the probability of a detection with a polarizer in place on its trajectory is less than or equal to the detection probability with the polarizer removed (Clauser and Horne, 1974).

In practice, this new assumption is equivalent to the following four inequalities

$$p(a, \lambda) \le p(\infty, \lambda)$$

$$p(a', \lambda) \le p(\infty, \lambda)$$

$$q(b, \lambda) \le q(\infty, \lambda) \tag{87}$$

$$q(b', \lambda) \le q(\infty, \lambda)$$

where the symbol ∞ indicates that the polarizer has been removed. The new relations (87) allow one to use equations (83) and (84) with

$$X = p(\infty, \lambda), \qquad Y = q(\infty, \lambda) \tag{88}$$

Substituting equations (88) and (84) into equation (83), multiplying by $\rho(\lambda)$, and integrating over λ, one obtains

$$-D_0 \le D(a, b) - D(a, b') + D(a', b) + D(a', b')$$

$$- D(a', \infty) - D(\infty, b) \le 0 \tag{89}$$

where D_0 denotes the same physical quantity as in Section 2.2 and the meaning of the new symbols is obvious.

This is an inequality of the strong type, deduced with the help of additional assumptions and much stronger than the weak-type inequality (86). In fact, inequality (89) contains only double-detection probabilities and the quantum-mechanical predictions can be shown to violate it for suitable choices of the polarizer's axes. Since the ratio of double-detection probabilities coincides with the corresponding ratio of detection rates, relation (89) can easily be shown to coincide with relation (58). From this observation it follows that all the results deduced in the CHSH approach from (58) are valid also in the present Clauser and Horne (CH) approach.

2.6. An Infinite Set of Inequalities (1980)

The restrictions imposed by local realism are not fully expressed by Bell's inequality, and an inequality can be deduced for an arbitrary linear combination of correlation functions (Garuccio and Selleri, 1980). In fact, given the numerical coefficients c_{ij}, real but otherwise arbitrary, and the correlation functions $P(a_i, b_j)$, with $i = 1, \ldots, n$ and $j = 1, \ldots, m$, local realism implies that

$$\sum_{i=1}^{n} \sum_{j=1}^{m} c_{ij} P(a_i, b_j) \leq M_0 \tag{90}$$

where

$$M_0 = \underset{\xi, \eta}{\text{Max}} \left(\sum_{i=1}^{n} \sum_{j=1}^{m} c_{ij} \xi_i \eta_j \right) \tag{91}$$

where, among all possible choices of the sign factors $\xi_i = \pm 1 (i = 1, \ldots, n)$ and $\eta_j = \pm 1 (j = 1, \ldots, m)$, one must take the one giving the maximum value to the quantity within parentheses in equation (91).

The whole story of this development, with the various methods of proof, and with the contributions made by different authors, is told in Chapter 3. Here, only four points will be enumerated:

1. Bell's inequality is a particular case of equation (90) with $m = n = 2$, with three c_{ij}'s equaling $+1$, and with the fourth one equaling -1.
2. All the physical restrictions of the set of inequalities with $n = m = 2$ are given by Bell's inequality.
3. An inequality is "trivial" (i.e., it does not provide physical restrictions) if the c_{ij}'s have factorizable signs, that is, if they can be written

$$c_{ij} = |c_{ij}| \sigma_{ij}$$

where

$$\sigma_{ij} = \mu_i \nu_j$$

with $\mu_i = \pm 1$ and $\nu_j = \pm 1$ $(i = 1, \ldots, n; j = 1, \ldots, m)$.

4. There are inequalities providing physical restrictions on the $P(a_i, b_j)$ which cannot be deduced from any Bell's inequality. The so-called "superinequalities" are of this type.

5. Recently Lepore (1988) could show that the set of inequalities (90) and (91) is not the most general one and that physically meaningful inequalities can be deduced from local realism for arbitrary linear combinations *of joint probabilities*. The set of inequalities (90) and (91) is recovered as a particular case of a more general set of inequalities.

Interesting consequences of local realism were found by Garg and Mermin (1982), who were able to deduce Bell-type inequalities for two spin-*j* particles (with arbitrary *j*). They could show that the singlet state for two particles with spin *j* leads to violations of local realism for arbitrarily large values of *j* up to and beyond the threshold of classical physics. But in the classical domain it is always possible to assign *a priori* well defined values to all observable quantities. The result of Garg and Mermin does not augur well for the coherence and rationality of the existing quantum theory, which seems to extend its "magic" predictions to include the macroscopic domain, where classical physics had successfully banished all "magic" approaches.

2.7. Rapisarda's Experiment (1981)

Up to the present section, formulations of the EPR paradox for experiments in which a one-way polarizer is put on the path of each photon were considered. However this is not a very convenient configuration, since the dichotomic choice is between the transmission of the photon and its lack of transmission (i.e., absorption or reflection). Now, of course, an absorption cannot be detected and a considerable amount of information is therefore lost inside the polarizer where the photon is absorbed. A better experiment would be one in which a truly binary choice is made, and where the two alternatives are both detected.

In 1981 Garuccio and Rapisarda (GR) studied an experiment in which a piece of calcite, monitored by two detectors put on the ordinary and on the extraordinary ray, was used as analyzer for each of the two photons. While the theoretical approach of Garuccio and Rapisarda was similar to that of Clauser and Horne, with the variable λ and with factorizable probabilities, an important difference is that they dealt with four simultaneously-measurable coincidence rates. Denoting a photon detection on the ordinary ray and on the extraordinary ray by $+$ and $-$, respectively, one has instead of equation (82):

$$D(a_{\pm}, b_{\pm}) = \int d\lambda \, \rho(\lambda) p(a_{\pm}, \lambda) q(b_{\pm}, \lambda) \qquad (92)$$

where $p(a_+, \lambda)$ is the probability that the photon α emerges and is detected in the ordinary beam when the axis of the calcite has orientation a, and so on.

Garuccio and Rapisarda proposed an entirely new definition of correlation function, based on all the available experimental information, and wrote

$$E(a, b) \equiv \frac{D(a_+, b_+) - D(a_+, b_-) - D(a_-, b_+) + D(a_-, b_-)}{D(a_+, b_+) + D(a_+, b_-) + D(a_-, b_+) + D(a_-, b_-)} \quad (93)$$

Substituting equation (92) into the latter expression one gets

$$E(a, b) = \frac{\int d\lambda\, \rho(\lambda) f(a, \lambda) g(b, \lambda)}{\int d\lambda\, \rho(\lambda) F(a, \lambda) G(b, \lambda)} \quad (94)$$

where

$$f(a, \lambda) \equiv p(a_+, \lambda) - p(a_-, \lambda)$$

$$g(b, \lambda) \equiv q(b_+, \lambda) - q(b_-, \lambda)$$

$$\quad (95)$$

$$F(a, \lambda) \equiv p(a_+, \lambda) + p(a_-, \lambda)$$

$$G(b, \lambda) \equiv q(b_+, \lambda) + q(b_-, \lambda)$$

The problem with equation (94) is that no inequality violated by quantum theory can be obtained from it for experiments that are actually feasible. Therefore, also in this case, one introduces an additional assumption, which can be formulated as follows

> For every photon in the state λ, the sum of the detection probabilities in the "ordinary" and in the "extraordinary" beams emerging from a two-way polarizer does not depend on the polarizer's orientation (Garuccio and Rapisarda, 1981).

The practical implications are that the function F does not depend on a, the function G does not depend on b, and the denominator of equation (94) does not depend on either a or b. A better notation is then

$$F(\lambda) = p(a_+, \lambda) + p(a_-, \lambda)$$

$$G(\lambda) = q(b_+, \lambda) + q(b_-, \lambda) \quad (96)$$

$$H_0 = \int d\lambda\, \rho(\lambda) F(\lambda) G(\lambda)$$

It will next be shown that the previous simplifications allow one to obtain a new inequality of the strong type, violated by quantum mechanics for real experiments. It is not difficult to show that

$$|E(a, b) - E(a, b') + E(a', b) + E(a', b')|$$

$$\leq H_0^{-1} \int d\rho(\lambda) \left[|f(a, \lambda)| |g(b, \lambda) - g(b', \lambda)| \right.$$

$$\left. + |f(a', \lambda)| |g(b, \lambda) + g(b', \lambda)| \right]$$

whence, using the obvious inequalities

$$|f(a, \lambda)| \leq F(\lambda), \qquad |f(a', \lambda)| \leq F(\lambda)$$

$$|g(b, \lambda)| \leq G(\lambda), \qquad |g(b', \lambda)| \leq G(\lambda)$$

one obtains

$$|E(a, b) - E(a, b') + E(a', b) + E(a', b')| \leq 2 \qquad (97)$$

since any two numbers g and g' satisfying $|g| \leq G$ and $|g'| \leq G$ must also satisfy: $|g - g'| + |g + g'| \leq 2G$. Garuccio and Rapisarda could show that the quantum-mechanical predictions violate equation (97) by as much as 50%. The quantum-mechanical expression for $E(a, b)$ will be given in Section 2.10. The Rapisarda experiment was carried out by Aspect, Grangier, and Roger (1982). An improved version is now underway in Catania (see Chapter 8).

2.8. Is Factorizability General Enough?

The factorizability condition, equation (82), has been proposed by Clauser and Horne as the most general possible formulation of local realism at the probabilistic level. Their idea was that λ specified the state of a single pair of correlated quantum objects, so that $p(a, \lambda)$ and $q(b, \lambda)$ were the probabilities of a certain behavior of a given *object*. One can then say that the Clauser–Horne idea was based on *objective probabilities for individual systems*. Although most people are unable to see anything philosophically dangerous in this notion, it is a fact that the history of probability calculus has developed without it. The realistic definition of probability is based not on individuals, but on statistical ensembles.

The EPR paradox is not a standard and well-established notion, and ought to be formulated starting only from generally accepted ideas. Quite apart from this general criticism, some difficulties have been found which cast doubts on the general validity of the CH formula.

1. Deterministic models have been found that are factorizable as such, but that lose factorizability as soon as they become probabilistic models owing to averaging over one of the hidden variables (Suppes and Zanotti, 1976; Selleri and Tarozzi, 1980; Garuccio and Rapisarda, 1981).
2. It has been shown that a probabilistic model, assumed factorizable in the CH sense in n variables, loses factorizability as soon as it is averaged over one of these variables (Selleri, 1987).
3. A concrete physical model has been constructed in the macroscopic domain, based entirely on local and realistic ideas, which is factorizable neither directly, nor in any conceivable indirect way (the example of the identical twins: Liddy, 1983; Selleri, 1987).
4. A numerical example of probabilities satisfying the inequalities of Bell's type cannot be written in terms of factorizable probabilities, no matter how many hidden variables are introduced (Garg and Mermin, 1982).

In the present section only the last problem will be reviewed. The situation discussed is similar to the one of equation (92), with the difference that three parameters a_i and three parameters b_j will be used. The notation is simpler if one writes

$$D(a_{i\pm}, b_{k\pm}) \equiv D_{ik}(\sigma, \tau) \tag{98}$$

where $\sigma = \pm$ and $\tau = \pm$ are sign factors. The model proposed by Garg and Mermin (1982) is

$$D_{ik}(\sigma, \tau) = \tfrac{1}{4}[1 - c(\sigma + \tau) + A_{ik}(\sigma\tau)] \tag{99}$$

where $i, k = 1, 2, 3$; $0 < c \leq \tfrac{1}{3}$; $A_{11} = A_{22} = 1$ and $A_{ik} = -\tfrac{1}{3}$ in all other cases. It is very easy to show that all inequalities of the type (86) that can be written with the nine quantities (99) reduce to

$$|A_{ik} - A_{il} + A_{jk} + A_{jl}| \leq 2 \tag{100}$$

so that they are always satisfied. We note that the coefficient c has disappeared from the locality condition (100). Assuming factorizability, one has

$$D_{ik}(\sigma, \tau) = \int d\rho(\lambda)\, p_i(\sigma, \lambda) q_k(\tau, \lambda) \tag{101}$$

where

$$0 \leq p_i(\sigma, \lambda) \leq 1, \qquad 0 \leq q_k(\tau, \lambda) \leq 1 \tag{102}$$

and

$$\int_\Lambda d\rho(\lambda) = 1 \tag{103}$$

Since in all cases it follows from equation (99) that

$$D_{ik}(+, +) + D_{ik}(+, -) + D_{ik}(-, +) + D_{ik}(-, -) = 1 \tag{104}$$

one has for all $\lambda \in \Lambda$ and all $i, k = 1, 2, 3$:

$$p_i(+, \lambda) + p_i(-, \lambda) = 1 = q_k(+, \lambda) + q_k(-, \lambda) \tag{105}$$

Of the twelve probabilities entering into equation (102) only six are thus seen to be independent. Their number can be further reduced to four; in fact from equation (99) one sees that both $D_{11}(+, -)$ and $D_{22}(+, -)$ vanish. By using equations (101) and (105) one then gets

$$q_1(+, \lambda) = p_1(+, \lambda) = 1 \text{ or } 0$$
$$q_2(+, \lambda) = p_2(+, \lambda) = 1 \text{ or } 0 \tag{106}$$

Let Λ_1 (Λ_2) be the region of Λ where $p_1(+, \lambda) = 1$ [$p_2(+, \lambda) = 1$], and outside which it vanishes. Obviously the product $p_1(+, \lambda)q_2(+, \lambda)$ equals unity in $\Lambda_1 \cap \Lambda_2$ and vanishes outside. Therefore

$$D_{12}(+, +) = \int_{\Lambda_1 \cap \Lambda_2} d\rho(\lambda) = \tfrac{1}{6} - c/2 \tag{107}$$

where the numerical value was taken from the Garg–Mermin formula (99). Consider next the integral J below, *a priori* of unclear physical meaning:

$$J = \int_{\Lambda_1 \cup \Lambda_2} d\rho(\lambda) \, q_3(-, \lambda) \tag{108}$$

The following four relations allow one to get a lower limit for J:

$$\int_{\Lambda_1 \cup \Lambda_2} = \int_{\Lambda_1} + \int_{\Lambda_2} - \int_{\Lambda_1 \cap \Lambda_2} \tag{109}$$

$$\int_{\Lambda_1} d\rho(\lambda) \, q_3(-, \lambda) = \int_\Lambda d\rho(\lambda) \, p_1(+, \lambda) q_3(-, \lambda) = \tfrac{1}{3} \tag{110}$$

$$\int_{\Lambda_2} d\rho(\lambda)\, q_3(-,\lambda) = \tfrac{1}{3} \tag{111}$$

$$\int_{\Lambda_1 \cap \Lambda_2} d\rho(\lambda)\, q_3(-,\lambda) \leq \tfrac{1}{6} - c/2 \tag{112}$$

Of these, relation (109) is of intuitive validity, relation (110) [(111)] is a consequence of the definition of Λ_1 [Λ_2] and of the value of $D_{13}(+,-)$ [$D_{23}(+,-)$] deducible from equation (99), and relation (112) is a consequence of equation (107).

By using the relations (109) to (112) one easily gets

$$J \geq \tfrac{1}{2} + c/2 \tag{113}$$

But the right-hand side of equation (113) equals the single probability $q_3(-)$, since it follows from equation (99) that

$$q_3(-) = \int_\Lambda d\rho(\lambda)\, q_3(-,\lambda) = D_{13}(+,-) + D_{13}(-,-) = \tfrac{1}{2} + c/2 \tag{114}$$

From the definition of J one sees that it cannot be larger than $q_3(-)$. Therefore, comparing equations (113) and (114), one sees that it must be $J = q_3(-)$. This result entails two conclusions:

1. $q_3(-,\lambda) = 1$, for all $\lambda \in \Lambda_1 \cap \Lambda_2$
2. $q_3(-,\lambda) = 0$, for all $\lambda \in \tilde{\Lambda}$ if $\tilde{\Lambda} = \Lambda - \Lambda_1 \cup \Lambda_2$

Therefore

$$q_3(+,\lambda) = 1, \qquad \text{for all } \lambda \in \tilde{\Lambda} \tag{115}$$

Identical reasoning can be invoked for $p_3(-,\lambda)$ and $p_3(+,\lambda)$, with the result

$$p_3(+,\lambda) = 1 \qquad \text{for all } \lambda \in \tilde{\Lambda} \tag{116}$$

From equations (115) and (116) it follows that $p_3(+,\lambda)q_3(+,\lambda) = 1$ for all $\lambda \in \tilde{\Lambda}$. Therefore

$$D_{33}(+,+) = \int_\Lambda d\rho(\lambda)\, p_3(+,\lambda)q_3(+,\lambda) \geq \int_{\tilde{\Lambda}} d\rho(\lambda)\, p_3(+,\lambda)q_3(+,\lambda)$$

$$= \int_{\tilde{\Lambda}} d\rho(\lambda) = \tfrac{1}{2} - c/2 \tag{117}$$

The latter inequality is, however, in disagreement with the value of $D_{33}(+, +)$ deduced from the Garg–Mermin formula (99). Therefore the latter formula is incompatible with the Clauser–Horne factorizability condition.

The four difficulties listed above are strong enough to make one feel somewhat uncertain about the generality of the Clauser–Horne formula. A safer definition of probabilistic local realism should therefore be looked for, with the possibility, in case of lack of success, of finding a reconciliation between local realism and the quantum-mechanical predictions in the yet unexplored region of local-realistic situations which do not satisfy the CH formula.

Unfortunately this possibility does not exist, as shown in Chapter 6, in which a fully general formulation of probabilistic local realism is developed that leads to the validity of Bell's inequality and of the other inequalities previously deduced from CH factorizability.

It has been shown by Lepore (1988) that the Garg–Mermin model equation (99) is nonlocal in spite of its satisfying all the inequalities of the set of equations (90) and (91). A satisfactory solution of the factorizability problem has been found by Garuccio, Lepore, and Selleri (1988): It is not factorizability itself that breaks down, but the assumed independence of $\rho(\lambda)$ and Λ on the considered experimental parameters a_1, $a_2, \ldots, b_1, b_2, \ldots$.

2.9. The EPR Paradox in Particle Physics (1981–1987)

Chapters 4 and 5 of the present book deal in an excellent way with some very interesting processes in particle physics which allow a formulation of the EPR paradox. The quantum-mechanical predictions for these processes disagree with some expectations which can be obtained from local realism. It is therefore possible to carry out experimental investigations of the paradox in particle physics.

The processes considered are:

1. Decay of the $J^P = 0^-$ meson η_c of mass 2980 MeV/c² into a Λ-hyperon plus $\bar{\Lambda}$-antihyperon pair, with subsequent decays $\Lambda \to p + \pi^-$ and $\bar{\Lambda} \to \bar{p} + \pi^+$. If $\hat{\mathbf{a}}$ and $\hat{\mathbf{b}}$ are unit vectors in the directions of the emitted π^- and π^+ mesons in the Λ and $\bar{\Lambda}$ rest frames, respectively, the suitably normalized decay rate, summed over p and $\bar{p}$ spins, is given by

$$r_0(\hat{\mathbf{a}}, \hat{\mathbf{b}}) = 1 + \alpha^2 \hat{\mathbf{a}} \cdot \hat{\mathbf{b}} \tag{118}$$

where $\alpha = -0.642 \pm 0.013$ is the Λ-decay asymmetry parameter. The latter quantum-mechanical prediction disagrees by as much as 10% from a limit deducible from local realism (Törnqvist, 1981).

2. Decay of the $J^P = 1^-$ meson J/Ψ of the mass 3097 MeV/c^2 into the same $\Lambda\bar{\Lambda}$ channel discussed above for the η_c. If the J/Ψ is polarized along the direction $\hat{n}$, the decay rate summed over p and $\bar{p}$ spins, is given by

$$r_1(\hat{a}, \hat{b}) = 1 - \alpha^2 \hat{a} \cdot \hat{b}' \tag{119}$$

where

$$\hat{b}' \propto \hat{b} - \hat{n}(2\hat{b} \cdot \hat{n}) \tag{120}$$

The latter quantum-mechanical prediction disagrees with expectations deduced from local realism.

3. Decay of a $J^{PC} = 1^{--}$ state into $K^0\bar{K}^0$. If charge conjugation-parity (CP) conservation is assumed and t_a (t_b) is the proper time of the kaon moving to the left (right) and $t = t_b - t_a$, then the probability of a double $\bar{K}^0$ observation at times t_a and t_b is given by

$$\omega(t_a, t_b) = \tfrac{1}{8} \exp(-\gamma_s t_a)[1 + \exp(-\gamma_s t)$$

$$- 2 \exp(-\gamma_s t/2) \cos \Delta t] \tag{121}$$

where γ_s is the total decay rate for the short-lived kaon, and Δ is the $K_{long} - K_{short}$ mass difference. The previous quantum-mechanical prediction disagrees by up to 12% from an upper limit rigorously deducible from local realism and given by the right-hand side of equation (121) without the term proportional to $\cos \Delta t$ (Selleri, 1983).

4. Decay of the $Y(4s)$ vector meson into a pair of neutral pseudoscalar mesons $B^0\bar{B}^0$. The formalism for treating this $B^0\bar{B}^0$ system is exactly the same as that for the $K^0\bar{K}^0$ system. However, the B^0-$\bar{B}^0$ oscillations are still in doubt and it is therefore not clear that this particular test can be carried out (Datta and Home, 1986).

5. Quantum-mechanical treatment of K^0-$\bar{K}^0$ mixing in double kaon decays of a spin-1 resonance with inclusion of CP violation. The predicted number of K^0 or $\bar{K}^0$ observed in one hemisphere seems to depend on the position of an absorber in the other hemisphere, in evident contradiction to the locality condition (Datta, Home, and Raychaudhuri, 1987).

It would be very interesting to have experimental evidence for the above five processes, as a starting point for a deeper understanding of the relationship between local realism and particle physics. One should be careful, however, not to take too literally the theoretical predictions "deduced" from local realism. Additional assumptions of the type introduced in the analysis of the experiments with optical photon pairs have often been made. There is, in particular, little doubt that the quantum-mechanical predictions (118) and (119) cannot be reproduced by local realistic models. A model of the Kasday type (see the final part of Section 1.6), in particular, is certainly possible. By and large, it can be said that the interesting perspectives that are opened by strong interactions into providing high-efficiency detectors, and therefore in allowing one to avoid some additional assumptions, are still to be investigated.

2.10. Experiments with Pairs of Atomic Photons (1972–1987)

In the first five sections of this chapter sets of α and β particles were considered and assumed to be "well-behaved" in the sense that all α (β) particles were taken to propagate toward a well-defined region R_α (R_β), where an observer O_α (O_β) was supposed to use some instruments in order to perform measurements on them. This is like saying that in experiments with pairs of optical photons the "source" is defined as incorporating the lenses which define the right solid angles.

The above assumption/definition does not cause any trouble in practice, because experimentalists know which fraction of photon pairs is well-behaved in the previous sense. As a consequence of this, our probabilities were all defined without factors representing the fractions of particles α and β actually arriving on their respective measuring apparatuses.

Actual experiments on the EPR paradox have almost always been carried out with photons. The quantum-mechanical treatment of photon polarization is similar to that of spin-$\frac{1}{2}$ in one important respect: both observables are dichotomic. The absence of a photon mass has the practical effect of eliminating from the theoretical scheme the longitudinal polarizations. Only linear-polarization states perpendicular to the direction of propagation are therefore left, similarly to the case of classical electromagnetic waves whose transverse nature is well known. Considering also states of circular polarization, one can define

$|R\rangle$, a single-photon state with right-handed circular polarization
$|L\rangle$, a single-photon state with left-handed circular polarization
$|x\rangle$, a single-photon state with linear polarization along the x-axis
$|y\rangle$, a single-photon state with linear polarization along the y-axis

These two sets of states are not unrelated. Elementary quantum theory gives

$$|R\rangle = (|x\rangle + i|y\rangle)/\sqrt{2}$$

$$|L\rangle = (|x\rangle - i|y\rangle))/\sqrt{2}$$

(122)

if the photon propagates in the positive z direction. The existence of dichotomic observables for photons has the practical effect that Bell-type and CHSH-type inequalities can be formulated also for pairs of photons. There are situations where quantum theory describes the polarization of two correlated photons with nonfactorizable state vectors, analogous to the singlet state of two spin-$\frac{1}{2}$ objects, which imply violations of local realism.

In the case of photons, the parity quantum number plays an important role and it is necessary to distinguish, for example, the $J^P = 0^+$ from the $J^P = 0^-$ states, represented respectively by the state vectors

$$|0^+\rangle = (|R_\alpha\rangle|R_\beta\rangle + |L_\alpha\rangle|L_\beta\rangle))/\sqrt{2}$$

$$|0^-\rangle = (|R_\alpha\rangle|R_\beta\rangle - |L_\alpha\rangle|L_\beta\rangle))/\sqrt{2}$$

(123)

These states can also be expressed in terms of linear polarizations and one obtains

$$|0^+\rangle = (|x_\alpha\rangle|x_\beta\rangle + |y_\alpha\rangle|y_\beta\rangle)/\sqrt{2}$$

$$|0^-\rangle = (|x_\alpha\rangle|y_\beta\rangle - |y_\alpha\rangle|x_\beta\rangle))/\sqrt{2}$$

(124)

The basis states with respect to which the linear polarization is expressed are arbitrary. Using the rotated x'- and y'-axes one obtains results identical to equations (124) for both states, with x' and y' in place of x and y. This property is due to the invariance under rotations around the z-axis of the zero-angular-momentum states.

All the inequalities of the Bell type and of the CHSH type found in the previous sections clearly apply also to photon pairs, since they were deduced from the dichotomic nature of the measured quantities, besides, of course, from locality and realism. In order to check that the quantum-mechanical predictions often violate those inequalities we will carefully present the theoretical formulas for the most important probabilities and correlation functions introduced in previous sections. This will also allow us to stress again the very important distinction between inequalities of the Bell type and strong inequalities.

The most widely-used cascade is the $(J = 0) \rightarrow (J = 1) \rightarrow (J = 0)$ cascade of calcium. The quantum-mechanical predictions following from the state $|0^+\rangle$, which applies to this case, are, for the double-transmission probabilities:

$$\omega(a_+, b_+) = \tfrac{1}{4}[\varepsilon_+^1 \varepsilon_+^2 + \varepsilon_-^1 \varepsilon_-^2 F_1(\theta) \cos 2(a - b)]$$

$$\omega(a_+, \infty_+) = \tfrac{1}{2}\varepsilon_+^1$$

$$\omega(\infty_+, b_+) = \tfrac{1}{2}\varepsilon_+^2 \qquad\qquad (125)$$

$$\omega(\infty_+, \infty_+) = 1$$

These relations give the correlation function $P(a, b)$ through equation (53):

$$P(a, b) = (1 - \varepsilon_+^1)(1 - \varepsilon_+^2) + \varepsilon_-^1 \varepsilon_-^2 F_1(\theta) \cos 2(a - b) \qquad (126)$$

In these relations $F_1(\theta)$ is a function of the half-angle θ subtended by the primary lenses representing a depolarization due to noncollinearity of the two photons and

$$\varepsilon_\pm^1 = \varepsilon_M^1 \pm \varepsilon_m^1, \qquad \varepsilon_\pm^2 = \varepsilon_M^2 \pm \varepsilon_m^2 \qquad (127)$$

Here ε_M^1 (ε_m^1) is the transmittance of the first polarizer for light polarized parallel (perpendicular) to the polarizer axis; and a similar notation has been used for the second polarizer. All these transmittances are usually very near to the ideal case, with ε_M^i close to unity and ε_m^i close to zero ($i = 1, 2$). Also, the depolarization factor F_1 is usually very close to unity, so that $P(a, b)$, as given by equation (126), violates Bell's inequality (47). However, as already stressed, the trouble is that transmission probabilities are not measurable, so that Bell's inequality (47) cannot be tested.

If the CHSH additional assumption of Section 2.2 is made, the double-detection probability D_0 becomes a crucial quantity, which is assumed independent of the presence and of the orientation of the polarizers. Quantum theory predicts

$$D_0 = \eta_1 \eta_2 \qquad (128)$$

where η_1 (η_2) is the quantum efficiency of the first (second) photomultiplier. In the experiments performed, η_1 and η_2 were of the order of 10%, so that D_0 was of the order of 10^{-2}. The latter quantity relates the double-transmission probabilities ω to the measurable double-transmission *and* double-detection probabilities D, *once the CHSH additional assumption has been made.*

In the usual quantum theory, the CHSH assumption does not need to be explicitly made, its validity being always taken for granted. The quantum-mechanical expressions for the D probabilities defined in equation (54) are

$$D(a, b) = \tfrac{1}{4}[\varepsilon_+^1\varepsilon_+^2 + \varepsilon_-^1\varepsilon_-^2 F_1(\theta) \cos 2(a - b)]\eta_1\eta_2$$

$$D(a, \infty) = (\varepsilon_+^1/2)\eta_1\eta_2$$

$$D(\infty, b) = (\varepsilon_+^2/2)\eta_1\eta_2 \qquad (129)$$

$$D(\infty, \infty) = \eta_1\eta_2$$

These double-detection and double-transmission probabilities are obviously proportional to the respective coincidence rates R [see equations (55)], the proportionality factor being N_0, the number of photon pairs entering per second into the right solid angles defined by the optical apparatuses. The inequality (58) can thus also be written

$$-1 \leqslant \frac{D(a, b)}{D_0} - \frac{D(a, b')}{D_0} + \frac{D(a', b)}{D_0} + \frac{D(a', b')}{D_0}$$

$$- \frac{D(a', \infty)}{D_0} - \frac{D(\infty, b)}{D_0} \leqslant 0 \qquad (130)$$

and it can easily be shown to be violated by the quantum-mechanical predictions (128) and (129). Experimentally, it has been found to be violated. One should remember that the inequality (130), as well as the inequality (58), is a consequence of local realism *and* of the additional assumption. Its violation can only mean that one of these tenets is wrong, but it cannot say which one. It is, for example, possible to build explicit local realistic models that do not satisfy the CHSH additional assumption, and that violate the inequality (130). This will be discussed in Section 3.7.

We note that the inequality (130) essentially coincides with the inequality (89) deduced with the help of the CH additional assumption. If instead, the inhomogeneous inequality (86), which was deduced only from local realism, is considered, one can see that the quantum-theoretical predictions for the single-photon transmission and detection probabilities are

$$p(a') = (\varepsilon_+^1/2)\eta_1, \qquad q(b) = (\varepsilon_+^2/2)\eta_2 \qquad (131)$$

Owing to the presence of a single η-factor these probabilities are an order of magnitude larger than the double-transmission and double-detection probabilities of equations (129). This implies that the inequality (86) is never violated.

We now come to Rapisarda-type experiments with two-way polarizers. The quantum-mechanical predictions for the D probabilities defining $E(a, b)$ [see equation (93)] are

$$D(a_+, b_+) = \tfrac{1}{4}[T_+^1 T_+^2 + T_-^1 T_-^2 F_1(\theta) \cos 2(a - b)]\eta_1 \eta_2$$

$$D(a_+, b_-) = \tfrac{1}{4}[T_+^1 R_+^2 - T_-^1 R_-^2 F_1(\theta) \cos 2(a - b)]\eta_1 \eta_2$$

$$D(a_-, b_+) = \tfrac{1}{4}[R_+^1 T_+^2 - R_-^1 T_-^2 F_1(\theta) \cos 2(a - b)]\eta_1 \eta_2 \qquad (132)$$

$$D(a_-, b_-) = \tfrac{1}{4}[R_+^1 R_+^2 + R_-^1 R_-^2 F_1(\theta) \cos 2(a - b)]\eta_1 \eta_2$$

where

$$T_+^i = T_\parallel^i + T_\perp^i, \qquad T_-^i = T_\parallel^i - T_\perp^i \qquad (133)$$

and

$$R_+^i = R_\parallel^i + R_\perp^i, \qquad R_-^i = R_\parallel^i - R_\perp^i \qquad (134)$$

($i = 1, 2$). The T and R parameters are transmittances defined in the following way. There are two prisms, denoted by the index $i = 1, 2$ above. From each prism two beams are emitted, a reflected one and a transmitted one. $T_\parallel$ ($T_\perp$) denotes the prism transmittance along the transmitted path for incoming light polarized parallel (perpendicular) to the transmitted-channel polarization plane; and $R_\parallel$ ($R_\perp$) denotes the prism transmittance along the reflected path for incoming light polarized parallel (perpendicular) to the reflected-channel polarization plane.

A recent measurement (Falciglia *et al.*, 1983) gave, for example,

$$T_\parallel = 0.9095 \pm 0.0023, \qquad T_\perp = 0.0044 \pm 0.0002$$

$$R_\parallel = 0.7625 \pm 0.0024, \qquad R_\perp = 0.0041 \pm 0.0003$$

Insertion of equations (132) into equation (93) gives

$$E(a, b) = \frac{f + g \cos 2(a - b)}{f' + g' \cos 2(a - b)} \qquad (135)$$

where

$$f = (T_+^1 - R_+^1)(T_+^2 - R_+^2), \qquad g = (T_-^1 + R_-^1)(T_-^2 + R_-^2)$$

$$f' = (T_+^1 + R_+^1)(T_+^2 + R_+^2), \qquad g' = (T_-^1 - R_-^1)(T_-^2 - R_-^2) \qquad (136)$$

Garuccio and Rapisarda (1981) showed that the prediction (135) violates the inequality (97). However, it should once more be remembered that the inequality (97) was deduced by means of the GR additional assumption and that local-realistic models exist for which that assumption is not valid.

Experiments with pairs of atomic photons were actually carried out by Freedman and Clauser (1972), Holt and Pipkin (1973), Clauser (1976), Fry and Thompson (1976), Aspect, Grangier, and Roger (1981 and 1982), Aspect, Dalibard, and Roger (1982) and Perrie, Duncan, Beyer, and Klein-poppen (1985). In all cases but one very good agreement with quantum theory was found and inequalities of the strong type were found to be violated. The story of these experiments is told in detail in Chapter 7.

Important experiments remain to be done in at least three areas, even in the case of atomic cascades:

1. Only two experiments have been reported which tried to measure *circular* polarizations of the photon pairs and in both cases strange effects were reported which were attributed to distortions generated in the $\lambda/4$ plates. In particular, no violations of the strong inequalities were found. In view of the great importance of the EPR paradox it is vital that these measurements be repeated.
2. A very important effect which has never been checked should present itself when the atomic source of photon pairs is inserted in a strong enough magnetic field. A kind of phase transition should take place as soon as the field is switched on, with a jump from a correlation function violating the strong inequalities to one respecting it. See Chapter 8 for more details.
3. The "variable-probability" models discussed in the Section 3.7 lead to correlation functions very similar to the quantum-mechanical ones, but with a small extra term proportional to $\cos 4(a - b)$. It is very important that this effect be looked for.

3. Attempted Solutions of the Paradox

> So the quantum, fiery creative force of modern physics, has burst forth in eruption after eruption and for all we know the next may be the greatest of all.
>
> (J. A. WHEELER, 1980)

3.1. Unbroken Wholeness

According to Bohm, the essential new feature implied by quantum theory is nonlocality: a system cannot be analyzed in parts whose basic properties do not depend on the state of the whole system, no matter how "well-separated" in space these parts might appear to be. He believes, with Hiley, that the well-known experiments on Bell's inequality reveal, in an especially clear way, the nonlocal nature of quantum phenomena (Bohm and Hiley, 1978). However, nonlocality, "is involved in an essential way in every manifestation of a many-body system, as treated by Schrödinger's equation in a $3N$-dimensional configuration space" (Bohm and Hiley, 1978, p. 94).

In the case of two particles with mass m Schrödinger's equation is

$$i\hbar \frac{\partial \Psi}{\partial t} = -\frac{\hbar^2}{2m} [\nabla_1^2 + \nabla_2^2] \Psi + V\Psi \tag{137}$$

where

$$\Psi = \Psi(\mathbf{x}_1, \mathbf{x}_2, t) \tag{138}$$

is the wave function of the two particles, $V(\mathbf{x}_1, \mathbf{x}_2)$ is the potential acting on them, and ∇_1^2 and ∇_2^2 refer to particles 1 and 2, respectively. Writing

$$\Psi = R \exp(iS/\hbar) \tag{139}$$

and

$$P = R^2 = \Psi^*\Psi \tag{140}$$

one can obtain from equation (137)

$$\frac{\partial P}{\partial t} + \boldsymbol{\nabla}_1 \cdot \left(P \frac{\boldsymbol{\nabla}_1 S}{m} \right) + \boldsymbol{\nabla}_2 \cdot \left(P \frac{\boldsymbol{\nabla}_2 S}{m} \right) = 0 \tag{141}$$

and

$$\frac{\partial S}{\partial t} + \frac{(\boldsymbol{\nabla}_1 S)^2}{2m} + \frac{(\boldsymbol{\nabla}_2 S)^2}{2m} + V + Q = 0 \tag{142}$$

where

$$Q = Q(\mathbf{x}_1, \mathbf{x}_2, t) = -\frac{\hbar^2}{2m} \left(\frac{\nabla_1^2 R}{R} + \frac{\nabla_2^2 R}{R} \right) \tag{143}$$

Evidently equation (141) describes the conservation of probability with density $P = \Psi^*\Psi$ in the configuration space of the two particles. Equation (142) is instead a Hamilton–Jacobi equation for the system of the two particles, acted on, not only by the classical potential V, but also by the quantum potential $Q(\mathbf{x}_1, \mathbf{x}_2, t)$. There are two strikingly new features of this quantum potential:

1. In general, it does not produce a vanishing interaction between the two particles as $|\mathbf{x}_1 - \mathbf{x}_2| \to \infty$.
2. It cannot be expressed as a universally determined function of the coordinates $\mathbf{x}_1$ and $\mathbf{x}_2$. Rather, it depends on $\Psi(\mathbf{x}_1, \mathbf{x}_2)$ and therefore on the "quantum state" *of the system as a whole.*

It is the latter feature that brings out the nonlocal nature of quantum phenomena. Bohm has also suggested a sort of ontological model of quantum nonlocality which tries to provide a general framework in which the new phenomena might look less unnatural. He has done this by introducing the notion of "unbroken wholeness" that characterizes two correlated quantum systems (see, for example, Bohm, 1987). He considers the interesting example of a hologram and stresses that the different parts of the object are not in correspondence with different parts of its hologram, but rather that each of the latter parts, individually, is somehow expressing the whole object. Accordingly, if one illuminates only part of the hologram, one gets information about the entire object, even if less-detailed and from fewer angles. Similarly Bohm thinks that what appears to us as two separated quantum objects might in actuality only be a manifestation of a truly-interconnected wholeness. The hologram of two spheres, for instance, stores the information of each ball over the entire hologram. It can therefore be said that *in the hologram* the two spheres are really, in a way, amalgamated and impossible to separate. Bohm views this as an example of the true physical situation giving rise to the EPR paradox: In space there is only an "unbroken wholeness," which sometimes can give rise to manifestations which *appear* as two separate objects.

3.2. Superluminal Connections in Dirac's Aether

A second possible solution of the EPR paradox is provided by the nonlocal model of Vigier and collaborators (Vigier, 1979). They adopt the idea, first presented by Dirac (1952), that the *aether*, with suitable properties, is no longer ruled out by special relativity, especially if the probabilistic nature of quantum phenomena is taken into account. In this approach, it is assumed that the velocity distribution of the particles constituting the aether has a constant value over the hyperboloid

$$v_0^2 - v_1^2 - v_2^2 - v_3^2 = 1$$

In such a case, in fact, the velocity distribution looks the same to all observers and the aether does not produce any physical effect on moving bodies. In Vigier's model this aether-like physical vacuum is made of extended *rigid* particles which can support, within their interiors, signals with superluminal velocity. The statistical properties of quantum objects reflect, then, nothing but the real random fluctuations of the ether.

In this theory there are also (quantum) waves which propagate as real physical collective excitations (i.e., as density waves) on the top of the foregoing Dirac aether. In this way, information originating on the boundary of the ψ wave (such as the opening or closing of a slit in the double-slit experiment, or the observation of one of the two particles forming an EPR pair) reacts with superluminal velocity (*via the quantum potential*) on the particle motions which propagate with subluminal group velocities along the flow lines of the quantum-mechanical ψ waves.

In Vigier's opinion the existence of superluminal propagations does not necessarily imply a breakdown of causality, if "causality" is defined as follows:

1. The possibility of solving the two-particle problem in the forward (or backward) time direction as a Cauchy problem.
2. The time-like nature of all particle trajectories.
3. The invariance of the formalism under the Poincaré group of transformations.

We consider the following objection that can be raised against a theory in which superluminal connections are introduced: There are two particles propagating in two widely-separated regions of space R_1 (on the Earth) and R_2 (in the Andromeda galaxy) and forming an EPR pair. Their propagation takes place according to precise deterministic equations containing nonlocal potentials like, for instance, Bohm's potential Q of equation (143). Each particle "knows" instantaneously what the other particle is doing, and reacts accordingly. It seems, therefore, obvious to conclude that the switching on and off of a magnetic field in R_1 must have instantaneous consequences on the particle located in R_2, because of the superluminal physical connection. The experiment can then be set up in such a way that the second particle enters a detector D_1 (a detector D_2) if the magnetic field in R_1 is off (is on). Therefore the observer in Andromeda can instantaneously learn what his fellow observer is doing on Earth. Using ensembles of correlated EPR pairs it then becomes possible to transmit instantaneous information from R_1 to R_2.

The problem of causality in general and the previous objection in particular have been discussed by Cufaro Petroni (1985) in a very clear way. His answer to the objection is that we live in a completely-deterministic world and that it does not make any sense to consider "modifications" of its properties, such as the one introduced before, through the switching on and off of a magnetic field:

> In a completely deterministic world there is no possible "modification": The world IS and we cannot intervene from exterior to its tissue in order to modify it, because we are IN the world (Cufaro Petroni, 1985).

A signal always needs a free will that is external to the physical process considered and that, at a given time, decides to modify the regular evolution of the process in order to send a message. But if instead we assume, with Cufaro Petroni, that the particles of the human brain are connected to all physical processes, obey the same equations, and therefore behave in an unique and strictly correlated way, then there are no "signals" at all.

3.3. Nonlocal Weak Realism

Stapp believes that the quantum-mechanical predictions for the situation dealt with in the EPR paradox have been accurately confirmed "under experimental conditions essentially equivalent to those needed for the EPR argument." Hence he concludes that the world we live in is nonlocal. However, he does not believe that the results obtained by Bell (1965) and by Clauser and Horne (1974) are sufficient for establishing the need for nonlocality since, in his opinion, these authors made very strong assumptions about microscopic reality that are not compatible with orthodox quantum thinking. The refutation of these "strong" assumptions of realism does not imply, however, any retreat to idealism or subjectivism. It is, in fact, possible to substitute them with an "informal" Copenhagen interpretation of quantum phenomena.

In this way Stapp distinguishes a *strict* Copenhagen interpretation, in which nothing at all is said about any reality other than our observations, from an *informal* interpretation, in which one accepts the common sense idea of a *macroscopic* reality that exists independently of our observations and can be described, at least approximately, with the concepts of classical physics. This "informal" interpretation is partly related, at least by Stapp, to Heisenberg's idea of a transition from the "possible" to the "actual" taking place during the act of measurement. Stapp's microworld is a "sea of microlevel potentialities," that become "well-defined" physical properties only by interacting with an experimental apparatus.

A model theory proposed by Stapp (1987) contains certain "hidden variables" λ which represent all the deterministic and stochastic quantities that characterize the unified organic world and which are *not* used to provide the basis for a Clauser–Horne factorization structure of probabilities. They do not reflect ideas of separation, localization, or microscopic structure.

Stapp writes $\lambda = (\lambda', \lambda'')$, where λ' is strictly predetermined, and λ'' is any stochastic variable.

Furthermore, in this theory it is assumed that every act of measurement involves a *choice*. This choice "picks the actual from among what had previously been mere possibilities: the choice renders fixed and settled something that had prior to the choice been undetermined." A "choice" variable Z is also introduced and written $Z = (x, y)$, where x and y represent the choices of experiment in the regions R_α and R_β, respectively, where two correlated observations of the EPR type are made. The "choices" x and y are treated as independent free variables. Each of them can assume an infinite number of different values.

Suppose there are two observables A and A' that can be measured in R_α and another two, B and B', that can be measured in R_β. *The choice variable picks one observable before an act of measurement is made.* More precisely, the chosen observable is, in R_α,

$$A \quad \text{if } x \in X, \quad A' \quad \text{if } x \in X' \tag{144}$$

where $X \cup X'$ is the set of possible values of x. Furthermore, in R_β the chosen observable is

$$B, \quad \text{if } y \in Y, \quad B', \quad \text{if } y \in Y' \tag{145}$$

where $Y \cup Y'$ is the set of possible values of y.

Depending on the values of x and y, there are four possible experiments that can be chosen to be performed in R_α and R_β, corresponding to the four pairs of observables

$$(A, B), \quad (A, B'), \quad (A', B), \quad (A', B') \tag{146}$$

Now, the results of the measurements of whatever observables have been chosen are assumed to be

$$r_\alpha(x, y, \lambda) \quad \text{in } R_\alpha, \quad r_\beta(x, y, \lambda) \quad \text{in } R_\beta \tag{147}$$

in a general nonlocal theory, while r_α does not depend on y and r_β does not depend on x if instead locality is assumed.

Stapp could easily prove that the local choice contradicts the empirical predictions of quantum theory and concluded:

> ... neither determinism, nor counterfactual definiteness, nor any idea of reality
> incompatible with orthodox quantum thinking need be assumed in order to

prove the incompatibility of the empirical predictions of quantum theory with
the EPR idea that no influence can propagate faster than light.

The remark about the absence of a "counterfactual definiteness" is, of
course, justified by the important fact that, in Stapp's theory, the fixing of
x and y and λ fixes the value *only of the observable that is actually measured.*
The values of the other three observables remain, instead, completely
indefinite.

3.4. Actions of the Future on the Past

A solution of the EPR paradox based on the idea that it is possible to
modify past events by means of retroactions from the future was first
proposed by Costa de Beauregard (1977). He noted that twice in classical
physics contradictions were discovered between *fact-like* irreversible processes and *law-like* reversibility of the physical theory: (1) When Boltzmann
used statistical mechanics for deducing the Second Law of Thermodynamics: the paradox inherent in extracting time asymmetry from a theory
like Newtonian mechanics that is intrinsically time-symmetric was exposed
in specific forms by Loschmidt and Zermelo; and (2) When the principle
of retarded waves was used in physical optics and in classical electrodynamics in order to exclude one half of the mathematically permissible
solutions of the wave equations.

Costa de Beauregard's idea is that a careful examination of the world
in which we live is bound to lead to the conclusion that retroactions in time
do play a role and should not be discarded in the formulations of our
theories. One way to see this is to remember that for Aristotle, creator of
the concept, *information* was not only knowledge, as is intended today, but
it was *also*, symmetrically, an *organizing power.* The examples given were
the craftman's or the artist's work, and also biological ontogenesis. A second
way to see a final cause at work is to consider modern cybernetics which,
surprisingly, came to rediscover the two faces of Aristotle's information. In
computers and other information-processing machines the chain

$$\text{Information} \xrightarrow{(1)} \text{negentropy} \xrightarrow{(2)} \text{information}$$

means that a concept is coded and sent as a message, before being decoded
and received. Negentropy is of course entropy with a minus sign. Step (2)
above is the *learning transition,* where information shows up as gain in
knowledge, while Step (1) is the *willing transition,* where information shows
up as an organizing power.

In the theoretical framework (*de jure*) there is a complete symmetry between the two transitions. In spite of this there is a dissymmetry in practice (*de facto*) because irreversibility is generated by misprints in the coding: noise along the line, mistakes in decoding, and so on.

The relationship between the variation of negentropy ΔN and the variation of information ΔI is

$$\Delta N = k \cdot \ln 2 \Delta I$$

If N and I are both expressed in "practical" units, it turns out that the factor multiplying ΔI is very small, of the order of 10^{-16}. Therefore, Costa de Beauregard concludes that it is very difficult to produce important increases of negentropy (decreases of entropy) by increasing the information. *Vice versa*, even a very small increase of negentropy can give rise to a large gain of information. If one lets $k \rightarrow 0$, one obtains a situation where gaining knowledge is absolutely costless, but producing order is utterly impossible. In this limit, consciousness is made totally passive: it registers what is going on outside itself, and that is all.

If the roots of Costa de Beauregard's conceptions go deep into classical physics, it is in quantum theory that he thinks the most important effects of retroaction can be seen. Again, he stresses, the theory is completely time-symmetrical, but only until the idea of the *collapse of the wave function* is introduced. At this point quantum theory commits itself to the philosophy of retarded waves. In Costa de Beauregard's opinion this happens because "the Copenhagen school has forgotten the hidden face of Aristotle's information."

It is precisely in the situations envisaged by the EPR paradox that this "hidden face" shows up again. In order to understand the essence of the EPR paradox, Costa de Beauregard considers the mathematical apparatus of quantum theory and concludes that the problem, today, is only that of *tailoring the wording of the EPR situation after the mathematics.* In his opinion there has in fact been, in our century, an irreversible victory of formalism over modelism.

From this starting point he deduces that when an EPR pair, for instance two photons described by one of the state vectors of equation (124), is measured by two observers in two regions separated by a space-like distance, then it is precisely the act of observation that produces *in the past* of the measurement process, the right physical properties of the photon pair. Each observer is thus considered capable of *telediction* plus *teleaction*, by taking, so to say, a relay in the past, or more precisely, in the source that emitted the two photons.

The conclusion that one can draw from this theory is that the element of reality introduced in the formulation of the EPR paradox can be accepted as real, but that it is viewed as created by *actually performed* acts of observation, and as propagating backward in time with one of the two correlated quantum objects, from the region of measurement to the source.

In particular, there can be no question of associating elements of reality with observables that are not concretely measured, as was done originally by Einstein, Podolsky, and Rosen, and later by Bell[12] and other authors. In this sense the solution of the EPR paradox proposed by Costa de Beauregard is similar to that of Bohr.

Several other authors have proposed propagations toward the past as a solution of the EPR paradox. In chronological order one can list: Stapp (1975), Davidon (1976), Rayski (1979), Rietdijk (1980), Cramer (1980), Sutherland (1983).

3.5. The Nonergodic Interpretation

The nonergodic interpretation of quantum mechanics assumes that a sequence of quantum objects, even if separated by large time intervals from one another, do not behave independently in their interaction with the measuring apparatus. The basic idea is that these objects may essentially interact with each other, by means of memory effects in an hypothetical medium filling the space crossed by them, on their way toward the measuring instruments.

Let us consider, for instance, the double-slit experiment. The previous type of indirect interaction is such that a particle passing through a slit knows if the other slit is open, because this information is recorded in the medium filling the space between the two screens. Those particles, which came previously from the second slit, modified the physical properties of space, and gave rise to the storage of the relative information. Obviously, interference can happen only after a sufficiently large number of particles have crossed the apparatus and conditioned the medium. In this way particles interfere, with other particles, but only indirectly, through the medium (Buonomano, (1980, 1987)).

More generally, we consider a quantum experiment repeated a large number of times, every repetition being called a "run." Let R represent the number of runs, and N the number of quantum objects in every run, assumed constant for simplicity. Let λ_{rn} represent the state of the nth particle in the rth run, and s_{rn} represent the state of the experimental apparatus just before interacting with the nth particle of the rth run. The result of the measurement, A_{rn}, is assumed to be completely fixed once λ_{rn} and s_{rn} are given. Therefore

$$A_{rn} = A(\lambda_{rn}, s_{rn}) \tag{148}$$

Starting from these numbers two types of averages are possible:

$$\bar{A}_r = \frac{1}{N} \sum_{n=1}^{N} A_{rn}, \qquad \bar{A}_n = \frac{1}{R} \sum_{r=1}^{R} A_{rn} \qquad (149)$$

where $\bar{A}_r$ is called the *run average* and $\bar{A}_n$ is the *ensemble average* at "time" n. Buonomano observed that it is always implicitly or explicitly assumed that

$$\bar{A}_r = \bar{A}_n \qquad (150)$$

(the *ergodic assumption*) but that such an assumption should really be checked with suitably designed experiments. In order to do so it must be made clear that the only way to avoid the medium polarization effects is to keep the runs distant in time from one another, and eventually also to keep them in different regions of space where no experiments have been carried out previously. Thus the ensemble average for $n = 1$:

$$\bar{A}_{n=1}$$

should represent events collected in conditions where the medium does *not* act on the particles (there are no memory effects for $n = 1$, since no previous particles entered the apparatus in any of the runs considered!). Therefore $\bar{A}_{n=1}$ should describe a situation in which no quantum phenomenon appears and classical physics holds unreservedly. Instead, $\bar{A}_n$ for large n, and $\bar{A}_r$ for all r describe quantum-mechanical situations. The case of $\bar{A}_n$ for not-too-large values of n, but with $n \neq 1$, represents mixed situations where a transition between classical and quantum physics is taking place.

This nonergodic interpretation of quantum mechanics can, in principle, solve the EPR paradox, because it can explain the apparent violations of local realism as due to nonergodic effects within a strictly local theory. Let us consider, in fact, the left-hand side of a polarization–correlation experiment and divide the space between polarizer and source into M cells, numbering them from left to right. Thus the polarizer is in cell 1 and the source in cell M. We assume that the state of the cell m depends on the previous state of the neighboring cells. It follows that after one photon has passed the state of cell 2 depends on the state of the polarizer. After two photons have passed, cell 3 depends on the state of the polarizer, as so on. Then, after $n \geq M$ photons have passed, cell M, that is the source, depends on the state of the polarizer.

If the right-hand side of the polarization–correlation experiment is treated in the same manner, one obtains a situation in which the source produces pairs of photons in a state dependent on the configuration of the analyzing–detecting apparatus. As is well known, no Bell-type inequality can be obtained in such a case, and the EPR paradox does not exist.

3.6. Negative Probabilities

The idea of negative probabilities has been entertained in different times by physicists such as Dirac and Feynman. In 1942 Dirac expressed the opinion that

> Negative energies and probabilities should not be considered as nonsense. They are well-defined concepts mathematically, like a negative sum of money, since the equations which express the important properties of energies and probabilities can still be used when they are negative. Thus negative energies and probabilities should be considered simply as things which do not appear in experimental results (Dirac, 1942).

More recently Feynman (1982) has stated that the only difference between a probabilistic classical world and the quantum world "is that somehow or other it appears as if the probabilities would have to go negative . . . ."

Following these ideas, a "negative-probability solution" of the EPR paradox has been proposed by Mückenheim (1982). In order to understand the logical possibility of solving the EPR paradox by extending the range of variation of probabilities, we should remember that in the proofs of Bell's inequality the implicit assumption is always made that probabilities (and frequencies in ensembles) are positive and not larger than one. For example, in Wigner's proof of Bell's inequality the probabilities $\omega(s, s'; t, t')$ were introduced [see equation (65)] which were, by definition, positive and not larger than unity. Similarly, the proof based on factorizable probabilities used, in an essential way, the inequalities

$$0 \leq x, x', y, y' \leq 1$$

where x, x', y, and y' were later to be identified with probabilities. In both examples, if these conditions are relaxed, Bell's inequality no longer has any validity.

In view of these considerations, it is perhaps not surprising that Mückenheim could build a negative-probability *local* hidden-variable model that reproduces all the predictions of quantum theory for the "singlet" state of two spin-$\frac{1}{2}$ particles.

The two particles have spin vectors **S** for the first one and $-$**S** for the second one, where **S** is assumed to have a random distribution over the sphere of radius $(\sqrt{3}/2)\hbar$, in a statistical ensemble of such pairs. The length $(\sqrt{3}/2)\hbar$ is chosen, of course, in such a way as to reproduce the quantum-mechanical eigenvalue of $\mathbf{S}^2$, which is $\frac{3}{4}\hbar^2$. If $\hat{\mathbf{a}}$ is a unit vector, the projection of **S** over $\hat{\mathbf{a}}$ satisfies

$$-(\sqrt{3}/2)\hbar \leq \mathbf{S} \cdot \hat{\mathbf{a}} \leq +(\sqrt{3}/2)\hbar \qquad (151)$$

Next Mückenheim assumes that the probabilities, $\omega(\hat{\mathbf{a}}_+, \mathbf{S})$ and $\omega(\hat{\mathbf{a}}_-, \mathbf{S})$, of measuring $\mathbf{S} \cdot \hat{\mathbf{a}}$ and finding the positive and the negative eigenvalue, respectively, are linear functions of $\mathbf{S} \cdot \hat{\mathbf{a}}$, and that their expressions satisfying

$$\omega(\hat{\mathbf{a}}_+, \mathbf{S}) + \omega(\hat{\mathbf{a}}_-, \mathbf{S}) = 1 \tag{152}$$

are given by

$$\omega(\hat{\mathbf{a}}_+, \mathbf{S}) = 0.5 + \mathbf{S} \cdot \hat{\mathbf{a}}/\hbar \qquad \text{and} \qquad \omega(\hat{\mathbf{a}}_-, \mathbf{S}) = 0.5 - \mathbf{S} \cdot \hat{\mathbf{a}}/\hbar \tag{153}$$

Obviously, these probabilities can assume negative values because of equation (151).

In the case of an EPR pair, one can consider the case of correlated spin measurements along $\hat{\mathbf{a}}$ and $\hat{\mathbf{b}}$ for the first and the second particle, respectively. The correlation function is given by

$$P(\hat{\mathbf{a}}, \hat{\mathbf{b}}) = \frac{\hbar^2}{16\pi} \int d\Omega \, [\omega\hat{\mathbf{a}}_+, \mathbf{S}) - \omega(\hat{\mathbf{a}}_-, \mathbf{S})][\omega(\hat{\mathbf{b}}_+, -\mathbf{S}) - \omega(\hat{\mathbf{b}}_-, \mathbf{S})]$$

Substituting equations (153) into the previous expression and carrying out the integration one obtains

$$P(\hat{\mathbf{a}}, \hat{\mathbf{b}}) = -(\hbar^2/4)\hat{\mathbf{a}} \cdot \hat{\mathbf{b}}$$

which coincides with the quantum-mechanical correlation function for the singlet state. A local model is thus able to reproduce the quantum-mechanical violations of Bell's inequality, if negative probabilities are introduced.

It has also been shown that the introduction of complex probabilities into the EPR paradox can reconcile locality with the quantum-mechanical predictions (Ivanovic, 1978).

3.7. Variable Probabilities

The idea of "variable probabilities" as a solution of the EPR paradox starts from the evidence provided by the experiments performed with atomic photon pairs and assumes that the inequalities of the strong type (deduced from local realism *and* from additional assumptions) are violated. This is probably a correct assumption, even though there is a debate going on regarding the role of rescattering in the atomic source (see: Sanz and Sanchez Gomez, 1987 and the bibliography quoted therein).

The point of view adopted with this line of research is that not local realism but the additional assumptions should be blamed for the failure of the strong inequalities. One must then study local models of reality in which the logical negation of the additional assumptions is explicitly taken as true. The interesting models should thus imply the simultaneous validity of the following three statements:

1. *Given that a pair of photons emerge from two regions of space where two polarizers can be located, the probability of their joint detection from two photomultipliers depends on the presence and/or on the orientation of the polarizers (the CHSH property).*

2. *For a photon in the state λ, the probability of a detection with a polarizer in place on its trajectory can be larger than the detection probability with the polarizer removed (the CH property).*

3. *For a photon in the state λ, the sum of the detection probabilities in the "ordinary" and in the "extraordinary" beams emerging from a two-way polarizer depends on the polarizer's orientation (the GR property).*

A detailed survey of results and problems concerning this line of research is provided in Chapters 15, 16, 17, and 19, so the present comments will be minimal.

From a general point of view, one can maintain that local realism cannot be proved wrong by experiments designed for testing the strong inequalities. Only if weak inequalities could be tested, could a crucial confrontation between quantum theory and local realism finally take place. This appears unlikely in the foreseeable future as far as experiments with pairs of atomic photons are concerned. The situation is, however, better for some proposed particle-physics experiments and for experiments with pair of atoms, since detectors operate, in these cases, nearer to the ideal behavior.

Even in the case of low-efficiency detectors there are interesting investigations to be carried out, for example, by replacing the usual additional assumptions (the CHSH, CH, and GR assumptions) with more physical restrictions. After all, it is unlikely that the considerable disagreement between quantum theory and local realism for high-efficiency detectors becomes perfect agreement for low-efficiency detectors! For instance, it would be interesting to study the use of symmetrical functions for describing the detection processes of the two photons, since it has been shown by Caser (1984) that the quantum-theoretical predictions cannot, in such a case, agree with the factorizable probabilities of Clauser and Horne.

It is interesting to recall that the idea of variable probabilities presents itself as a natural consequence of probabilistic local realism, as shown in

Chapter 6. Also for this reason, it would be very interesting to carry out the experiments mentioned at the end of Section 2.10, which were:

1. Insertion of the atomic source of photon pairs in a magnetic field of about 200 to 300 gauss.
2. Use of $\lambda/4$ plates for systematic measurements of circular polarizations.
3. Search for small terms proportional to $\cos 4(a - b)$ in the correlation function $P(a, b)$.

Literature Cited and Bibliography

Aspect, A., P. Grangier, and G. Roger, 1981, *Phys. Rev. Lett.* **47**, 460.

Aspect, A., P. Grangier, and G. Roger, 1982, *Phys. Rev. Lett.* **49**, 91.

Aspect, A., J. Dalibard, and G. Roger, 1982, *Phys. Rev. Lett.* **49**, 1804.

Bell, J. S., 1965, *Physics* **1**, 195.

Bohm, D., 1951, *Quantum Theory*, Prentice-Hall, Englewood Cliffs.

Bohm, D., 1988, in: *Microphysical Reality and Quantum Formalism* (A. van der Merwe *et al.*, eds.), Reidel, Dordrecht.

Bohm, D., and Y. Aharonov, 1957, *Phys. Rev.* **108**, 1070.

Bohm, D., and B. J. Hiley, 1978, *Found. Phys.* **8**, 93.

Bohr, N., 1935, *Phys. Rev.* **48**, 696.

Buonomano, V., 1980, *Nuovo Cim. B* **57**, 146.

Buonomano, V., 1988, in: *Microphysical Reality and Quantum Formalism* (A. van der Merwe *et al.*, eds.), Reidel, Dordrecht.

Capasso, V., D. Fortunato, and F. Selleri, 1973, *Int. J. Theor. Phys.* **7**, 319.

Caser, S., 1984, *Phys. Lett. A* **102**, 152.

Clauser, J. F., 1976, *Phys. Rev. Lett.* **37**, 1223.

Clauser, J. F., 1976 *Nuovo Cim. B* **33**, 740.

Clauser, J. F., M. A. Horne, A. Shimony, and R. A. Holt, 1969, *Phys. Rev. Lett.* **23**, 880.

Clauser, J. F., and M. A. Horne, 1974 *Phys. Rev. D* **10**, 526.

Costa de Beauregard, O., 1977, *Nuovo Cim. B* **42**, 41.

Cramer, J. G., 1980, *Phys. Rev. D* **22**, 362.

Cufaro Petroni, N., 1985, in: *Open Questions in Quantum Physics* (G. Tarozzi and A. van der Merwe, eds.), Reidel, Dordrecht.

Datta, A., and D. Home, 1986, *Phys. Lett. A* **119**, 3.

Datta, A., D. Home, and A. Raychaudhuri, 1987, *Phys. Lett.*, A **123**, 4.

Davidon, W. C., 1976, *Nuovo Cim. B* **36**, 34.

de Broglie, L., 1974, *C. R. Acad. Sci. Paris*, **278**, 721.

Dirac, P. A. M., 1941, *Proc. R. Soc. London Ser. A* **180**, 1.

Dirac, P. A. M., 1952, *Nature* **169**, 702.

Einstein, A., B. Podolski and N. Rosen, 1935, *Phys. Rev.* **47**, 777.

Falciglia, F., A. Garuccio, *et al.*, 1983, *Lett. Nuovo Cim.* **37**, 66.

Falciglia, F., A. Garuccio, *et al.*, 1983, *Lett. Nuovo Cim.* **38**, 52.

Ferrero, M., and E. Santos, 1986, *Phys. Lett. A* **116**, 356.

Feynman, R. P., 1982, *Int. J. Theor. Phys.* **21**, 467.

Fortunato, D., 1976, *Lett. Nuovo Cim.* **15**, 289.

Freedman, S. J., and J. F. Clauser, 1972, *Phys. Rev. Lett.* **28**, 938.

Fry, E. S., and R. C. Thompson, 1976, *Phys. Rev. Lett.* **37**, 465.

Furry, W. H., 1936, *Phys. Rev.* **49**, 393; **49**, 476.

Garg, A., and N. D. Mermin, 1982, *Phys. Rev. Lett.* **49**, 901; **49**, 1220.

Garuccio, A., and V. Rapisarda, 1981, *Nuovo Cim. A* **65**, 269.

Garuccio, A., and V. Rapisarda, 1981, *Lett. Nuovo Cim.* **30**, 443.

Garuccio, A., and F. Selleri, 1980, *Found. Phys.* **10**, 209.

Garuccio, A., and F. Selleri, 1984, *Phys. Lett.* **A103**, 99.

Garuccio, A., V. L. Lepore, and F. Selleri, 1988, University of Bari preprint.

Ghirardi, G. C., A. Rimini and T. Weber, 1980, *Lett. Nuovo Cim.* **27**, 293.

Haji-Hassan, T., A. J. Duncan, *et al.*, 1987, *Phys. Lett., A* **123**, 110.

Holt, R. A., and F. M. Pipkin, 1974, University of Harvard, preprint.

Ivanovic, I. D., 1978, *Lett. Nuovo Cim.* **22**, 14.

Jauch, J. M., 1971, in: *Foundations of Quantum Mechanics* (B. d'Espagnat, ed.), Italian Physical Society, Course IL, Academic Press, New York.

Kasday, L., 1971, in: *Foundations of Quantum Mechanics* (B. d'Espagnat, ed.), Italian Physical Society, Course IL, Academic Press, New York.

Lepore, V. L., 1988, *Found. Phys.* (submitted).

Liddy, D. E., 1983, *J. Phys. A* **16**, 2703.

Marshall, T. W., E. Santos and F. Selleri, 1983, *Phys. Lett. A* **98**, 5.

Mückenheim, W., 1982, *Lett. Nuovo Cim.* **35**, 300.

Perrie, W., A. J. Duncan, H. J. Beyer, and H. Kleinpoppen, 1985, *Phys. Rev. Lett.* **54**, 1790.

Piccioni, O., and W. Mehlhop, 1988, in: *Microphysical Reality and Quantum Formalism* (A. van der Merwe *et al.*, eds.), Reidel, Dordrecht.

Rayski, J., 1979, *Found. Phys.* **9**, 217 (1981).

Rietdijk, C. W., 1980, *Found. Phys.* **10**, 403; *Found. Phys.* **11**, 783.

Sanz, A. L., and J. L. Sanchez-Gomez, 1987, *Europhys. Lett.* **3**, 519.

Schrödinger, E., 1935, *Proc. Camb. Phil. Soc.* **31**, 555.

Selleri, F., 1983, *Lett. Nuovo Cim.* **36**, 521.

Selleri, F., 1988, in: *Microphysical Reality and Quantum Formalism* (A. van der Merwe *et al.*, eds.), Reidel, Dordrecht.

Selleri, F., and G. Tarozzi, 1980, *Lett. Nuovo Cim.* **29**, 533.

Stapp, H. P., 1975, *Nuovo Cim. B* **29**, 270.

Stapp, H. P., 1977, *Nuovo Cim. B* **40**, 191.

Stapp, H. P., 1988, in: *Microphysical Reality and Quantum Formalism* (A. van der Merwe *et al.*, eds.), Reidel, Dordrecht.

Suppes, P., and M. Zanotti, 1976, in: *Logic and Probability in Quantum Mechanics*, Reidel, Dordrecht.

Sutherland, R. I., 1983, *Int. J. Theor. Phys.* **22**, 377.

Törnqvist, N., 1981, *Found. Phys.* **11**, 171.

Vigier, J. P., 1979, *Let. Nuovo Cim.* **24**, 258 and 265.

Wigner, E. P., 1970, *Am. J. Phys.* **38**, 1005.

Wu, C. S., and I. Shaknov, 1950, *Phys. Rev.* **77**, 136.

2

Are Faster-Than-Light Influences
Necessary?

HENRY P. STAPP

1. Faster-Than-Light Influences and Signals

The question of whether influences act instantaneously over finite distances is as old as modern science itself. Newton, when he proposed his universal law of gravitation, was asked how the postulated force was transmitted. He declined to frame a hypothesis regarding the mechanism, but declared that anyone who believed that the force could act over a finite distance without an intervening medium had a mind not fit for the contemplation of such matters. But in spite of Newton's conviction, no significant progress was made on the question of action-at-a-distance for two centuries. Then Maxwell propounded his theory for the analogous problem of electric and magnetic forces. This theory entailed the existence of light, and correctly predicted its velocity. It also entailed that no electric or magnetic influence of a sufficiently tangible kind could be transmitted faster than light. During the present century Einstein, generalizing this result, formulated the principle that no "signal" could propagate faster than light.

A *signal* is a special kind of influence. For our purpose it is enough to identify as a particular type of signal an influence that can be initiated by human choice, which controls a faraway response. For example, the choice of whether or not to depress a telegraph key controls, under appropriate conditions, whether or not a device will sound at the other end of the telegraph line.

HENRY P. STAPP • Lawrence Berkeley Laboratory, Berkeley, California 94720, United States.

The human choice and the response it controls can each be localized in a corresponding space–time region. A faster-than-light signal is a signal such that no point in the region of the response can be reached from any point in the region of the choice, without moving faster than light. Relativity theory postulates the nonexistence of faster-than-light signals, but does not necessarily impose an analogous requirement upon all other conceivable kinds of influences.

2. The Spin-Correlation Experiment

Einstein, Podolsky, and Rosen[1] argued in their famous 1935 paper that quantum theory did not provide a complete description of physical reality. Their argument was based on the analysis of a complicated experimental situation. David Bohm[2] later clarified the situation, by introducing a simpler experimental setup that exhibited all the essential features. Bohm's "spin version" of the EPR experimental arrangement is the basis of the present considerations. It is described in the introductory part of this book, and need not be further discussed here.

One point should, however, be emphasized. My starting point, like that of Einstein, Podolsky, and Rosen, is the assumption that the predictions of quantum theory, for the experiments under consideration, are valid. Some other authors start, instead, from the experimental data. Then questions concerning the counter efficiencies and the geometric details of those particular experiments that have already been performed become relevant. But here we start directly from the predictions of quantum theory. These predictions are, for the experiments under consideration, expressions of the core ideas of quantum theory: the possibility that they are seriously incorrect appears to me to be extremely unlikely.

One further stipulation should be made: in the experiment I am considering, the particles in the two initial beams of identical spin-$\frac{1}{2}$ particles initially scatter near the center of a spherical array of counters. This array has two escape holes that allow some pairs of particles, which have scattered at 90°, to escape. These escaping pairs i are numbered from 1 to n by fast electronics. The geometric arrangement is such that one particle from each pair i will enter a deflection device in a space–time region R_1, and the other particle from the pair i will enter a deflection device in a space–time region R_2. Detecting and recording devices are arranged so as to record in R_1, for each i from 1 to n, either $r_{1i} = +1$ or $r_{1i} = -1$, according to whether the particle from pair i is deflected "up" or "down" in R_1, relative to the preferred direction D_1 of the deflection device in R_1. The numbers $r_{2i} = \pm 1$ are similarly defined and recorded in R_2. A choice is made in R_1 between two alternative possible preferred directions, D_1' or D_1'', of the device in

R_1; and a choice is made in R_2 between two alternative possible preferred directions, D_2' or D_2'', of the device in R_2. The two regions R_1 and R_2 are space-like separated, which means that the information about the choice of setting made by the experimenter in R_1 does not have time to get to region R_2 before the results $r_{2i} = \pm 1$ are recorded there, and *vice versa*, without traveling faster than light.

The choices of the experimenters in R_1 and R_2 are considered, for the purposes of this analysis, to be two independent free parameters. This does not mean that these choices are, necessarily, literally free and nonpredetermined. It only means that one is allowed, within the specific context of the analysis of the implications of the quantum-theoretical predictions for these particular experiments, to treat the choices of the two experimenters as two independent free variables. These predictions are extracted from a quantum-theoretical representation of the state of the two particles. The mathematical formalism used in this calculation has no representation at all of the processes of making choices that are going on in the brains of experimenters: it involves only the states of various mechanical devices during those periods in which the particles are in, or near, these devices, and not how the devices came to be in those particular states. The EPR analysis is, in this respect, identical to that of quantum theory itself, which also treats the choices of the experimenters as independent free variables, within the context of the study of these experiments.

For our purposes the important prediction of quantum theory pertains to the correlation parameter defined by

$$c(r_1, r_2) = \frac{1}{n} \sum_{i=1}^{n} r_{1i} r_{2i} \tag{1}$$

Since each r_{1i} and r_{2i} is, according to the definitions given earlier, either $+1$ or -1, each term in the above sum is also either $+1$ or -1. Thus the largest and smallest possible values of this sum are n and $-n$. Consequently, c must lie between $+1$ and -1.

The relevant prediction of quantum theory is that if n is very large, then the value of c, computed according to equation (1), will be very close to

$$\bar{c}(D_1, D_2) = -\cos \theta(D_1, D_2) \tag{2}$$

where $\theta(D_1, D_2)$ is the angle between the preferred directions, D_1 and D_2, of the deflection devices in R_1 and R_2, respectively.

An important special case is that in which the directions of D_1 and D_2 are the same. Then the angle $\theta(D_1, D_2)$ is zero, and $\bar{c}(D_1, D_2)$ is -1. The only way in which the value of c, computed according to equation (1), can be -1 is for every term in the sum to be -1. This means that, for every value of i, the signs of r_{1i} and r_{2i} must be opposite: if the deflection in R_1 is "up" then the deflection in R_2 is "down," and *vice versa*. This means that if the directions of D_1 and D_2 are the same, then the deflections in R_1 and R_2 are perfectly "anticorrelated": an "up" deflection in one region is (almost) invariably accompanied by a "down" deflection in the other region, and *vice versa*.

In the spin version of the EPR argument the alternative possible directions, D_1' and D_1'', differ by 90°, but D_1' is the same as D_2', and D_1'' is the same as D_2''.

Einstein, Podolsky, and Rosen constructed a simple looking but actually rather subtle argument for the incompleteness of the quantum-mechanical description of physical reality. Before describing the EPR argument I shall describe a naive argument that appears to lead to the same conclusion.

3. The Naive Argument

Suppose that the choices of directions in R_1 and R_2 were such that D_1 and D_2 were the same. Then the deflections in R_1 and R_2 would be perfectly anticorrelated, as discussed above: each deflection "up" in one region would be paired with a deflection "down" in the other region, and *vice versa*.

There is a natural way to explain this perfect anticorrelation: for each pair i, the decisions as to whether the deflections will be "up" or "down" in each of the two regions R_1 and R_2 are already fixed at the time and location of the initial collision between the two particles of this pair. The information about these decisions can then be carried by the particles into the regions R_1 and R_2 where the deflections occur. In this way the perfect anticorrelation is understood in a completely natural way without requiring any faster-than-light transfer of information.

There is an alternative way of understanding the perfect anticorrelation. In this second scheme the decision as to whether the particle is deflected "up" or "down" in R_1 is made only during the processes of deflection, detection, and registration in R_1. In this case the information regarding this choice made in R_1 cannot get to R_2 without traveling faster than light. And the analogous statement holds also for the choice of result made in R_2: the information about this choice cannot get to R_1 without traveling faster than light. Thus there is, in this second scheme, no way to understand the existence of the perfect anticorrelation without allowing faster-than-light transfer of information. If one rules out such transfers then one also rules out the possibility that the choices of results are fixed in R_1 and R_2. One is led to the conclusion that the choices of results that will eventually appear in R_1 and R_2 must be determined by information contained in the intersection of the backward light cones from R_1 and R_2 (see Figure 1).

This natural solution leads, however, to a problem. In the experimental situation under consideration here the choice between the directions D_1' and D_1'' is not made until a time long after the original collision has taken place. And the same is true of the choice between D_2' and D_2''. Thus the information about which experiments will eventually be performed in the two regions, R_1 and R_2, is not available in the intersection of the backward light cones from R_1 and R_2. (Here it has been assumed that the information about the choice of experiment performed in either region can propagate only forward in time.) Consequently, the information residing in the intersection of the backward light cones from R_1 and R_2 must fix the results of *both* of the then-existing possibilities for the experiment that will eventually be chosen in each region.

This latter conclusion entails that the quantum-theoretical description is incomplete. For this conclusion amounts to admitting the predetermination of the results of several experiments, only one of which can actually be performed. And these alternative possibilities are, according to the quantum formalism, incompatible possibilities. Therefore quantum theory has no way to represent, simultaneously, a well-defined result for all of these alternative possible measurements. So if these various results were, in fact, simultaneously well-defined, then the quantum-theoretical description, being unable to represent all this information, would necessarily be incomplete. This is the naive form of the argument for the incompleteness of the quantum-theoretical description.

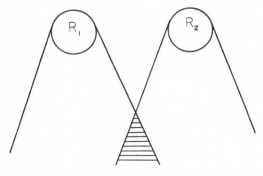

Figure 1. The naive argument. If information travels no faster than light, then information sufficient to determine the results in R_1 and R_2 must be contained in the shaded region, which is the intersection of the backward light cones from R_1 and R_2, in order to explain the exact anticorrelation of results in R_1 and R_2. But information about which experiment is performed in R_1 and R_2 is confined to the forward light cones from these regions, and hence is not present in the shaded region. Thus information sufficient to determine the results of all four possible measurements must be contained in the shaded region. This entails counterfactual definiteness, and hence the incompleteness of the quantum-theoretical description.

4. Orthodox Responses to the Naive Argument

To understand the response of the orthodox quantum theorist to the naive argument, it is necessary to recognize that the orthodox Copenhagen interpretation has two levels, here called the "strict" and "informal" levels. A principal element of both is that the quantum formalism is to be regarded as merely a set of rules for calculating connections between certain kinds of observations. In the words of Bohr:

> Strictly speaking, the mathematical formalism of quantum theory... merely offers rules of calculation for the deduction of expectations pertaining to observations obtained under well-defined conditions specified by classical physical concepts[3]

and

> ... the formalism does not allow pictorial representation along accustomed lines, but aims directly at establishing relations between observations obtained under well-defined conditions.[4]

The attitude that demands rigorous adherence to this point of view, and admits no discussion at all of what is "happening," is here called the "strict" interpretation.

Bohr's words are, however, fully compatible with the idea that our observations are observations of things that are actually "happening" in the external world, on the macroscopic level. But Bohr carefully avoided making specific ontological commitments about these "happenings." Heisenberg, on the other hand, was more forthcoming. He speaks of transitions from the "possible" to the "actual" and says:

> If we want to describe what happens in an atomic event, we have to realize that the word 'happens' can apply only to the observations, not to the state of affairs between observations. It applies to the physical, not the psychical act of observation, and we may say that the transition from the 'possible' to the 'actual' takes place as soon as the interaction of the object with the measuring device has come into play; it is not connected with the act of registration of the result in the mind of the observer. The discontinuous change in the probability function, however, takes place with the act of registration, because it is the discontinuous change in our knowledge in the instant of registration that has its image in the discontinuous change in the probability function.[5]

He also speaks of the probability function as representing "tendencies" or "potentia" for these actual happenings or events, which take place when the interaction of the object with the measuring device has come into play.

This idea, that the transition from the "possible" to the "actual" takes place when the interaction of the object with the measuring device comes into play, leads to the second scheme described above for understanding the existence of the strict anticorrelations. The problem with that second scheme was that it required faster-than-light transfer of information.

Heisenberg deals with this question in his 1929 book, *The Physical Principles of the Quantum Theory*. He discusses there the simpler situation in which a photon wave packet strikes a half-silvered mirror, and divides into two packets that move into separated regions. He then says:

> ... if now an experiment yields the result that the photon is, say, in the reflected part of the packet, then the probability of finding the photon in the other part of the packet immediately becomes zero. The experiment at the position of the reflected packet then exerts a kind of action (reduction of the wave packet) at the distant point occupied by the transmitted packet, and one sees that this action is propagated with a velocity greater than light. However, it is also obvious that this kind of action can never be utilized to transmit a signal so that it is not in conflict with the postulates of the theory of relativity.[6]

If, in accordance with Heisenberg's ideas, the probability function is regarded as representing tendencies for the macroscopic happenings, such as firings of particle counters, then the sudden "reduction of the wave packet" represents an abrupt change in tendencies at the distant point, and hence an immediate physical influence of some sort. The anticorrelation of the results in R_1 and R_2 that occur in the EPR-Bohm experiments can then easily be explained—and reconciled with the idea that the result of the experiment is not fixed until the interaction of the object and the device has come into play—by exploiting the abrupt change in tendencies in the distant region. The naive argument for the incompleteness of quantum theory is thereby dissolved, by considerations that were a standard part of quantum-theoretical thinking as early as 1929. However, this way of thinking admits the existence of faster-than-light influences. But the information that is transmitted faster than light pertains to nature's selections, rather than the experimenter's choices.

Heisenberg's approach admits the existence of faster-than-light influences, but not faster-than-light signals. However, the more usual strategy is to retreat to the strict interpretation, and simply refuse to discuss what is happening beyond what we do and what appears. Then the naive argument loses its force, because the quantum theorist simply refuses to recognize the categories of things upon which the argument is based.

5. The EPR Argument

The EPR argument is a modification of the naive argument. It is designed to invade the seemingly-impregnable position of the strict interpretation. The EPR argument meets the quantum theorist on his own ground, and on his own terms. For only in this way can the argument carry weight in the minds of these theorists.

The quantum theorist's terms are the acceptance of nothing other than: (1) the freedom of the experimenters to choose the experiments they will perform, and (2) the validity of the predictions of quantum theory.

Restriction imposed by locality requirements can be entertained, but restrictions on influences must be confined to the influences of human choices: human choices can be assumed to be localized in the regions in which they are made, and it can be assumed that no such choice made in one region can influence anything in a second region if the second region can be reached from the first only by traveling faster than light.

The aim of the EPR argument is to answer (in the negative) the question posed by the title of their paper: Can quantum-mechanical description of physical reality be considered complete? Thus they must give meaning to the words "physical reality." They do this by introducing their famous criterion of physical reality:

> If, without in any way disturbing a system, we can predict with certainty (i.e., with probability unity) the value of a physical quantity, then there exists an element of physical reality corresponding to that physical quantity.

Einstein, Podolsky, and Rosen discuss this criterion of physical reality, and argue that it accords with the ideas of both classical physics and quantum theory.

Let the two alternative possible physical quantities pertaining to region R_1 be denoted by r_1' and r_1'' respectively. And let the two alternative possible physical quantities pertaining to region R_2 be denoted by r_2' and r_2'' respectively. The prime and double prime relate to the superscripts on D_1' and D_1'', and on D_2' and D_2''. Thus each possible value for r_1' is a set of numbers $r_{1i}' = \pm 1$, etc., and the equalities $D_1' = D_2'$ and $D_1'' = D_2''$ lead, through equations (1) and (2), to

$$r_{1i}' = -r_{2i}' \qquad \text{for all } i \text{ if } D_1 = D_1' \text{ and } D_2 = D_2' \qquad (3)$$

and

$$r_{1i}'' = -r_{2i}'' \qquad \text{for all } i \text{ if } D_1 = D_1'' \text{ and } D_2 = D_2'' \qquad (4)$$

These equations represent, algebraically, the strict anticorrelations that were discussed above.

Einstein, Podolsky, and Rosen's argument (with appropriate replacements of symbols) is this:

> by measuring either r_1' or r_1'' we are in a position to predict with certainty [by using either equation (3) or equation (4)], and without in any way disturbing the system in R_2, either the value of r_2' or the value of r_2''. In accordance with our criterion of physical reality, in the first case we must consider r_2' as being an element of physical reality, in the second case r_2'' is an element of physical reality. Thus either r_2' or r_2'' is an element of physical reality depending on

whether we measure r_1' or r_1'' in region R_1. But maintaining that either r_2' alone or r_2'' alone is an element of physical reality, depending on what we measure in R_1, would make "the reality of r_2' and r_2'' depend upon the process of measurement carried out in region R_1. No reasonable definition of reality could be expected to permit this.[1]

Thus r_2' and r_2'' must be simultaneous elements of physical reality. This immediately entails, for reasons already explained, the incompleteness of the quantum-theoretical description of physical reality.

6. Bohr's Rebuttal

Rosenfeld[7] has described the reaction in Copenhagen that the EPR paper evoked. Bohr's initial attempts at an answer were not satisfactory, but after six weeks of effort his reply was completed. This reply,[8] however, was addressed mainly to the question of the *consistency* of the quantum-theoretical description, in the experimental situation discussed by Einstein, Podolsky, and Rosen, and not to their argument itself, which questioned not the consistency but rather the *completeness* of the quantum-theoretical description.

Bohr's rebuttal to the EPR argument itself was this:

From out point of view we now see that the wording of the above-mentioned criterion of physical reality proposed by Einstein, Podolsky, and Rosen contains an ambiguity as regards the meaning of the expression "without in any way disturbing a system." Of course there is in a case like that just considered no question of a mechanical disturbance of the system under investigation during the last critical stage of the measuring procedure. But even at this stage there is essentially the question of *an influence on the very conditions which define the possible types of predictions regarding the future behavior of the system.* Since these conditions constitute an inherent element of the description of any phenomenon to which the term "physical reality" can be properly attached, we see that the argumentation of the mentioned authors does not justify their conclusion that quantum-mechanical description is essentially incomplete.

The point of this rebuttal was to tie "physical reality" to what can be predicted about a system, and then to maintain that, since our predictions pertaining to region R_2 depend upon what we do in R_1, the physical reality in R_2 is disturbed by what we do in R_1.

The subtlety of Bohr's response testifies to the strength of the EPR argument: Bohr evidently found no simple, adequate reply. In the end he denied the EPR locality assumption that what we do in one region leaves the physical reality in the other region undisturbed. Heisenberg's approach also denies this assumption: he accepts the existence of faster-than-light actions that are not faster-than-light signals. The fact that the responses of both Bohr and Heisenberg effectively reject the EPR locality assumption suggests that what we have here is some subtle sort of faster-than-light connection. This is exactly what a deeper analysis, based on the work of Bell, appears to show.

7. Bell's Theorem

The problem of faster-than-light influences remained dormant in the minds of most physicists until it was stirred up in 1965 by a paper written by John Bell[9] who began his paper with a brief account of the EPR argument:

> Since we can predict in advance the result of measuring any chosen component of σ_2 by previously measuring the same component of σ_1 it follows that the result of any such measurement must actually be predetermined. Since the initial quantum mechanical wave function does not determine the result of an individual measurement, this predetermination implies the possibility of a more complete specification of the state. Let this more complete specification be effected by means of a set of parameters λ.

This version of the EPR argument introduces many elements that are not present in the carefully sculpted EPR argument itself: "in advance," "previously," "predetermined," and "parameters λ." Bell put these extra ideas together to form the idea of a deterministic hidden-variable theory. This theory he subjected to a locality requirement, which demanded that the results that would appear in each region, under either of the conditions that might be set up there, must be independent of the choice made by the experimenter in the other region, which is space-like separated from the first. He then showed that no such local deterministic hidden-variable theory could reproduce all the statistical predictions of quantum theory for spin-correlation experiments of the kind we have been discussing. A key innovation was to consider not just the predictions associated with settings of D_1 and D_2 at 0° and 90°, but to consider also some other appropriately chosen settings.

This result of Bell's did not immediately appear to have any great significance for the question of faster-than-light influences in nature, for the assumptions of determinism and of hidden variables seemed doubtful: orthodox quantum thinking explicitly rejects both of these ideas. However, both of these extra assumptions can, as we shall see, be stripped away.

8. Failure of Local Microrealism

Bell's theorem has been extended by Clauser *et al.*[10] to a broader class of hidden-variable theories, which accommodate stochastic elements. Much of the work of these authors is concerned with experimental tests, and hence with problems connected, for instance, to counter efficiencies. These considerations do not concern us, since we are accepting the validity of the quantum predictions.

Locality conditions for these stochastic hidden-variable theories were introduced by invoking semiclassical ideas[11,12] at the microscopic level. These ideas suggested a certain hidden-variable factorized form for the probabilities of coincidence counts.

In an effort to express in general terms the assumptions that underlie this proposed hidden-variable factorization property, Clauser and Shimony[12] have considered the concept of realism:

> Realism is a philosophical view according to which external reality is assumed to exist and have definite properties whether or not they are observed by somebody.

In the consideration of quantum theory it is necessary to distinguish macrorealism from microrealism. The Copenhagen interpretation of quantum theory is certainly compatible with macrorealism: it is compatible with the idea that our observations are observations of a macroscopic external reality created by myriads of macroevents of the kind discussed by Heisenberg. Of course, the strict Copenhagen interpretation enjoins us not to clutter quantum theory with superfluous ontological suppositions about the precise nature of these happenings. But it certainly allows their existence. Thus the general assumption of macrorealism does not take us outside the strict Copenhagen interpretation.

However, the ideas that underlie the justification of the hidden-variable factorization property of Clauser *et al.* are ideas about a microscopic level of reality that is totally alien to orthodox quantum-theoretical thinking. Theories that satisfy this hidden-variable factorization property should perhaps be called local-microrealistic theories, instead of local-realistic (or objective) theories, to emphasize the fact that they express certain ideas about the character of reality at the microscopic level that go far beyond the simple idea that external reality exists and has some well-defined (macroscopic) properties whether or not they are observed by anybody.

Clauser and Shimony have noted that an assumption of physical realism underlies the EPR argument. However, the EPR reality assumption is expressed by general principles that were designed to be compatible with orthodox quantum thinking and is thus totally different in character from the semiclassical ideas about a local microscopic space–time structure that underlie the hidden-variable factorization properties used by Clauser *et al.*

9. Failure of EPR Local Realism

The logical form of the EPR argument is this:

$$QM + (LOC + REALITY) \rightarrow CFD \qquad (5a)$$

and

$$\text{CFD} \to \text{QM IS INCOMPLETE} \qquad (5b)$$

where CFD stands for counterfactual definiteness. That is, from the assumption that the predictions of quantum theory are valid, and certain combined assumptions about locality and physical reality, Einstein, Podolsky, and Rosen conclude that the results of some unperformed (and mutually incompatible) experiments must be simultaneously well defined. This first conclusion, CFD, immediately entails, as noted by EPR, that the quantum-mechanical description is incomplete.

Simple arithmetic shows, as will be discussed presently, that[13]

$$\text{CFD} + \text{LOC} \to -\text{QM} \qquad (6)$$

That is, counterfactual definiteness plus locality entails the nonvalidity of the predictions of quantum theory. The combination of this result with the first part of the EPR argument, (5a), entails

$$\text{QM} + (\text{LOC} + \text{REALITY}) + \text{LOC} \to -\text{QM} \qquad (7a)$$

and hence, equivalently,

$$\text{QM} \to -(\text{LOC} + \text{REALITY})_{\text{EPR}} \qquad (7b)$$

where

$$(\text{LOC} + \text{REALITY})_{\text{EPR}} \equiv (\text{LOC} + \text{REALITY}) + \text{LOC} \qquad (7c)$$

The LOC that occurs in (6), which applies within a context in which CFD holds, is not identical to the LOC that occurs in the combined assumption (LOC + REALITY) that occurs in (5a). But it expresses, within this CFD context, the same basic EPR locality idea that nothing in R_i can be disturbed by what the experimenters do in R_j ($j \neq i$). This justifies the notation of (7c).

The result (7b), which is based on (6), invalidates the EPR argument, for it shows that its general assumptions are mutually incompatible. This purely logical argument eliminates the need for Bohr's epistemological rebuttal. It also yields a nonlocality result potentially far more interesting than the result of Bell, for it says that any theory that reproduces the predictions of quantum theory cannot satisfy the relatively weak locality and reality requirements that went into the EPR argument.

A key ingredient here is the one symbolized by (6). The meaning of this result is as follows: The CFD conclusion of the EPR argument, (5a), says that the results of the two alternative possible experiments that might be performed in R_2 are simultaneously well-defined, and the same conclusion holds for the results of the two alternative possible experiments in R_1. (A slight elaboration of the EPR argument is needed when three different angles 0°, 90°, and 135° are used, instead of only two). Since the results in all four alternative possible combinations of experimental conditions are then simultaneously well-defined, we may construct a table that shows these values. One conceivable possibility is shown in Table 1. Here $n = 8$, and the value of c, calculated according to equation (1), is shown. Also shown is the predicted value $\bar{c}$, calculated according to equation (2), for the following choices of the azimuthal angles that define the possible directions of D_1 and D_2:

$$D_1' \sim \theta_1' = 0°$$

$$D_1'' \sim \theta_1'' = 135°$$

$$D_2' \sim \theta_2' = 0°$$

$$D_2'' \sim \theta_2'' = 90°$$

Once counterfactual definiteness is established, the EPR locality idea can be formulated as the requirement that what would happen in either region, under either of the two alternative possible conditions that might be set up in that region, does not depend upon which of the two alternative possible experiments is chosen by the experimenters in the other region. This means that the set of results r_1' in R_1 does not depend upon the choice between D_2' and D_2'' made in R_2, and so on.

Table 1. Conceivable Set of Possibilities for the Results of the Four Alternative Possible Experiments[a]

i	(D_1', D_2')		(D_1', D_2'')		(D_1'', D_2')		(D_1'', D_2'')	
	r_1'	r_2'	r_1'	r_2''	r_1''	r_2'	r_1''	r_2''
1	+1	−1	+1	+1	−1	−1	−1	+1
2	−1	+1	−1	+1	+1	+1	+1	+1
3	−1	+1	−1	−1	−1	+1	−1	−1
4	+1	−1	+1	−1	−1	−1	−1	−1
5	+1	−1	+1	+1	−1	−1	−1	+1
6	−1	+1	−1	+1	+1	+1	+1	+1
7	+1	−1	+1	−1	−1	−1	−1	−1
8	−1	+1	−1	−1	+1	+1	+1	−1
	$c = -1$		$c = 0$		$c = 0.75$		$c = 0.25$	
	$\bar{c} = -1$		$\bar{c} = 0$		$\bar{c} = 0.707$		$\bar{c} = -0.707$	

[a] $c = (1/n) \sum_{i=1}^{n} r_{1i} r_{2i}$; $\bar{c}(\theta_1, \theta_2) = -\cos(\theta_1 - \theta_2)$.

In the first three pairs of columns the values have been arranged so that the value of c is close to the value $\bar{c}$ predicted by quantum theory. But the fourth case then shows a large disagreement. It is in fact easy to show[13,14] that this is always the case, for all values of n, provided the angles θ_1', θ_1'', θ_2', and θ_2'' are selected in the way shown: for this choice of these angles there is no conceivable possible arrangement of $r_{1i}' = \pm 1$, $r_{1i}'' = \pm 1$, $r_{2i}' = \pm 1$, and $r_{2i}'' = \pm 1$ that satisfies both the locality conditions and the quantum-theoretical predictions.

According to (7b), we may conclude from the mathematical result stated above that the assumptions that characterize EPR local realism are invalid. However, the significance of this conclusion is not totally clear. This is because the EPR assumptions of locality and reality are expressed in a manner not suited to our present aim, which is very different from that of Einstein, Podolsky, and Rosen. They wished to say something about "physical reality," and hence had to build their argument around a definition, or at least a criterion, of physical reality. And they wished to prove counterfactual definiteness in order to establish the incompleteness of the quantum-theoretical description. We are not interested in defining "physical reality," or in proving either counterfactual definiteness or the incompleteness of the quantum-theoretical description. Rather, we wish to clarify the result suggested by the independent considerations of Heisenberg and Bohr, namely, that the quantum aspects of nature are tied up to some subtle sort of faster-than-light connection. We shall need, therefore, to reformulate the results of this section in a way that circumvents the assumptions about "physical reality" that are not germane to our purpose. First, however, we shall introduce our criterion for the existence of an influence.

10. Criterion for the Existence of an Influence

In discussing the question of "influence" we are in a position similar to that of Einstein, Podolsky, and Rosen in their discussion of "physical reality": almost any symbol one writes, or word one uses, can, from the point of view of the strict orthodox interpretation, prejudice the issue. No models or words suggesting determinism or counterfactual definiteness can be invoked. One must base the considerations on general principles that are reasonable in their own right. The problem for Einstein, Podolsky, and Rosen was to set forth a reasonable criterion for "physical reality." Our problem is to set forth a reasonable criterion for the existence of an "influence."

Consider a theory that has a variable y, and an independent variable x. The idea that, within the structure imposed by this theory, the choice of the value of x does not influence y does not mean that within this structure the value of y must *necessarily* remain unchanged if the value of x is changed. For y might depend upon many things, and some of these, such as random variables, might not necessarily stay the same if x were changed. However, the idea that the choice of the value of x does not influence y does entail that, for each choice of the values of the other independent variables, the value of x can, within the constraints imposed by the theory, be varied over its domain without the value of y changing: the random variables *could* be left undisturbed. That is, in terms of values, if $\bar{x}$ is the set of all independent variables other than x, then for each value of $\bar{x}$ there is a value $y(\bar{x})$ such that the theory allows y to be held fixed at the value $y(\bar{x})$ as x varies over its entire domain (see Figure 2). If no such value $y(\bar{x})$ exists, then the theory forces y to vary as x is varied, and it cannot be said that, within the theory, the choice of x has no influence on y.

11. The Existence of Faster-Than-Light Influences

One principal aim here is to avoid the use of CFD. So we begin by specifying what CFD is, in the context of the specific situation under consideration here. CFD: "Regardless of which of the four alternative possible measurements is performed, the results of all four possible measurements are determinate." In more detail, "Regardless of which of the four alternative possible measurements is performed, nature, according to some underlying theoretical conception, fixes a quartet of values (r_1, r_2, r_3, r_4), in which r_m can be identified as the value that would be obtained as the result of the measurement if the measurement m were performed; i.e., all four values r_m are fixed or determined within nature, according to some underlying conception, even though only one of these values can be revealed by actual measurement."

> *Remark 1.* We distinguish here between physical theories, such as quantum theory and classical physics, and some perhaps less-completely-defined theoretical conception of the nature to which our physical theories are supposed to refer.
>
> *Remark 2.* No significance is supposed to be attached to tense, i.e., to the distinction between is, was, or will be performed: CFD is supposed to mean that all four values eventually become fixed, even though only at most one of the four alternative possible measurements can ever be performed.
>
> *Remark 3.* This CFD property can be decomposed into two parts, one referring to the unperformed measurements (strict CFD), and one referring to the performed experiment (definite result). Here we take CFD to be their symmetrically-stated combination.

Remark 4. This CFD property is the property that is supposed to be proved by the Naive Argument, and by the EPR argument.

Remark 5. Given CFD one can immediately deduce a contradiction with the predictions of quantum theory from the result[13] discussed in connection with Figure 2.

This CFD property may be contrasted with the property of "Unique Results." UR: "For each of the four alternative possible measurements m, if m is performed then nature must select some unique value for the result of this measurement m, and will never fix any values for the results that the remaining three measurements would have had if they had been performed."

The property UR is coordinated with quantum theory in the following way: QT: "For each of the four alternative possible measurements m, if m is performed then the unique value r that nature must, according to UR, select will, with probability greater than $1 - \varepsilon$, lie in a set $Q_m(\varepsilon)$. This set can, for any $\varepsilon > 0$, however small, be taken to be the set

$$Q_m(\varepsilon) = \{r;\ C(r) - \bar{C}m| < 0.01\} \tag{8}$$

by taking n, the number of pairs, sufficiently large."

Let S_m be the set of 4^n conceivable possible values of the result of measurement m. On the basis of UR we may, for any quartet (r_1, r_2, r_3, r_4) in $S_1 \otimes S_2 \otimes S_3 \otimes S_4$, contemplate the *conceivable possibility* that:

1. If we perform measurement $m = 1$, then nature will select the value r_1 for the result of this measurement, and will select no values for the results of the unperformed measurements $m \neq 1$.
2. If we perform measurement $m = 2$, then nature will select the value r_2 for the result of this measurement, and will select no values for the results of the unperformed measurements $m \neq 2$.
3. If we perform measurement $m = 3$, then nature will select the value r_3 for the result of this measurement, and will select no values for the results of the unperformed measurements $m \neq 3$.

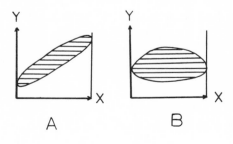

Figure 2. Conditions for influence. The shaded region indicates the region allowed by the theory, for some set of values of the independent variables other than x. If for some set of values of the independent variables other than x it is not possible to vary x over its entire domain without y changing (as in A, but not in B), then y is influenced by the choice of x.

4. If we perform measurement $m = 4$, then nature will select the value r_4 for the result of this measurement, and will not select any values for the results of the unperformed measurements $m \neq 4$.

This conceivable possibility has a certain similarity to CFD: both refer to a quartet (r_1, r_2, r_3, r_4) in $S_1 \otimes S_2 \otimes S_3 \otimes S_4$. But they are logically different. The existence of the conceivable possibility corresponding to the quartet (r_1, r_2, r_3, r_4) does not entail or rest upon the assumption that nature has selected, or in any way determined, this quartet of values: the conceivable possibility is defined by a conjunction of assertions, each of which applies only under the condition that the appropriate measurement is performed, and these four measurements cannot be performed simultaneously.

The CFD property contradicts quantum thinking, but the UR property does not. In fact, UR is completely in line with the quantum-theoretical idea that the values of the results of unperformed experiments are not physically well-defined. It is also in line with Heisenberg's idea that a transition from "possible" to "actual" takes place when the interaction between the quantum object and the measuring device comes into play. However, it goes slightly beyond the strict Copenhagen interpretation, which makes no explicit assumption that nature selects the unique value that appears to us. For example, the Copenhagen interpretation probably does not strictly contradict the many-worlds ontology, in which nature selects no unique values. (According to the many-worlds ontology, the uniqueness of the value that we perceive is a consequence of a limitation of human faculties, rather than a consequence of any singling-out of a unique value by nature herself).

Since no faster-than-light action occurs in the many-worlds ontology, any argument for the existence of faster-than-light actions must be predicated, in part, on an assumption that excludes the many-worlds ontology. In the present case this assumption is UR.

The EPR locality idea is this: nothing in R_i can be disturbed or influenced by what we can freely choose to do in the space-like-separated region R_j. To implement this idea we must deal, conceptually, with comparisons involving alternative possible conditions of measurement. Our aim is to introduce the necessary comparisons by using UR, and the framework of conceivable possibilities, in such a way as to completely avoid the CFD idea that nature fixes the quartet of results (r_1, r_2, r_3, r_4).

The argument proceeds in two steps. First a concrete, local non-CFD model will be considered; then its abstract essence will be extracted.

The concrete, local non-CFD model is constructed as follows: we suppose for any measurement m, if m is performed then nature will construct, in some local but nonpredetermined manner, a local mechanism that will pick a value for the result of measurement m. We suppose further that precisely the same set of constituents is used to construct this mechanism in all four alternative possible cases, so that it is physically impossible for mechanisms corresponding to any two of the four alternative possible measurements to exist simultaneously. Since the mechanism is constructed in a nondeterministic manner there are no plans or traces or other indications of what the mechanism would have been if some other choice of measurement had been made. Thus this concrete local model is manifestly non-CFD.

We consider now the process of the construction of the mechanism and, in view of the nondeterministic character of this process, the various conceivable possible mechanisms that might eventually be constructed. For each of these conceivable possible local mechanisms we can consider the part lying in $V^-(R_i)$, which is the backward light cone from R_i. According to the locality idea that causal influences propagate only forward in time and no faster than light, it is only this part of the mechanism lying in $V^-(R_i)$ that can have any influence upon the output in R_i. Similarly, it is only the part of the mechanism in $V^+(R_j)$, the forward light cone from R_j, that can have been influenced by the human input in R_j.

Owing to the space-like separation of R_1 and R_2 these two cones are, for $i \neq j$, disjoint:

$$V^-(R_i) \cap V^+(R_j) = \varnothing \qquad (i \neq j) \tag{9}$$

Thus, for any of the possible mechanisms that might be constructed, the parts that can influence the output in R_i cannot have been influenced by the input in the space-like separate region R_j.

We are concerned with these mechanisms only insofar as they can influence the outputs in R_1 or R_2. Thus the mechanisms can be considered to be confined to the region $V^-(R_1) \cup V^-(R_2)$. The part of this region that can contain parts of the mechanisms that can have been influenced by the inputs in R_1 and R_2 is

$$[V^-(R_1) \cup V^-(R_2)] \cap [V^+(R_1) \cup V^+(R_2)] \tag{10}$$

Again owing to the space-like separation of R_1 and R_2, this region consists of two disjoint regions, $V^-(R_1) \cap V^+(R_1)$ and $V^-(R_2) \cap V^+(R_2)$, such that the part of the mechanism in the first of these regions cannot influence the output in R_2 and cannot have been influenced by the input in R_2, and the part in the second region cannot influence the output in R_1 and cannot have been influenced by the input in R_1. Thus each of the conceivable possible mechanisms has two disjoint parts such that all of the influence of the input in R_1 is confined to one of these parts, $V^-(R_1) \cap V^+(R_1)$, and all of the influence of the input in R_2 is confined to the second part, $V^-(R_2) \cap V^+(R_2)$. Since the influences of the inputs in R_1 and R_2 are, within the mechanism that fixes the outputs in R_1 and R_2, wholly confined to disjoint parts of the mechanism, the influences of these two inputs must act independently upon the mechanism: neither input can influence the influence of the other.

According to our criterion of influence, the part of the mechanism that is not influenced by the input in R_1 must be *allowed* to remain unaltered as the input in R_1 is varied. This means that, for either condition that might be set up in R_2, the output in R_2, though nonpredetermined, and in general dependent upon the input in R_2, must be allowed to remain unaltered as the input in R_1 is varied. Similarly, for either condition that might be set up in R_1, the output in R_1, though nonpredetermined, and in general dependent upon the input in R_1, must be allowed to remain unaltered as the input in R_2 is varied. Physically, the requirement is that there be, conjunctively, no influence in either direction. Thus both conditions of noninfluence may be applied together.

The condition that, under either condition that might be set up in R_1 the nonpredetermined output that appears there be unaltered as the input in R_2 is varied restricts the allowed conceivable possibilities discussed above: the allowed quartets (r_1, r_2, r_3, r_4) are restricted to those in which the output in R_1 is independent of the input in R_2. The other condition of noninfluence imposes a similar condition. These two conditions together restrict the allowed quartets (r_1, r_2, r_3, r_4) to those in which the output in R_1 is independent of the input in R_2 and the output in R_2 is independent of the input in R_1.

Remark 6. This restriction on the allowed quartets is the same as the one that arises from locality in the case where CFD holds. Thus the mathematical proof used in that case can be carried over to show the incompatibility of locality and the predictions of quantum theory also in the present case.

Remark 7. Although the mathematical aspect of the incompatibility of locality and quantum theory is the same here as in the case where CFD holds, the physical basis is different. The present argument is based on a model that is, generally, in line with orthodox quantum thinking in that it explicitly excludes CFD.

Remark 8. The model does not entail or suggest that the "particles" exist in any objective or separate sense, or can be separated from the macroscopic devices.

Remark 9. The localizations involved in the argument are not microscopic: they can be millimeters or centimeters, or, in principle, even meters or kilometers.

The concrete model has several functions. The first is to cast in solid form the crucial property that under any one of the four alternative possible conditions nature must select a unique value for the result of the measurement that corresponds to the condition. This idea of a selection of unique value is essential, for it is the possible influence of the input in R_j upon this unique value appearing in R_i that is the subject of the analysis: in a many-worlds ontology there is no analogous uniquely defined value upon which to base the argument.

The second function of the concrete model is to exclude from the outset any possibility of satisfying CFD. The occurrence in the argument of a hidden assumption of CFD would eviscerate any claim of a violation of locality, since a violation of CFD would be far more plausible.

These first two functions played by the concrete model are simply to put in a visualizable form the content of UR.

A third function of the concrete model is to provide a concrete structure for describing the assumed faster-than-light limitation on causal influences. However, this concrete structure can be bypassed by formulating the locality condition directly as the requirement that, for each of the two possible values of i, and for each of the two possible inputs in R_i, the selection pertaining to the output in R_i that, according to UR, nature must make cannot be influenced by the input made in the space-like separated region R_j.

A fourth function of the concrete model is to establish the independence of the influences of the two inputs. However, this is a consequence of the disjointedness and space-like separation of the regions that can both be influenced by these two inputs and have a bearing on the results in R_1 and R_2. Abstractly, the independence of these two influences is expressed as the lack of any physical meaning to the order in which the events in R_1 and R_2 take place.

A fifth function of the concrete model is simply to allow one to think more concretely about the various alternative possibilities, as various alternative possible mechanisms. But there is no reason not to think abstractly, in the equivalent way, simply about the various alternative possibilities themselves. So, although the concrete model is perhaps helpful as an aid to thinking, it does not play any essential role that goes beyond the combination of UR and the idea that causal influences can propagate only forward in time, and no faster than light.

12. Analysis of a Counterclaim

A recent article[15] contains a purported proof that quantum theory is fully compatible with the demand that there be no faster-than-light influences of any kind:

$$QT \in \mathscr{L} \tag{11}$$

Here $\mathscr{L}$ is the set of theories that are "fully compatible with the demand that there be no faster-than-light influence of any kind". What is actually proved in Reference 15 is that quantum theory satisfies a certain "locality condition" that we here call KLC:

$$QT \in \mathscr{K} \tag{12}$$

where $\mathscr{K}$ is the set of theories that is consistent with KLC. What is needed to complete the proof of the claimed result (11) from the proved result (12) is that

$$\mathscr{K} \subset \mathscr{L} \tag{13}$$

But we shall exhibit a model theory (MT) that is contained in $\mathscr{K}$ but not in $\mathscr{L}$. This shows that KLC is too weak: it does not ensure full compatibility with the demand that there be no faster-than-light influences of any kind.

The condition KLC, restricted to our special situation, is this: for each of the four possible values of the pair (X_1, X_2), and for each pair of values (r', r'_2) that satisfies the statistical predictions of the theory under the condition that $[M_1(X_1), M_2(X_2)]$ is performed, there is some pair of values (r''_1, r''_2) such that: (1) (r'_1, r''_2) satisfies the statistical predictions of the theory under the condition that $[M_1(X_1), M_2(-X_2)]$ is performed, and (2) (r''_1, r'_2) satisfies the statistical predictions of the theory under the condition that $[M_1(-X_2), M_2(X_2)]$ is performed. The conditions for the four possible values of (X_1, X_2) are imposed *disjunctively*: the values r'_1, r'_2, r''_1, r''_2 occurring for each of the four alternative possible values of the pair (X_1, X_2) are allowed to be independently chosen quantities. This disjunctive form is to be contrasted with the conjunctive form obtained in Section 11.

Let us consider the model theory defined by (with $x_i = \pm 1$)

$$r_{1i} = \lambda_i \left[\frac{|X_1 + X_2| + (X_1 - X_2)}{2} \right] \tag{14a}$$

and

$$r_{2i} = \lambda_i \left[\frac{|X_1 + X_2| + |X_1 - X_2|}{2} \right] \tag{14b}$$

where each $\lambda_i = \pm 1$ is a random variable, with probability $\frac{1}{2}$ assigned to each of its two possible values. This model is blatantly nonlocal: the r_{1i} depend on X_2, and the r_{2i} depend on X_1. The observable averages are easily computed. Owing to the random variables λ_i the average values of r_1 and r_2 are zero:

$$\langle \bar{r}_2 \rangle = \left\langle \frac{1}{N} \sum_{i=1}^{M} r_{1i} \right\rangle = 0 \tag{15a}$$

and

$$\langle \bar{r}_2 \rangle = \left\langle \frac{1}{N} \sum_{i=1}^{N} r_{1i} \right\rangle = 0 \tag{15b}$$

The predicted correlation function $\bar{c}(X_1, X_2)$ is computed from equation (1), and its values are

$$\bar{c}(+1, +1) = +1 \tag{16a}$$

$$\bar{c}(+1, -1) = +1 \tag{16b}$$

$$\bar{c}(-1, +1) = -1 \tag{16c}$$

$$\bar{c}(-1, -1) = +1 \tag{16d}$$

The nonlocal character of any theory that satisfies equations (16) is easy to see. Consider first the two equations (16a) and (16c). Under the experimental conditions pertaining to equation (16a) the results appearing in R_1 and R_2 are perfectly correlated: the value of r_{1i} is always equal to the value of r_{2i}. But under the experimental conditions pertaining to equation (16c) the results appearing in R_1 and R_2 are perfectly anticorrelated: the value of r_{1i} is always equal to the negative of the value of r_{2i}. If one assumes that the results r_{2i} appearing in R_2 are undisturbed by what is done in R_1, then one can conclude that the two possible measurements in R_1 measure exactly the same thing, apart from a minus sign. That is, the two measurements in R_1 are related in the same way as the measurements performed by two Stern–Gerlach devices that are oriented in exactly opposite directions.

If we could find in nature two different possible measurement procedures that yielded correlation functions of the form (16a) and (16c), respectively, relative to a measurement performed in R_2, and if we could assume that the choice between the two measurement procedures in R_1 necessarily had no effect upon the results r_{2i} appearing in R_2, then we could certainly conclude that the two different possible measurements in R_1 were measuring exactly the same thing, apart from a minus sign.

But let us now change the experiment performed in R_2. Then we find from equations (16b) and (16d), by means of the same argument as before, that the same two measurements in R_1 are measuring exactly the same thing, *with no sign change*. Thus the two measurements in R_1 measure either exactly the same thing, or exactly the same thing with a reversed sign. And which of these two cases holds depends upon which experiment is performed in the other region.

This state of affairs is manifestly incompatible with the idea that there are no faster-than-light influences of any kind. Yet it is easy to show that the predictions (15) and (16) entail KLC. For to verify KLC it is sufficient to show that for any set of values r_{1i} satisfying (15a) [resp., values r_{2i} satisfying (15b)] there is some set of values r_{2i} [resp., r_{1i}] that satisfy both (15b) [resp., (15a)] and the appropriate correlation value from (16)]. But the two conditions (15) say that the set of r_{1i}'s must be half + and half −, and the same must be true for the set of r_{2i}'s. But then any correlation in the allowed range $1 \geq \bar{c} \geq -1$ can be readily constructed by making an appropriate matching of the +1's and −1's from the two sets.

It has therefore been shown that expression (13) is false. Hence the result (11) claimed to be proved in Reference 15 does not follow from the result (12) that is proved there.

ACKNOWLEDGMENT. I thank L. Ballentine, D. Bedford, J. Cushing, M. Nitschke, E. Ruhnau, and A. Sudbery for discussions and correspondence that contributed significantly to the form of this paper. This work was supported by the Director, Office of Energy Research, Office of High Energy and Nuclear Physics, Division of High Energy Physics of the U.S. Department of Energy under Contract DE-AC03-76SF00095.

References

1. A. Einstein, B. Podolsky, and N. Rosen, *Phys. Rev.* **47**, 777 (1935).
2. D. Bohm and Y. Aharonov, *Phys. Rev.* **103**, 1070 (1957).
3. N. Bohr, *Essays 1958–1962 on Atomic Physics and Human Knowledge*, p. 60, Wiley, New York (1963).
4. N. Bohr, *Atomic Physics and Human Knowledge*, p. 71, Wiley, New York (1958).
5. W. Heisenberg, *Physics and Philosophy*, Chap. 3, Harper and Rowe, New York (1958).
6. W. Heisenberg, *The Physical Principles of the Quantum Theory*, p. 39, Dover, New York (1930).
7. L. Rosenfeld, in: *Quantum Theory and Measurement* (J. A. Wheeler and W. H. Zurek, eds.), p. 142, Princeton Univ. Press, Princeton, N.J. (1983).
8. N. Bohr, *Phys. Rev.* **48**, 696 (1935).
9. J. S. Bell, *Physics* **1**, 195 (1964).
10. J. F. Clauser, M. A. Horne, A. Shimony, and R. A. Holt, *Phys. Rev.* **23**, 880 (1969).
11. J. F. Clauser and M. A. Horne, *Phys. Rev.* **10D**, 526 (1974).
12. J. F. Clauser and A. Shimony, *Rep. Prog. Phys.* **41**, 1881 (1978).
13. H. P. Stapp, *Phys. Rev.* **30**, 1303 (1971).
14. H. P. Stapp, *Am. J. Phys.* **53**, 306 (1985).
15. K. Kraus, in: *Symposium on the Foundations of Modern Physics* (P. Lahti and P. Mittelstaedt, eds.), World Scientific, Singapore (1985).

All the Inequalities of Einstein Locality

Augusto Garuccio

1. Introduction

The Einstein, Podolosky, and Rosen paradox[1] proved the existence of an incompatibility among three hypotheses: (1) quantum mechanics is correct; (2) quantum mechanics is complete; and (3) "elements of reality" exist associated with the atomic system that determine the result of a measurement eventually performed.

This paradox opened an as yet unsettled debate about which one of the three hypotheses should be discarded. For instance, Einstein proposed to admit that quantum mechanics is not complete, while Bohr[2] considered it unnecessary to suppose that "elements of reality" exist.

An important step forward in this argument was taken in 1965 by Bell,[3] who found an inequality that is violated by quantum mechanics but satisfied by every theory satisfying the third hypothesis of the EPR paradox.

This third hypothesis is usually known as Einstein locality and consists of the assumption that the results of measurements on atomic systems are determined by "elements of reality" (sometimes called hidden variables), which are associated with the systems being measured and, eventually, also with the measuring apparatuses, and which remain unaffected by measurements on other distant atomic systems. This determination can be either deterministic[4,5] in the true philosophical sense, or probabilistic,[6-8] in the sense that only the probabilities of the different outcomes of correlated measurements are fixed by the hidden variables.

AUGUSTO GARUCCIO • Department of Physics, University of Bari, 70216 Bari, Italy.

These two different approaches to the Einstein locality give rise to two different mathematical formulations, which we present in this first section.

Let us consider, therefore, two measurements in two space–time regions $R(1)$ and $R(2)$ with a space-like separation. In $R(1)$ the first observer measures the dichotomic observable $A(a)$, dependent on the instrument parameter a, while in $R(2)$ the second observer measures the similar observable $B(b)$. The measurements are performed on correlated systems, for instance, on two photons produced in the same atomic cascade. The only possible values of dichotomic observables are assumed to be ± 1.

In the deterministic approach (DA), the hidden variable λ determines the result of every measurement:

$$A(a, \lambda) = \pm 1 \qquad B(b, \lambda) = \pm 1$$

and the correlation function of the two measurements is

$$P(a, b) = \int d\lambda \, \rho(\lambda) A(a, \lambda) B(b, \lambda)$$

where $\rho(\lambda)$ is the normalized probability density.

In the probabilistic approach (PA), one introduces the probabilities $p_{\pm}(a, \lambda)$ that the results of measurements of $A(a)$ give ± 1, respectively, and the analogous probabilities $q_{\pm}(b, \lambda)$ for $B(b)$. Therefore, the correlation function becomes

$$P(a, b) = \int d\lambda \, \rho(\lambda) p(a, \lambda) q(b, \lambda)$$

where

$$p(a, \lambda) = p_{+}(a, \lambda) - p_{-}(a, \lambda)$$

and

$$q(b, \lambda) = q_{+}(b, \lambda) - q_{-}(b, \lambda)$$

It has been shown[9-11] that the deterministic approach and the probabilistic one are equivalent, in the sense that inequalities for linear combinations of correlation functions deducible from the former are true also in the latter, and *vice versa*.

The aim of this paper is to deduce systematically from Einstein locality and in the case of dichotomic observables,* all the possible inequalities which it can generate and we shall consider in turn, the deterministic and probabilistic approaches, following the historical development of the subject.

In Section 2 we will review the methods and the results based on the deterministic approach. In Section 3 we will present a general method for deducing all the possible inequalities of Einstein locality. The comparison between these inequalities and Bell's inequality will be developed in Section 4, where it will be shown that the physical content of Einstein locality is not fully expressed by Bell's inequality. In Section 5 new and more stringent inequalities for linear combination of joint probabilities are discussed.

2. The Deterministic Approach

2.1. First Method

The first general method for deducing inequalities from Einstein locality was introduced by Selleri,[14] and is based on the assumptions of deterministic local hidden-variable theory, and perfect total anticorrelation between the two measurements,

$$A(a, \lambda) = -B(a, \lambda) \tag{1}$$

If one considers the instrument parameters $a_1, a_2, \ldots, a_n$ (n odd), one can always write

$$\left[\sum_{i=1}^{n} \eta_i A(a_i, \lambda) \right]^2 \geq 1 \tag{2}$$

(where η_i are factors equal to ± 1 and can be chosen arbitrarily), because the quantity within square brackets can assume only the values n, $n - 2, \ldots, 1, -1, \ldots, 2 - n, -n$. Developing the square, applying the operation $\int d\lambda \, \rho(\lambda)$, and using condition (1), one obtains

$$\sum_{i>j} \eta_i \eta_j P(a_i, a_j) \leq \frac{n - 1}{2} \tag{3}$$

---

* For Bell-type inequalities deduced with multivalued observables refer to Baracca *et al.*[12] and Mermin.[13]

On the left-hand side (lhs) of equation (3) there are $\frac{1}{2}(n^2 - n)$ correlation functions, and 2^{n-1} different ways to choose the sign factors η_i in equation (2). In the case $n = 3$, it is possible to deduce the original Bell inequality, the Gutkowski–Masotto inequality (15) and the Clauser *et al.* inequality.[4]

2.1.1. Generalization to Arbitrary Coefficients

A generalization of Selleri's method can be obtained[16] by considering the inequality

$$[\alpha A(a, \lambda) + \beta A(b, \lambda) + \gamma A(c, \lambda)]^2 \geq \min(\pm\alpha \pm \beta \pm \gamma)^2 \qquad (4)$$

which is obviously always true since $A(x, \lambda) = \pm 1$ and the minimum on the right-hand side (rhs) is taken over all the possible sign choices.

We assume that the real parameters α, β, and γ are positive. Without loss of generality we can also assume that

$$\alpha \geq \beta \geq \gamma > 0 \qquad (5)$$

since the ordering of the three terms on the lhs of the inequality (4) is arbitrary, and the possibility that some of the coefficients α, β, and γ are equal to zero leads only to trivial inequalities.

It is easy to show that equation (5) implies

$$\min(\pm\alpha \pm \beta \pm \gamma)^2 = (-\alpha + \beta + \gamma)^2 \qquad (6)$$

Therefore, if one carries out the squares in equation (4), recalls that $A^2(a, \lambda) = A^2(b, \lambda) = A^2(c, \lambda) = 1$, divides by two, uses equations (1) and (6), multiplies by $\rho(\lambda)$, and integrates over λ, one obtains

$$-\alpha\beta P(a, b) - \alpha\gamma P(a, c) - \beta\gamma P(b, c) \geq -\alpha\beta - \alpha\gamma + \beta\gamma \qquad (7)$$

Dividing the previous inequality by $-\alpha\beta < 0$ and putting

$$x = \gamma/\beta \qquad \text{and} \qquad y = \gamma/\alpha \qquad (8)$$

one obtains

$$1 + x - y \geq P(a, b) + xP(a, c) + yP(b, c) \qquad (9)$$

which is our basic generalized inequality.

It is not difficult to prove that this inequality is the strongest possible one for three correlation functions. We note that from equation (5) it follows that

$$1 \geq x \geq y > 0 \qquad (10)$$

The generalized system of inequalities for three correlation functions, which can be obtained by repeating, with minor changes, the previous reasoning, is

$$1 + x - y \geqslant P(a, b) + xP(a, c) + yP(b, c)$$

$$1 + x - y \geqslant P(a, b) - xP(a, c) - yP(b, c)$$

$$1 + x - y \geqslant -P(a, b) + xP(a, c) - yP(b, c)$$

$$1 + x - y \geqslant -P(a, b) - xP(a, c) + yP(b, c)$$

(11)

It is interesting to see for which values of x and y the inequality (9) is violated by the quantum-mechanical correlation function for the singlet state

$$P_0(\theta) = -\cos \theta$$

The rhs R of the inequality (9) becomes

$$R_0 = -\cos(\theta_1 + \theta_2) - x \cos \theta_1 - y \cos \theta_2 \tag{12}$$

The partial derivatives of R_0 with respect to θ_1 and θ_2 can be shown to vanish only if

$$\cos \theta_2 = \frac{x^2 - y^2(1 + x^2)}{2xy^2} \tag{13}$$

$$\cos \theta_1 = \pm \frac{y^2 - x^2(1 + y^2)}{2x^2 y} \tag{14}$$

$$\cos(\theta_1 + \theta_2) = \pm \frac{x^2 y^2 - x^2 - y^2}{2xy} \tag{15}$$

Calling η and ξ the ± 1 factors on the rhs of equations (14) and (15), respectively, and substituting equations (13), (14), and (15) into equation (12), one obtains

$$R_0^{max} = \frac{1}{2xy} [x^2(\eta + \xi - 1) + y^2(\eta - \xi + 1) + x^2 y^2(-\eta + \xi + 1)]$$

It is not difficult to show that the largest value of R_0 is obtained, for all possible values of x and y, by taking $\eta = \xi = +1$.

Therefore the plus sign should be adopted in equations (14) and (15) and

$$R_0^{\max} = \frac{x^2 + y^2 + x^2 y^2}{2xy} \qquad (16)$$

The previous results hold for all those values of x and y for which the three cosines of equations (13), (14), and (15) lie within the physical region. It is easy to show that this results in the conditions

$$
\begin{aligned}
y/(1+y) &\le x \le y/(1-y) \qquad &\text{if } y < \tfrac{1}{2} \\
y/(1+y) &\le x \le 1 \qquad &\text{if } \tfrac{1}{2} \le y \le 1
\end{aligned}
\qquad (17)
$$

The low limit is obviously always satisfied, since in our case $x \ge y$. In Figure 1 the region in which these conditions are satisfied is shown as a hatched area. Within this region, the generalized inequality (9) is always violated [except on the borderline $x = y/(1 - y)$]. In fact the relation

$$R_0^{\max} > 1 + x - y$$

can be easily be transformed into

$$[x(1 - y) - y]^2 > 0$$

which is always true, except on the line $x = y/(1 - y)$, where the lhs vanishes.

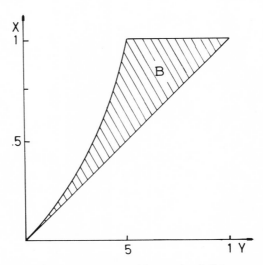

Figure 1. The region B of the x, y plane in which equation (12) has a maximum for physical values of the angles θ_1 and θ_2.

Outside the B region of Fiugre 1 the partial derivatives, with respect θ_1 and θ_2, of equation (12) never vanish in the physical region. This means that the maximum value of R_0 must be looked for on the boundary, that is, for

$$\theta_1 = 0, \pi \qquad \text{and} \qquad \theta_2 = 0, \pi \qquad (18)$$

Here, all the cosines on the rhs of equation (12) assume the values ± 1. Therefore it is not difficult to check that R_0 on the boundary can never be larger than $1 + x - y$. As a consequence, on the boundary, equation (18), the inequality (9) is never violated. It follows that for values of x and y lying *outside* the B region of Figure 1 the inequality (9) is never violated.

We thus reach a strikingly simple conclusion: the inequality (9) is violated every time its rhs presents a maximum as a function of θ_1 and θ_2 (and therefore in the whole of region B of Figure 1).

It is, instead, never violated if the rhs of the inequality (9) assumes its maximum on the boundary of the θ_1, θ_2 region, that is, if the values of x and y are outside the B region of Figure 1.

It can be shown that the violation of inequality (9) reaches its maximum value only at the point $x = y = 1$, that is, in the case of the traditional Bell inequality for three correlation functions.

Coming to Bell's inequality for four correlation functions, we note that from the first two and the second two inequalities of system (11), one deduces, respectively,

$$1 + x - y \geqslant P(a, b) + |xP(a, c) + yP(b, c)|$$
$$1 + x - y \geqslant P(a, b) + |xP(a, c) - yP(b, c)| \qquad (19)$$

Performing the substitutions $x' \to x$, $y' \to y$, and $c' \to c$ in the second inequality, and summing the two together, one obtains

$$2 + x + x' - y - y' \geqslant |xP(a, c) + yP(b, c)| + |x'P(a, c') - y'P(b, c')| \qquad (20)$$

which generalizes Bell's inequality to arbitrary coefficients x, x', y, and y' such that

$$1 \geqslant x \geqslant y > 0 \qquad \text{and} \qquad 1 \geqslant x' \geqslant y' > 0 \qquad (21)$$

Substituting the quantum-mechanical result $P_0(\theta) = -\cos \theta$ on the right-hand sides, R_1 and R_2 respectively, of equations (19) and maximizing over the $\hat{c}$ direction, keeping the $\hat{a}$ and $\hat{b}$ directions fixed, one obtains

$$R_1 = -\cos \theta + (x^2 + y^2 - 2xy \cos \theta)^{1/2}$$
$$R_2 = \cos \theta + (x^2 + y^2 + 2x'y' \cos \theta)^{1/2} \qquad (22)$$

where θ is the angle between $\hat{a}$ and $\hat{b}$.

As can be seen in the inequality (20), one needs now to calculate R_1 plus R_2 and obtain its maximum value for varying θ. A straightforward calculation leads to

$$(R_1 + R_2)^{\max} = (xy + x'y')^{1/2} \left(\frac{x}{y} + \frac{y}{x} + \frac{x'}{y'} + \frac{y'}{x'} \right)^{1/2} \qquad (23)$$

A numerical calculation for variable parameters x, y, x', and y' satisfying the relations (21) shows that the maximum violation of inequality (20) occurs for the values

$$x = y = x' = y' = 1$$

of the numerical parameters, and only for these values. Therefore we conclude, once more, that Bell's inequality is the strongest one *of this set* of inequalities deduced from Einstein locality.

2.2. Second Method—The Roy and Singh Approach

A first generalization of the method discussed in Section 2.1 was introduced by Roy and Singh,[17] who substituted the results of measurement of $A(a, \lambda)$ by the expectation values $\bar{A}(a, \lambda)$, but assumed that

$$\bar{A}(a, \lambda) = -\bar{B}(a, \lambda) \qquad (24)$$

Let us note that this generalization is merely formal, because if equation (24) holds for all λ, then it is possible to find a distribution function $\rho'(\lambda')$ such that

$$\bar{A}(a, \lambda) = \int \rho'(\lambda')d\lambda' \, A(a, \lambda, \lambda')$$

$$= -\int_{\Lambda'} \rho'(\lambda')d\lambda' B(a, \lambda, \lambda') = -\bar{B}(a, \lambda)$$

and thus return to the deterministic case.

A second generalization of the method introduced by Roy and Singh does not assume the equality (24), but starts from the basic inequality $(\sum_{i=1}^{n} \eta_i A_i + \sum_{j=1}^{m} \eta_j B_j)^2 \geq 1$ for $n + m$ odd and $A_i = B_j = \pm 1$. By performing the square and using $A_i^2 = B_j^2 = 1$, it is possible to deduce the inequality

$\mathcal{B}(A_i, B_j) \geq 1 - n - m$, where $\mathcal{B}(A_i, B_j)$ is a bilinear form in A_i and B_j. Then, for the bilinear form $\mathcal{B}[\bar{A}(a_i, \lambda), \bar{B}(b_j, \lambda)]$, the following inequality holds:

$$\mathcal{B}[\bar{A}(a_i, \lambda), \bar{B}(b_j, \lambda)] \geq 1 - n - m \qquad (25)$$

since the bilinear form $\mathcal{B}$ reaches its maximum and minimum on the boundary, i.e., for $\bar{A}(a_i, \lambda) = A_i = \pm 1 = B_j = \bar{B}(b_j, \lambda)$. Integrating the inequality (25) over λ, one obtains on the lhs a linear combination of correlation functions $P(a_i, b_j)$ plus functions like $\int \rho(\lambda) \, d\lambda \, \bar{A}(a_i, \lambda) \bar{A}(a_k, \lambda)$, which do not have a direct physical interpretation. It is necessary to combine different inequalities in order to eliminate the meaningless functions.

2.2.1. Some Interesting Results

Using the previous method, Roy and Singh deduced three interesting inequalities that provide restrictions on $P(a_i, b_j)$ which are not implied by Bell's inequality.

Before presenting these inequalities let us note that given a linear combination

$$\sum_{i=1}^{n} \sum_{j=1}^{m} C_{ij} P(a_i, b_j)$$

the coefficients C_{ij} define an $n \times m$ matrix which can be taken to represent completely the original linear combination.

The first inequality is

$$\sum_{i=1}^{4} \sum_{j=1}^{5} C_{ij}^1 P(a_i, b_j) \leq 6 \qquad (26)$$

where

$$C_{ij}^1 = \begin{pmatrix} 1 & 1 & 1 & 0 & 1 \\ 1 & 1 & -1 & 1 & 0 \\ 1 & 1 & 0 & -1 & -1 \\ 1 & -1 & 0 & 0 & 0 \end{pmatrix}$$

The second one is

$$\sum_{i=1}^{4} \sum_{j=1}^{7} C_{ij}^2 P(a_i, b_j) \leq 8 \qquad (27)$$

where

$$C_{ij}^2 = \begin{pmatrix} 1 & 1 & 1 & 1 & 0 & 0 & 0 \\ 1 & -1 & 0 & 0 & 1 & 1 & 0 \\ 1 & 0 & -1 & 0 & -1 & 0 & 1 \\ 1 & 0 & 0 & -1 & 0 & -1 & -1 \end{pmatrix}$$

and the third one is

$$\sum_{i=1}^{6} \sum_{j=1}^{8} C_{ij}^3 P(a_i, b_j) \leq 16 \tag{28}$$

where

$$C_{ij}^3 = \begin{pmatrix} 1 & 1 & -1 & -1 & 1 & 1 & 1 & 1 \\ 1 & 1 & 1 & 1 & -1 & -1 & 1 & 1 \\ 1 & 1 & 1 & 1 & 1 & 1 & -1 & -1 \\ 1 & -1 & -1 & 1 & 1 & -1 & 1 & -1 \\ 1 & -1 & 1 & -1 & -1 & 1 & 1 & -1 \\ 1 & -1 & 1 & -1 & 1 & -1 & -1 & 1 \end{pmatrix}$$

It is easy to show that inequalities (26), (27), and (28) provide restrictions on $P(a_i, b_j)$ not implied by Bell's inequality. Let us suppose, for example, that

$$P(a_4, b_2) = P(a_2, b_3) = P(a_3, b_4) = P(a_3, b_5) = 0$$

and the remaining $P(a_i, b_j)$ occurring in the equality (26) are all equal to $\frac{2}{3}$; then all the Bell inequalities involving these $P(a_i, b_j)$ are obeyed, but the inequality (26) is violated.

We shall return to inequality (27) in Section 5.

2.3. Third Method

In a 1979 paper, Roy and Singh[18] proposed new method for deducing a larger set of inequalities.

The fundamental assumption is the perfect anticorrelation between the two expectation values:

$$\bar{A}(a_i, \lambda) = -\bar{B}(a_i, \lambda) = x_i(\lambda)$$

Then, given a number N of settings of a measuring device $(a_1, a_2, \ldots, a_N)$ and a number N of integers (positive, negative, or zero) $(n_1, n_2, \ldots, n_N)$, one obtains

$$\sum_{i<j} n_i n_j P(a_i, a_j) = -\frac{1}{2} \int d\lambda \, \rho(\lambda) \left\{ \left[\sum_{i=1}^{N} n_i x_i(\lambda) \right]^2 - \sum_{j=1}^{N} n_i^2 x_i^2(\lambda) \right\} \quad (29)$$

Since the expression in braces is linear in each $x_i(\lambda)$, its minimum is reached when all the $x_i(\lambda)$ lie on the boundary; hence

$$\sum_{i<j}^{N} n_i n_j P(a_i, a_j) \leq \begin{cases} \dfrac{1}{2}\left(\displaystyle\sum_{i=1}^{N} n_i^2 - 1 \right) & \text{if } \displaystyle\sum_{i=1}^{N} n_i = \text{odd} \\[2ex] \dfrac{1}{2} \displaystyle\sum_{i=1}^{N} n_i^2 & \text{if } \displaystyle\sum_{i=1}^{N} n_i = \text{even} \end{cases} \quad (30)$$

The inequalities (3) of Section 2.1 are special cases of the inequalities (30) with $n_i = 0$ or ± 1. For example, for $N = 6$, $n_1 = 2$ and $n_i = 1$ ($i = 2, \ldots, 6$), we obtain the following strong inequality:

$$\sum_{i,j=1}^{6} C_{ij} P(a_i, b_j) \leq 4$$

where

$$C_{ij} = \begin{pmatrix} 0 & 2 & 2 & 2 & 2 & 2 \\ 0 & 0 & 1 & 1 & 1 & 1 \\ 0 & 0 & 0 & 1 & 1 & 1 \\ 0 & 0 & 0 & 0 & 1 & 1 \\ 0 & 0 & 0 & 0 & 0 & 1 \\ 0 & 0 & 0 & 0 & 0 & 0 \end{pmatrix}$$

3. The Probabilistic Approach

3.1. General Method

In the usual probabilistic approach, the hidden variable λ determines the probability $p_\pm(a, \lambda) [q_\pm(b, \lambda)]$ that the result of the measurement of $A(a) [B(b)]$ gives ± 1, respectively (a more detailed and critical analysis of this definition of probability is given in Chapter 6).

If $\rho(\lambda) \geq 0$ is the probability density of the variable λ, the correlation function is

$$P(a, b) = \int d\lambda \, \rho(\lambda) p(a, \lambda) q(b, \lambda) \tag{31}$$

where

$$p(a, \lambda) = p_+(a, \lambda) - p_-(a, \lambda)$$
$$q(b, \lambda) = q_+(b, \lambda) - q_-(b, \lambda) \tag{32}$$

We wish to stress that λ, in the previous equation, is just a general symbol, which could cover several different "additional parameters." In fact, a given *local*-realistic theory of the type given by equations (31) and (32) above is specified by the following:

1. Number and nature of the additional parameters $\lambda_1, \lambda_2, \ldots$.
2. The functional dependence of $p(a, \lambda_1, \lambda_2, \ldots)$ and $q(b, \lambda_1, \lambda_2, \ldots)$ on these parameters.
3. The probability density $\rho(\lambda_1, \lambda_2, \ldots)$.

Two theories are different if they differ in any one of the three previous specifications. Denoting again by a single symbol the parameters $\lambda_1, \lambda_2, \ldots$, we can say, more simply, that two local-realistic theories are different if they are based on different functions $p(a, \lambda)$ and $q(b, \lambda)$ and/or different probability densities $\rho(\lambda)$.

The only interesting inequalities deduced from Einstein locality are those which hold true for all conceivable local-realistic theories of the type given by equations (31) and (32). Obviously, it is not possible today to say which one (if any) of the infinitely many theories based on Einstein locality is the correct one. Therefore, inequalities deduced from a particular theory (or from a particular set of theories) are not interesting.

The following lemma permits us to deduce inequalities which are true for all conceivable local-realistic theories.

Lemma. Given a real number M, the inequality

$$\sum_{ij} C_{ij} P(a_i, b_j) \leq M \tag{33}$$

can be true for all conceivable local-realistic theories if and only if the inequality

$$\sum_{ij} C_{ij} p(a_i, \lambda) q(b_j, \lambda) \leq M \tag{34}$$

is true for arbitrary values of λ and for arbitrary dependence of p and q on their arguments.

Proof. The inequality (34) is a consequence of inequality (33) since, among all the conceivable local-realistic theories, there are those in which the density function ρ is a delta function $\delta(\lambda - \lambda_0)$ and therefore

$$\sum_{ij} C_{ij} \int d\lambda \; \delta(\lambda - \lambda_0) p(a_i, \lambda) q(b_j, \lambda) \leq M$$

implies that

$$\sum_{ij} C_{ij} p(a_i, \lambda_0) q(b_j, \lambda_0) < M$$

where λ_0, being arbitrary, can assume any value. Conversely, if the inequality (34) is true for arbitrary λ and arbitrary dependence of $p(a_i, \lambda)$ and $q(b_j, \lambda)$ on their arguments, it is sufficient to multiply it by $\rho(\lambda)$ and integrate it in order to obtain inequality (33) as true for an arbitrary local-realistic theory. The proof is thus completed.

Of course, the previous lemma does not specify the value of M and we shall call an inequality of the type (33) "trivial" if

$$M \geq \sum_{ij} |C_{ij}|$$

In fact the lhs of inequality (33) cannot be larger than the rhs of the previous inequality since every correlation function $P(a_i, b_j)$ has, by definition, a modulus not exceeding one.

Our aim is therefore *the determination of nontrivial inequalities satisfied by all the conceivable local hidden-variables theories.*

Obviously, the most stringent inequality is found when M is taken equal to the maximum value of the lhs of the inequality (34):

$$M = M_0 \equiv \max \left\{ \sum_{ij} C_{ij} p(a_i, \lambda) q(b_j, \lambda) \right\} \tag{35}$$

If we are interested in theory-independent inequalities, then we must choose the maximum for all the conceivable dependences of p and q on λ. Among them, there is independence of λ, for which inequality (34) becomes

$$\sum_{ij} C_{ij} p(a_i) q(b_j) \leq M \tag{36}$$

The lhs of the inequality (36) is linear in $p(a)$ and $q(b)$, therefore its maximum M is found in principle on the boundary, namely, at one of the vertices of the hypercube C in the multidimensional space having $p(a_i)$ and $q(b_j)$ as Cartesian coordinates, i.e.,

$$M_0 = \max\left\{\sum_{ij} C_{ij}\xi_i\eta_j\right\} \tag{37}$$

where $\xi_i = \pm1$, $\eta_j = \pm1$, and the maximum is taken over all the possible choices of ξ_i and η_j.

It is now easy to show that M coincides with M_0. In fact, the lhs of the inequality (36) is limited, by any particular λ-dependence, to some curve or surface entirely within the hypercube C. The value of the lhs of inequality (36) itself depends only on the values of $p(a_i, \lambda)$ and $q(b_j, \lambda)$ for given coefficients C_{ij}, that is, to say, on the considered point P of the hypercube C with coordinates q_i and p_j, whatever the particular values of λ, a_i, and b_j which allow one to reach the point P. The largest value of the lhs of equation (35) is therefore, in all cases, at one of the vertices of hypercube C, where $p(a_i, \lambda) = \xi_i = \pm1$ and $q(b_j, \lambda) = \eta_j = \pm1$ (see Figure 2).

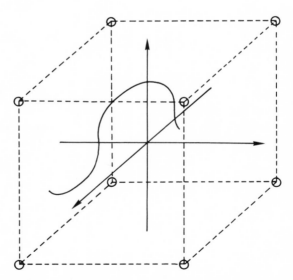

Figure 2. Given a hidden-variable theory, the functions $p(a_i, \lambda)$ and $q(b_j, \lambda)$ describe a surface or a line, inside the hypercube C, having $n + m$ dimensions, where n is the number of a_i and m the number of b_j. Since the lhs of inequality (33) is linear in $p(a_i, \lambda)$ and $q(b_j, \lambda)$, the maximum for all conceivable hidden-variable theories must lie on one of the vertices of C.

3.2. *Three Useful Theorems*

In this section we shall prove three theorems which narrow the set of inequalities of the type given by relations (33) and (37), which can be considered to be of physical interest.

Theorem 1. Every inequality whose coefficients C_{ij} have factorizable signs is trivial.

In fact, if

$$C_{ij} = |C_{ij}| \mu_i \nu_j \qquad (\mu_i = \pm 1, \quad \nu_j = \pm 1)$$

one has from equation (37)

$$M_0 = \max_{\xi_i \eta_j} \left\{ \sum_{ij} |C_{ij}| \mu_i \nu_j \xi_i \eta_j \right\} = \sum_{ij} |C_{ij}|$$

since it is possible to choose $\xi_i = \mu_i$ and $\eta_i = \nu_j$ for all i and j.

Theorem 2. If an argument a_i or an argument b_j appears only once, the inequality can be reduced to a more elementary one.

In fact, there is a one-to-one correspondence between experimental parameters a_i and sign factors ξ_i, and between b_j and η_j. As a consequence, if we suppose that a_1 enters only once then the sign factor ξ_1 enters only once, and

$$M_0 = \max_{\xi_i \eta_j} \left\{ C_{1l} \xi_1 \eta_l + \sum_{j=2}^{n} \sum_{j=1}^{m} C_{ij} \xi_i \eta_j \right\}$$

$$= |C_{1l}| + \max_{\xi_i \eta_j} \left\{ \sum_{i=2}^{n} \sum_{j=1}^{m} C_{ij} \xi_i \eta_j \right\}$$

since one can always choose η_l in such a way that $C_{1l} \xi_1 \eta_l = |C_{1l}|$. In this case the inequality

$$\sum_{ij} C_{ij} P(a_i, b_j) \leq M_0$$

can be reduced to the more elementary one

$$\sum_{i=2}^{n} \sum_{j=1}^{m} C_{ij} P(a_i, b_j) \leq M_0 - |C_{1l}|$$

Theorem 3. If the lhs of the inequality (33) can be split into two parts such that no argument a_i or b_j is common to two correlation functions belonging to each of these two parts, then the inequality deducible from Einstein locality can be reduced to two more elementary inequalities.

Proof. The correspondence between parameters a_i and b_j, and sign factors ξ_i and η_j, ensures that if the lhs can be split into two parts, with no argument a_i or b_j in common, then also the rhs can be split into two parts having no sign factor ξ_i or η_j in common. Hence, the original inequality can be written as the sum of two simpler one.

Before concluding this section it is important to stress that the "singlet" case, defined by $P(a_i, a_i) = -1$ for all a_i, requires special treatment. In fact, in this case, exactly the same technique that we used above leads to

$$\sum_{ij} C_{ij} P(a_i, a_j) \leq \max_{\xi} (-C_{ij} \xi_i \xi_j) \qquad (38)$$

The presence of two ξ factors gives new results with respect to the previous case and, in particular permits us to obtain all the results discussed in Section 2 that are deduced with the total anticorrelation assumption.

4. The Physical Content of Einstein-Locality Inequalities

4.1. First Results of a General Method

In this section we will discuss the results of the previous general method for deducing inequalities, and we will compare these inequalities with that of Bell.

From Section 2.2.1 we know that there are at least three inequalities which provide on $P(a, b)$ restrictions not implied by Bell's inequality.

Now, let us consider whether other inequalities exist that are simpler than the inequalities (26), (27), and (28), and not implied by Bell's inequality. In order to answer this question, let us use the previous method for deducing Einstein-locality inequalities.

It is easy to prove, using Theorem 2, that linear combinations of three correlation functions give only trivial inequalities.

Let us consider, then, the case of four correlation functions. By equation (37) the maximum for Einstein locality is obtained for

$$\sum_{i,j=1}^{2} C_{ij} P(a_i, b_j) \leq M_0 = \max \left\{ \sum_{i=1}^{2} \sum_{j=1}^{2} C_{ij} \xi_i \eta_i \right\}$$

$$= \max\left\{ \sum_{i=1}^{2} \xi_i \sum_{j=1}^{2} |C_{ij}| \sigma_{ij} \eta_j \right\}$$

$$= \max\left\{ \sum_{i=1}^{2} \left| \sum_{j=1}^{2} |C_{ij}| \sigma_{ij} \eta_j \right| \right\}$$

$$= \max\left\{ \left| |C_{11}| + |C_{12}|\xi \right| + \left| |C_{22}| + |C_{21}|\sigma\xi \right| \right\}$$

where σ_{ij} is the sign of C_{ij}, $\rho = \sigma_{11} \cdot \sigma_{12} \cdot \eta_1 \cdot \eta_2$, and $\sigma = \sigma_{11} \cdot \sigma_{12} \cdot \sigma_{21} \cdot \sigma_{22}$.

If $\sigma = +1$ (or, equivalently, the signs are factorizable) the result is trivial:

$$M_0 = \sum_{ij} |C_{ij}|$$

If, instead, $\sigma = -1$ then

$$M_0 = \sum_{ij} |C_{ij}| - 2 \min_{lm} |C_{lm}| \tag{39}$$

From equation (39) it is possible to deduce that, in the case of four correlation functions,

$$M_0 \geq \tfrac{1}{2} \sum_{ij} |C_{ij}|$$

where the equality holds only when C_{ij} is constant for all i and j, or, equivalently, for Bell's inequality; therefore we conclude that in the case of four correlations no inequality stronger than Bell's inequality exists. This result agrees completely with the one obtained in Section 2.1.1 using the deterministic approach.

In the case of five correlation functions, at least one of the elements a_i or b_j must appear only once and therefore, because of Theorem 2, all the inequalities are trivial.

The case $n = 6$ is more complicated and only the result will be presented here. One obtains

$$\sum_{i=1}^{2} \sum_{j=1}^{3} C_{ij} P(a_i, b_j) \leq M \tag{40}$$

where

$$M = \max_{\rho_1 \rho_2}\left\{ \left| |C_{11}| + |C_{12}|\rho_1 + |C_{13}|\rho_2 \right| + \left| |C_{21}| + |C_{22}|\rho_1\sigma_1 + |C_{23}|\rho_2\sigma_2 \right| \right\} \tag{41}$$

where ρ_1 and ρ_2 are sign factors to be chosen in such a way as to maximize M and

$$\sigma_1 = \sigma_{11}\sigma_{12}\sigma_{21}\sigma_{22}, \qquad \sigma_2 = \sigma_{11}\sigma_{13}\sigma_{21}\sigma_{23}$$

A particular application of inequality (40) is the following:

$$P(a_1, b_1) + P(a_2, b_2) - P(a_1, b_3) + P(a_2, b_1) + P(a_2, b_2) - P(a_2, b_3) \leq 2$$

$$(42)$$

as can easily be checked.

4.2. The Superinequalities

Before proceeding to study inequalities with larger numbers of correlation functions, it would be useful to introduce the following definition: Given an Einstein-locality inequality, an associated Bell inequality is a Bell inequality which contains correlation functions that also appear in the original inequality.

The following theorem gives a powerful method for analyzing Einstein-locality inequalities in order to single out those that provide restrictions on correlation functions not implied by Bell's inequality.

Theorem 4. Given a linear combination $L = \sum_{ij} C_{ij} P(a_i, b_j)$, if

$$\bar{M} \equiv \sum_{ij} |C_{ij}|$$

and if

$$M_0 = \max_{\xi_i \eta_j} \left\{ \sum_{ij} C_{ij} \xi_i \eta_j \right\}$$

is the maximum value of L allowed by Einstein locality, then the inequality $L \leq M_0$ implies the existence of physical restrictions not contained in any Bell inequality provided that

$$M_0 < \tfrac{1}{2} \bar{M} \tag{43}$$

Proof. Let us consider the $n \times m$ space in which the $P(a_i, b_j)$ are located on the axes, and the vector $\bar{P} = \{\bar{P}(a_i, b_j)\}$, which maximizes the linear combination L. The components of this vector all have, of course, modulus one and their signs are the same as those of the corresponding C_{ij}. If we consider the new vector

$$\{\bar{P}'(a_i, b_j)\} = \{\bar{P}(a_i, b_j)\} * 0.5 = \{0.5 * \bar{P}(a_i, b_j)\}$$

its components all satisfy the associated Bell inequalities (i.e., each $|\bar{P}'(a_i, b_j)| = 0.5$), but it results in

$$\sum_{ij}^{n,m} C_{ij} \bar{P}'(a_i, b_j) = \tfrac{1}{2} \bar{M} > M_0$$

We will call the Einstein-locality inequalities satisfying the condition (43) "superinequalities."

In what follows we will utilize the method given by Theorem 4 for studying the Einstein-locality inequalities, first in the case of three different directions for a and for b (3×3), and then in the case of four directions each (4×4).

Indeed equations (37) and (43) allows us to use a computational method to solve the problem; details of this method are given in the Appendix for the case 4×4.

4.2.1. The Case 3×3

The case of three values for a and three (or more) for b was analyzed by Garg,[20] who proved that the necessary and sufficient condition for the validity of Einstein locality is that every associated Bell inequality be satisfied.

Using the previous computational method, we can analyze the 3×3 linear combinations with the C_{ij} integer's in the range $\{-2, +2\}$; the result is that for all these 3×3 linear combinations, the maximum M_0 for Einstein locality is equal to or larger than one half of the maximum possible. This confirms the result of Garg.

4.2.2. The Case 4×4

An interesting result is obtained in the case of four different directions each, for a_i and b_j.

In this case the number of correlation functions is 16 and the number of different 4×4 matrices with integer coefficients C_{ij} in the range $\{-2, +2\}$ is 5^{16}. We analyzed, using the previous method, only 13,500,000 matrices (equivalent to 0.009% of the total) thereby obtaining 1050 superinequalities. Since the region analyzed has no special features, it is probably possible to generalize the result and conclude that an analysis of the complete set of 4×4 inequalities would give nearly 10^7 superinequalities of the stated type.

It is easy to prove that, given a 4×4 matrix, it is possible, by permuting or changing the signs of rows and/or columns, to obtain another 255 equivalent matrices, i.e., matrices with the same Einstein-locality maximum M_0. Therefore, we can conclude that the number of 4×4 independent linear combinations, leading to superinequalities, with coefficients in the range $\{-2, +2\}$, is nearly 45,000.

The following are some examples of these inequalities:

1.
$$\sum_{i,j}^{4,4} C_{ij}^1 P(a_i, b_j) \leq 11 \tag{44}$$

where

$$
C_{ij}^1 = \begin{pmatrix} 2 & 2 & 2 & 2 \\ 2 & 2 & 1 & -2 \\ -2 & 2 & 0 & 0 \\ 1 & 1 & -2 & 0 \end{pmatrix}
$$

2.
$$
\sum_{i,j}^{4,4} C_{ij}^2 P(a_i, b_j) \leq 10 \tag{45}
$$

where

$$
C_{ij}^2 = \begin{pmatrix} 2 & 2 & 2 & 0 \\ 2 & 1 & -1 & -2 \\ 2 & 1 & -1 & 2 \\ 2 & -2 & 0 & 0 \end{pmatrix}
$$

3.
$$
\sum_{i,j}^{4,4} C_{ij}^3 P(a_i, b_j) \leq 6 \tag{46}
$$

where

$$
C_{ij}^3 = \begin{pmatrix} 1 & 1 & 1 & -1 \\ -1 & 1 & 1 & -1 \\ 0 & 2 & -1 & 1 \\ 0 & 0 & 1 & 1 \end{pmatrix}
$$

This last inequality was discovered by Kemperman in 1984 in the course of studying other problems.

In order to clarify the content of Theorem 4, we shall analyze in detail the inequality (44). The maximum possible value of the lhs is obviously

$$
\bar{M}' = \sum_{i,j}^{4,4} |C_{ij}^1| = 23
$$

and is obtained for a suitable choice of $P(a_i, b_j)$:

$$
\{\bar{P}(a_i, b_j)\} = \begin{pmatrix} 1 & 1 & 1 & 1 \\ 1 & 1 & 1 & -1 \\ -1 & 1 & 0 & 0 \\ 1 & 1 & -1 & 0 \end{pmatrix} \tag{47}
$$

(We use a matrix representation for the vector $\{\bar{P}(a_i, b_j)\}$ in the 4×4 space.)

Starting from the set (47) it is possible to define a new set of correlation functions

$$\{\bar{P}'(a_i, b_j)\} = \begin{pmatrix} 0.5 & 0.5 & 0.5 & 0.5 \\ 0.5 & 0.5 & 0.5 & -0.5 \\ -0.5 & 0.5 & 0 & 0 \\ 0.5 & 0.5 & -0.5 & 0 \end{pmatrix} \tag{48}$$

Since all the $\bar{P}'(a_i, b_j)$ are within ± 0.5, all Bell inequalities containing the $\bar{P}'(a_i, b_j)$ are satisfied. Therefore, using only the Bell inequality, we could conclude that the set $\{\bar{P}'(a_i, b_j)\}$ describes a physical system compatible with Einstein locality. This is, however, not true since the set (48) introduced in inequality (44) gives

$$11.5 < 11$$

and therefore the inequality is violated and the set of correlation functions (48) cannot be obtained from a local theory.

Moreover, a hypersphere of center $\{\bar{P}'(a_i, b_j)\}$ and radius $R = \frac{1}{2}(0.5)$ exists such that all the sets of correlation functions inside this circle that satisfy Bell's inequality violate inequality (44).

It is possible to verify that the region of 4×4 space occupied by the $P(a_i, b_j)$ with these features is not unique; indeed the Kemperman inequality (46), for example, is violated for the following two sets of correlation functions:

$$\{P'(a_i, b_j)\} = 0.5 \begin{pmatrix} 1 & 1 & 1 & -1 \\ -1 & 1 & 1 & -1 \\ 0 & 1 & -1 & 1 \\ 0 & 0 & 1 & 1 \end{pmatrix}$$

and

$$\{P''(a_i, b_j)\} = \begin{pmatrix} 1 & 0.6 & 0.6 & -0.6 \\ 0.2 & 0.6 & 0.6 & -0.6 \\ 0.6 & 0.2 & 0.2 & -0.2 \\ 0.6 & 0.2 & 1 & -0.2 \end{pmatrix}$$

4.2.3. Conclusions

We conclude with three remarks:

1. We proved in the previous section that, in the case of 4×4 correlation functions, a large number of superinequalities exist. It is possible that only a finite number of these inequalities are independent and form a set which completely expresses Einstein locality. Further studies would answer this question.

2. The correlation functions analyzed in all the experimental tests are functions of the absolute value of the angle between two directions a_i and b_j, therefore they are symmetric with respect to the exchange of a_i and b_j. The matrix coefficients C_{ij} of inequalities stronger than Bell's inequality are, in general, nonsymmetric. Therefore, it is possible to suppose that there are symmetric matrices associated with the superinequalities.

    Let us consider now a superinequality defined by the 4×4 matrix C_{ij}; if the limit of Einstein locality is M_0, it is easy to see that the transposed matrix $\bar{C}_{ij} = C_{ji}$ defines a new superinequality with the same limit M_0. It is possible to define the symmetrical matrix associated with C_{ij} as

$$\hat{C}_{ij} = \tfrac{1}{2}(C_{ij} + \bar{C}_{ij})$$

For $\hat{C}_{ij}$, the limit of Einstein locality is

$$\hat{M} = \max_{\xi, \eta} \left\{ \sum_{ij}^{4} \tfrac{1}{2}(C_{ij} + \bar{C}_{ij}) \xi_i \eta_i \right\}$$

$$= \tfrac{1}{2} \max_{\eta} \left\{ \sum_i \left| \sum_j C_{ij}\eta_j + \sum_j C_{ji}\eta_j \right| \right\}$$

$$\leq \frac{1}{2} \left\{ \max_{\eta} \sum_i \left| \sum_j C_{ij}\eta_j \right| + \max_{\eta} \sum_i \left| \sum_j C_{ji}\eta_j \right| \right\}$$

$$= \tfrac{1}{2}\{M_0 + M_0\} = M_0$$

where the equality holds if C_{ij} and $\bar{C}_{ij}$ are maximized by the same choice of signs $\{\bar{\eta}_i\}$.

3. Theorem 4 only expresses a *sufficient* condition for the existence of an inequality stronger than Bell's inequality. For example, the inequality (27) of Section 2.2.1 has its rhs equal to one half of the possible maximum, but provides restrictions on $P(a_i, b_j)$ not implied by Bell's inequality.

5. New Inequalities for Joint Probabilities

5.1. A General Method for Joint Probabilities

New and more stringent inequalities have been deduced in 1987 by Lepore (22) for linear combinations of joint probabilities. The physical content of these inequalities is not deducible from any inequality discussed in the previous sections.

Let us consider M instrumental parameters $a_1, a_2, \ldots, a_m$ for the first measurement apparatus and n instrumental parameters $b_1, b_2, \ldots, b_n$ for the second measurement apparatus. Let

$$w_{hk}(a_i, b_j) = \int \rho(\lambda)\, d\lambda\, p_h(a_i, \lambda) q_h, (b_j, \lambda) \tag{49}$$

the joint probability of measuring $A(a_i)$ and obtaining h and measuring $B(b_j)$ and obtaining $k(h, k = +1)$. We can consider now the linear combination of joint probabilities

$$C = \sum_{\substack{hk \\ ij}} C_{ij}^{hk} w_{hk}(a_i, b_j) \tag{50}$$

where C_{ij}^{hk} are arbitrary $4mn$ real coefficients.

In order to deduce the inequalities

$$m_0 \leq C \leq M_0 \tag{51}$$

true for all conceivable local realistic theories, it is sufficient, using the lemma of Section 3.1, to prove that the inequality

$$m_0 \leq \sum_{\substack{hk \\ ij}} C_{ij}^{hk} P_h(a_i, \lambda) q_k(b_j, \lambda) \leq M_0 \tag{52}$$

is true for arbitrary values of λ and for arbitrary dependence of p_h and q_k from their arguments.

Using the relations

$$p_+(a_i, \lambda) + p_-(a_i, \lambda) = 1$$

$$q_+(b_j, \lambda) + q_-(b_j, \lambda) = 1$$

we can write

$$\sum_{\substack{hk \\ ij}} C_{ij}^{hk} p_h(a_i, h) q_k(b_j, \lambda)$$

$$= F(p_+(a_l, \lambda) \cdots p_+(a_m, \lambda), q_+(b_1, \lambda), \ldots, q_+(b_n, \lambda)) \qquad (53)$$

where F is a linear function of

$$p_+(a_2, \lambda), \ldots, p_+(a_m, \lambda), q_+(b_1, \lambda), \ldots, q_+(b_n, \lambda)$$

Obviously, the most stringent inequality is found when M_0 is taken equal to the maximum value (and m_0 equal to the minimum value) of function F. Since the linear function F is defined in the hypercube C in the multidimensional space having $P_+(a_i, \lambda)$ and $q_+(b_j, \lambda)$ as Cartesian coordinates, the maximum and the minimum is found in principle on the boundary, namely in one of the vertices of hypercube C.

Therefore, setting

$$m_0 = \min_{\substack{\xi_1, \ldots, \xi_m = 0,1 \\ \eta_1, \ldots, \eta_n = 0,1}} F(\xi_1, \ldots, \xi_m, \eta_1, \ldots, \eta_n) \qquad (54)$$

$$M_0 = \max_{\substack{\xi_1, \ldots, \xi_m = 0,1 \\ \eta_1, \ldots, \eta_n = 0,1}} F(\xi_1, \ldots, \xi_m, \eta_1, \ldots, \eta_n) \qquad (55)$$

we obtain the set of inequalities

$$m_0 \leq \sum_{\substack{hk \\ ij}} C_{ij}^{hk} w_{hk}(a_i, b_j) \leq M_0 \qquad (56)$$

For every choice of coefficient C_{ij}^{hk}, relation (56) provides the most stringent inequality that can be deduced from local realism.

5.2. A Particular Inequality

In order to prove the set of inequalities (52) is not equivalent to the set of inequalities of correlation function (37) and implies more stringent restriction for the Einstein locality, we will use the following model studied by Garg and Mermin in 1982.[23]

In this model the joint probabilities are

$$w_{hk}(a_i, b_j) = \tfrac{1}{4}[1 - C(h + k) + A_{ij}hk] \quad \begin{cases} i = 1, 2, 3 \\ j = 1, 2, 3 \end{cases} \qquad (57)$$

with

$$0 \le C \le \tfrac{1}{3} \qquad (58)$$

and

$$A_{11} = A_{22} = j$$
$$(59)$$
$$A_{ij} = -\tfrac{1}{3} \quad \text{for } i = 1, 2, 3; j = 1, 2, 3; (i, j) \ne (1, 1), (2, 2)$$

From equations (57) and (59) one has

$$P(a_i, b_j) = A_{\mu\nu}$$

and it is easy to prove that this model satisfies all inequalities deduced from Einstein locality for correlation functions.

Let us consider now the case of three directions a_1, a_2, a_3 for the first apparatus and three directions b_1, b_2, b_3 for the second; we can write the following linear combination

$$w_{++}(a_3, b_3) - w_{--}(a_2, b_2) + w_{-+}(a_1, b_1) + w_{+-}(a_2, b_2)$$

$$+ w_{--}(a_3, b_2) - w_{-+}(a_3, b_1) + 2w_{++}(a_1, b_2)$$

$$+ w_{--}(a_2, b_3) - w_{+-}(a_1, b_3)$$

By calculating the minimum value of the associated function F one gets min $F = 0$; therefore the following inequality holds

$$w_{++}(a_3, b_3) - w_{--}(a_2, b_2) + w_{-+}(a_1, b_2) + w_{+-}(a_2, b_2) + w_{--}(a_3, b_2)$$

$$- w_{-+}(a_3, b_1) + 2w_{++}(a_1, b_2) + w_{--}(a_2, b_2) - w_{+-}(a_1, b_3) \ge 0 \qquad (60)$$

If now we substitute equations (57) and (59) into equation (60), we obtain

$$-C \ge 0$$

and this contradicts equation (58). Hence the Garg–Mermin model satisfies all inequalities for correlation functions, but violates at least this particular inequality for joint probabilities: then the model cannot be reproduced by a local probabilistic theory.

We can therefore conclude that the set of inequalities (52) is the widest set of inequalities deduced from Einstein locality for linear combination of joint probabilities with real coefficients.

Appendix: The Computational Method for Deducing the Superinequalities in the Case 4×4

The maximum allowed by Einstein locality is given by

$$\max_{\eta} \sum_{i=1}^{4} \left| \sum_{j=1}^{4} C_{ij}\eta_j \right|$$

where $\eta_j = +1$. The number of possible choices of the four sign factors is 16, but the presence of modulus signs reduces the number of independent choices to eight. Therefore, we can define the following matrix:

$$\eta_j^l = \begin{pmatrix} 1 & -1 & 1 & 1 & 1 & -1 & -1 & -1 \\ 1 & 1 & -1 & 1 & 1 & -1 & 1 & 1 \\ 1 & 1 & 1 & -1 & 1 & 1 & -1 & 1 \\ 1 & 1 & 1 & 1 & -1 & 1 & 1 & -1 \end{pmatrix}$$

in which every column represents an independent choice of sign factors.

The problem of the calculating maximum is now reduced to computing the elements of the 4×8 matrix $|\sum_j C_{ij}\eta_j^l|$, summing the elements of each column, and finding the maximum sum ($\equiv M_0$).

If M_0 is less than or equal to $\frac{1}{2}\sum_{ij}|C_{ij}|$, then the inequality

$$\sum_{ij} C_{ij}P(a_i, b_j) \leq M_0$$

is a superinequality.

References

1. A. Einstein, B. Podolsky, and N. Rosen, *Phys. Rev.* **47**, 777 (1935).
2. N. Bohr, *Phys. Rev.* **48**, 696 (1935).
3. J. S. Bell, *Phys.* **1**, 195 (1965).
4. J. F. Clauser, M. A. Horne, A. Shimony, and R. A. Holt, *Phys. Rev. Lett.* **23**, 880 (1969).
5. F. Selleri, *Lett. Nuovo Cim.* **3**, 581 (1972).
6. J. S. Bell, in: *Rendiconti SIF* (B. d'Espagnat, ed.), IL Course, New York (1971).
7. E. P. Wigner, *Am. J. Phys.* **38**, 1005 (1970).
8. J. F. Clauser and M. A. Horne, *Phys. Rev. D* **10**, 526 (1974).
9. A. Garuccio and F. Selleri, *Lett. Nuovo Cim.* **23**, 555 (1978).
10. F. Selleri and G. Tarozzi, *Lett. Nuovo Cim.* **29**, 533 (1980).
11. A. Garuccio and V. A. Rapisarda, *Lett. Nuovo Cim.* **30**, 443 (1980).
12. A. Baracca, S. Bergia, R. Livi, and M. Restignoli, *Int. J. Theor. Phys.* **15**, 473 (1976).
13. N. D. Mermin, *Phys. Rev. D* **22**, 356 (1980).

14. F. Selleri, *Found. Phys.* **8**, 103 (1978).
15. D. Gutkowski and G. Masotto, *Nuovo Cim. B* **22** 121 (1974).
16. A. Garuccio, *Lett. Nuovo Cim.* **23**, 559 (1978).
17. S. M. Roy and V. Singh, *J. Phys. A* **11**, 167 (1978).
18. S. M. Roy and V. Singh, *J. Phys. A* **12**, 1003 (1979).
19. A. Garuccio and F. Selleri, *Found. Phys.* **10**, 209 (1980).
20. A. Garg, *Phys. Rev. D* **28**, 785 (1983).
21. A. Garg, Comment on "Systematic Derivation of all the Inequalities of Einstein Locality," preprint.
22. V. L. Lepore, New Inequalities from Local Realism, preprint (1987).
23. A. Garg and N. D. Mermin, *Phys. Rev. Lett.* **49**, 1220 (1982).

4

Einstein-Podolsky-Rosen Experiments Using the Decays of η_c or J/ψ into $\Lambda\bar{\Lambda} \to \pi^- p \pi^+ \bar{p}$

NILS A. TÖRNQVIST

1. Introduction

An important key to the resolution of the Einstein-Podolsky-Rosen (EPR) paradox[1] lies in the finding of new ways to test the nonlocal correlations predicted by quantum mechanics. In this paper we shall discuss a novel type of experiment, involving η_c and J/ψ decay, which has become experimentally feasible. Nowadays, a large number of J/ψ decays accumulate in the course of e^+e^- storage ring experiments. For the $\Lambda\bar{\Lambda}$ channel in particular, over 1000 decay events are seen in some current experiments, although the branching ratio[2] is only $(1.58 \pm 0.21) \times 10^{-3}$. Similar, although experimentally less feasible, reactions are $J/\psi \to \Sigma\bar{\Sigma} \to \pi N \pi \bar{N}$ and $e^+e^- \to \mu^+\mu^- \to e^+e^- +$ neutrinos.

Such reactions, where the spontaneous decay works as a spin analyzer, do not touch on the EPR paradox as generally as those where the direction of the spin analyzer can be chosen at will, by the external experimental setup. However, these reactions do test the quantum-mechanical correlations at macroscopic distances and they involve weak interactions. Therefore high-statistics experiments that observe these correlations would provide a valuable contribution to the verification of the nonlocal correlations predicted by quantum mechanics.

NILS A. TÖRNQVIST • Department of High Energy Physics, University of Helsinki, SF-00170 Helsinki 17, Finland.

In this chapter we first discuss, in Section 2, the nonrelativistic $\eta_c \to \Lambda\bar{\Lambda}$ and $J/\psi \to \Lambda\bar{\Lambda} \to \pi^- p \pi^+ \bar{p}$ decays. Then in Section 3 we look at the relativistic effects in J/ψ decay and in Section 4 we examine the Bell inequalities and present an instructive graph for displaying the domains separated by the Bell bounds on the one hand, and the bounds that follow from quantum mechanics on the other. The first experiment to test the quantum-mechanical correlations for the decays discussed in this chapter is currently being carried out,[3,4] and we review it in Section 5. Another experiment is also in progress.[5] (For the author's previous notes on the subject, see elsewhere.[6-8])

2. Nonrelativistic Resonance Decay to $\Lambda\bar{\Lambda}$

In this section we first recapitulate the well-known results of $\Lambda \to \pi N$ decay in the Λ c.m.s. Then we discuss the nonrelativistic situation for a spin-0 or spin-1 resonance which decays into $\Lambda\bar{\Lambda}$, gradually generalizing so as to make the physical interpretation as transparent as possible.

2.1. The $\Lambda \to \pi N$ Decay

Owing to parity violation, the Λ decay distribution depends on the Λ polarization, i.e., the decay works as a polarimeter. Denoting the S- and P-wave amplitudes by S and P, the transition matrix for the hyperon decay can be written (cf. Perkins[9])

$$M_a = S + P\boldsymbol{\sigma} \cdot \hat{\mathbf{a}} \tag{2.1}$$

where $\hat{\mathbf{a}}$ denotes the unit vector along the pion momentum in the Λ c.m.s. and $\boldsymbol{\sigma}$ the Pauli matrices. Forming $|\langle\chi_p|M_a|\chi_\Lambda\rangle|^2$ and summing over proton spins one obtains the decay rate for $\Lambda \to \pi N$:

$$R(\hat{\mathbf{a}}) \propto \text{Tr}[M_a\rho_\Lambda M_a^\dagger] \propto 1 + \alpha\hat{\mathbf{a}} \cdot \mathbf{P}_\Lambda \tag{2.2a}$$

where $\rho_\Lambda = |\chi_\Lambda\rangle\langle\chi_\Lambda| = \frac{1}{2}(1 + \mathbf{P}_\Lambda \cdot \boldsymbol{\sigma})$ is the Λ spin-density matrix, $\mathbf{P}_\Lambda$ is the Λ polarization, and α is the Λ decay asymmetry parameter given by

$$\alpha = \alpha_\Lambda = 2 \, \text{Re}(SP^*)/(|S|^2 + |P|^2)$$

which, for $\Lambda \to p\pi^-$, has the experimental value[10] -0.642 ± 0.013 (for $\Sigma^+ \to \pi^+ n$ it is near unity). For $\bar{\Lambda} \to \pi \bar{N}$ one gets, with our conventions ($|\chi^\dagger\rangle$ denotes a spinor which transforms as the conjugate representation), $\rho_{\bar{\Lambda}} = |\chi_{\bar{\Lambda}}^\dagger\rangle\langle\chi_{\bar{\Lambda}}^\dagger| = \frac{1}{2}(1 - \mathbf{P}_{\bar{\Lambda}} \cdot \boldsymbol{\sigma})$ and

$$R(\hat{\mathbf{a}}) \propto \text{Tr}[M_b\rho_{\bar{\Lambda}}M_b^\dagger] \propto 1 - \alpha\hat{\mathbf{a}} \cdot \mathbf{P}_{\bar{\Lambda}} \tag{2.2b}$$

Thus $\alpha_{\bar{\Lambda}} = -\alpha_{\Lambda}$ as required by CP invariance, a relation which has been experimentally tested.[11]

2.2. Resonance Decay into $\Lambda\bar{\Lambda} \rightarrow \pi^- p \pi^+ \bar{p}$

The decay matrix elements for η_c or $J/\psi \rightarrow \Lambda\bar{\Lambda} \rightarrow \pi^- p \pi^+ \bar{p}$ can be written

$$\sum_{ij} \langle \chi_p | M_a | \chi_{\Lambda_i} \rangle s_{ij} \langle \chi^\dagger_{\bar{\Lambda}_j} | M^\dagger_b | \chi^\dagger_{\bar{p}} \rangle$$

We denote the two-particle spin correlation $\sum |\chi_{\Lambda_i}\rangle s_{ij} \langle \chi^\dagger_{\bar{\Lambda}_j}|$ by $\hat{S}$. It is equal to the unit matrix for a singlet spin-0 particle (η_c) and $\mathbf{N} \cdot \boldsymbol{\sigma}$ for a spin-1 particle (J/ψ) with polarization $\mathbf{N}$. Forming the absolute square and summing over proton and antiproton spins gives the rate as a trace over Pauli spin matrices:

$$R(\hat{\mathbf{a}}, \hat{\mathbf{b}}) = \text{Tr}[\hat{S}|M_a|^2 \hat{S}^\dagger |M_b|^2] \propto \text{Tr}[\hat{S}(1 + \alpha\hat{\mathbf{a}} \cdot \boldsymbol{\sigma})\hat{S}^\dagger(1 + \alpha\hat{\mathbf{b}} \cdot \boldsymbol{\sigma})] \quad (2.3)$$

where $\hat{\mathbf{a}}$ is the unit vector along the π^- momentum in the Λ c.m.s. and $\hat{\mathbf{b}}$ is the corresponding quantity for $\bar{\Lambda}$. (The boosts to the rest systems in question are performed along the $\Lambda\bar{\Lambda}$-axis and the two rest frames are superimposed.)

2.3. The $\eta_c \rightarrow \Lambda\bar{\Lambda}$ Decay

For a spin-singlet initial-state (η_c) decay, one has simply $\hat{S} = 1$ and equation (2.3) takes the form

$$R(\hat{\mathbf{a}}, \hat{\mathbf{b}}) \propto 1 + \alpha^2 \hat{\mathbf{a}} \cdot \hat{\mathbf{b}} \quad (2.4)$$

This decay is a realization of the classic Bohm[12,13] variant of the EPR problem. Apart from the constant α^2 and the sign, the rate of equation (2.4) is equivalent to that obtained in measuring the spin correlation in the Bohm experiment, the directions of the pion momenta $\hat{\mathbf{a}}$ and $\hat{\mathbf{b}}$ replacing the spin-analyzing directions of the polarimeters.

It may be noted that the Λ decays as if it had a polarization $P_\Lambda = \alpha\hat{\mathbf{b}}$ "tagged" in the direction of the π^+ coming from the $\bar{\Lambda}$ [cf. equation (2.2a)]. Correspondingly, the $\bar{\Lambda}$ decays as if it had a tagged polarization $\mathbf{P}_{\bar{\Lambda}} = \alpha\hat{\mathbf{a}}$ (in the direction of the π^- coming from the Λ). This same conclusion also holds if complete spin measurements are done on the final proton or antiproton (see Törnqvist[6]).

In other words, Λs coming from a singlet $\Lambda\bar{\Lambda}$ state are indistinguishable from a Λ beam prepared to be "polarized" in a tagged direction $\hat{\mathbf{a}}$. This result is rather curious since, for individual events $\eta_c \to \Lambda\bar{\Lambda} \to \pi^- p \pi^+ \bar{p}$, the Λ can be thought of as polarized in the direction of the π^+, while the $\bar{\Lambda}$ is polarized in the direction of π^-. Knowledge of how one of the Λ decayed, or will decay (time ordering is not relevant here), tells an observer that the second Λ decayed, or will decay, as if it had a definite polarization. This is a practical demonstration of the conceptual peculiarities involved in the EPR problem.

Another remark worth noting is that if, through some (hidden or not) measuring process, the spin of the Λ are measured before their decay (along a random direction), the correlation between the pion momenta is reduced by a factor of three,[6] since now the spin is "measured" a second time during the decay and, clearly, the first measurement will alter the result of the second.

The η_c has been seen mainly in ψ or ψ' radiative decays ($\psi \to \eta_c\gamma$). Since the production of the η_c in this way is not easy, it is difficult to obtain sufficient statistics for a significant analysis. Producing the η_c through nearly-real $\gamma\gamma$ reactions ($e^+e^- \to e^+\gamma\gamma e^- \to e^+\eta_c e^-$) may make a detailed study feasible in the near future. We estimate that it would require of the order of 50 or more (all good ones) to allow a first look at the physics. To study the correlation experimentally, it is useful to plot the number of events against $\cos\Theta_{ab} = \hat{\mathbf{a}} \cdot \hat{\mathbf{b}}$ and to determine the slope (cf. Figure 1). In Figure 1 we also show bounds that can be obtained from Bell's inequalities supplemented by continuity arguments (details can be found elsewhere[6]). These are linear in the variable Θ_{ab}:

$$|E(\Theta_{ab})| \le 1 - 2\Theta_{ab}/\pi \tag{2.5}$$

where $E = (r-1)/\alpha^2$. With poor statistics, only the forward–backward asymmetry in this variable can be measured,[14] although even this gives a first insight into the physics. As can be seen, the Bell bound requires a smaller slope or smaller backward–forward asymmetry. This asymmetry $|(R-L)/(R+L)|$, where R and L stand for the integrals $\int_0^1 r(\Theta d\cos\Theta_{ab}$ and $\int_{-1}^0 r(\Theta)d\cos\Theta_{ab}$ respectively, is predicted by equation (2.11) to be $\alpha^2/2 = 0.21$ while equation (2.5) gives $(1 - 2/\pi)\alpha^2 = 0.15$.

With sufficient statistics it would be interesting to also analyze the space-like separated $\Lambda\bar{\Lambda}$ decays, since these are the most relevant to the EPR problem. For the η_c, a simple kinematic calculation shows that the events with space-like separated Λ or $\bar{\Lambda}$ decays satisfy $x_L/x_S < (1+\beta)/(1-\beta) = 4.94$ (for the J/ψ it is 5.53), where x_L and x_S are the longer and shorter decay lengths, respectively, and β is the velocity of the

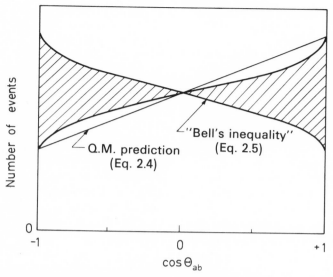

Figure 1. The expected distribution of the angle Θ_{ab} between the (boost-corrected) directions of the two pions produced in the reaction $\eta_c \to \Lambda\bar{\Lambda} \to \pi^- p\pi^+ \bar{p}$ predicted by quantum mechanics (solid upward-sloping straight line). If a random (hidden) polarization measurement is done on the Λ or $\bar{\Lambda}$ before the decay, the correlation (i.e., the slope) is reduced by a factor $\frac{1}{3}$. Bell's inequalities supplemented by continuity arguments give the shaded regions.[6]

Λ. Without experimental cuts in the fiducial volume the fraction of such events would be $\beta = 0.66$.

2.4. The Decay $J/\psi \to \Lambda\bar{\Lambda} \to \pi^- p\pi^+ \bar{p}$

This decay is, from the experimental point of view, much more interesting since nearly 10^7 J/ψ are seen in some current $e^+ e^- \to J/\psi$ experiments. This reaction is discussed elsewhere.[7] (An earlier discussion[6] was incomplete and contains some obvious misprints.) As we shall see, the decay of a vector state is, in some circumstances, equally interesting, from the point of view of the EPR paradox, as is the scalar state discussed above.

For the nonrelativistic decay of a spin-1 particle, polarized in the direction $\mathbf{N}$, $\hat{S} = \mathbf{N} \cdot \boldsymbol{\sigma}$. Then the spin dependence of the rate in equation (2.3) takes the form

$$R_N(\hat{\mathbf{a}}, \hat{\mathbf{b}}) \propto \mathrm{Tr}[(1 + \alpha\hat{\mathbf{a}} \cdot \boldsymbol{\sigma})\mathbf{N}^\dagger \cdot \boldsymbol{\sigma}(1 + \alpha\hat{\mathbf{b}} \cdot \boldsymbol{\sigma})\mathbf{N} \cdot \boldsymbol{\sigma}] \qquad (2.6a)$$

which, when normalized such that

$$r_N(\hat{\mathbf{a}}, \hat{\mathbf{b}}) = \frac{2R_N(\hat{\mathbf{a}}, \hat{\mathbf{b}})}{R_N(\hat{\mathbf{a}}, \hat{\mathbf{b}}) + R_N(\hat{\mathbf{a}}, -\hat{\mathbf{b}})} \qquad (2.7)$$

and evaluated for real N, reduces to

$$r_N(\hat{\mathbf{a}}, \hat{\mathbf{b}}) = 1 - \alpha^2 \left(\hat{\mathbf{a}} \cdot \hat{\mathbf{b}} - \frac{2\hat{\mathbf{a}} \cdot \mathbf{N} \hat{\mathbf{b}} \cdot \mathbf{N}}{N^2} \right) \tag{2.6b}$$

If we average over two orthogonal polarizations (which we shall choose to be in the x and y directions, although they are quite arbitrary at this stage) we simply get

$$\bar{r}_{xy}(\hat{\mathbf{a}}, \hat{\mathbf{b}}) = 1 - \alpha^2 \hat{a}_z \hat{b}_z \tag{2.8}$$

We specify that the x and y directions be transverse, i.e., a boost from the Λ rest system is done in the z direction. The z-axis will then be the $\Lambda\bar{\Lambda}$-axis in the reaction $e^+ e^- \to \Lambda\bar{\Lambda}$. (Later we shall choose x to be orthogonal to the $e^+ e^- \to \Lambda\bar{\Lambda}$ scattering plane, and y to be in the scattering plane.)

The case of $\mathbf{N} = \hat{\mathbf{e}}_z$ (longitudinal or helicity $\lambda = 0$) gives

$$r_z(\hat{\mathbf{a}}, \hat{\mathbf{b}}) = 1 - \alpha^2 (\hat{\mathbf{a}} \cdot \hat{\mathbf{b}} - 2\hat{a}_z \hat{b}_z) \tag{2.9}$$

while the average over all three polarizations gives

$$\bar{r}_{xyz}(\hat{\mathbf{a}}, \hat{\mathbf{b}}) = 1 - \frac{\alpha^2}{3} \hat{\mathbf{a}} \cdot \hat{\mathbf{b}} \tag{2.10}$$

From the point of view of EPR correlations, the simply-factorized form of equation (2.8) is uninteresting, as we show below. On the other hand, equations (2.6b), (2.9), and (2.10) are as interesting as the case of the decay of a spin-0 state:

$$r_0(\hat{\mathbf{a}}, \hat{\mathbf{b}}) = 1 + \alpha^2 \hat{\mathbf{a}} \cdot \hat{\mathbf{b}} \tag{2.11}$$

The sign of the the correlation between the pions in equations (2.6)–(2.11) is opposite to that between the Λ and $\bar{\Lambda}$ spins because, due to CP invariance, the decay parameters satisfy $\alpha_{\bar{\Lambda}} = -\alpha_{\Lambda}$, as noted above in connection with equation (2.2b).

The factor $\frac{1}{3}$ in equation (2.10) should not be confused with the same factor $\frac{1}{3}$ obtained when considering the possibility of a (hidden) spin measurement of the Λ or $\bar{\Lambda}$ before the decay (cf. the discussion above for η_c decay and elsewhere[6]). For the spin-1 case, a similar averaging over a random spin-measurement direction $\mathbf{X}$ implies that we should average a distribution of the form $(1 + \alpha \hat{\mathbf{a}} \cdot \mathbf{X})(1 + \alpha \hat{\mathbf{b}} \cdot \mathbf{X}_r)$ (where $\mathbf{X}_r$ is defined as in equation (2.12) below), which gives equation (2.6b), but with an extra factor of $\frac{1}{3}$ in the second term, similar to the spin-0 case. For the average over all three initial polarizations, the factor $\frac{1}{3}$ in equation (2.10) is similarly replaced by $\frac{1}{9}$ when a random spin measurement is performed before the decay. We note that in these situations the Λ spins are measured twice, first in the (hidden) spin measurement along X and then in the Λ decay. Clearly, the first measurement must disturb the result of the second, which explains why one gets two factors of $\frac{1}{3}$ (i.e., $\frac{1}{9}$) when we average over both the initial polarization and the additional spin measurement along a random direction.

The analogy between equations (2.6b) and (2.11) is obvious if one defines a new vector $\hat{\mathbf{b}}_r$ obtained from $\hat{\mathbf{b}}$ through a 180° rotation about $\mathbf{N}$:

$$\hat{\mathbf{b}}_r = -\hat{\mathbf{b}} + \frac{2\hat{\mathbf{b}} \cdot \mathbf{N}}{N^2} \mathbf{N} \tag{2.12}$$

There is an identical correlation between $\hat{\mathbf{a}}$ and $\hat{\mathbf{b}}_r$, as in the spin-0 case between $\hat{\mathbf{a}}$ and $\hat{\mathbf{b}}$

$$r_N(\hat{\mathbf{a}}, \hat{\mathbf{b}}) = 1 + \alpha^2 \hat{\mathbf{a}} \cdot \hat{\mathbf{b}}_r \tag{2.6c}$$

For $\hat{\mathbf{b}}_r = \pm\hat{\mathbf{a}}$ we have the maximal correlation or anticorrelation (i.e., if Λ is spin-up along $\hat{\mathbf{a}}$ then $\bar{\Lambda}$ is always spin-up along $-\hat{\mathbf{b}}_r$).

The helicity states $\lambda = \pm 1$ correspond to $\mathbf{N} = (\hat{\mathbf{e}}_x \pm i\hat{\mathbf{e}}_y)/\sqrt{2}$ in equation (2.6a) and give

$$R_{\lambda = \pm 1}(\hat{\mathbf{a}}, \hat{\mathbf{b}}) \propto (1 \mp \alpha\hat{a}_z)(1 \pm \alpha\hat{b}_z) \tag{2.13}$$

i.e., the rate simply factorizes. Therefore, in these cases, no interesting EPR-like correlations appear—for $\lambda = +1$ the Λ and $\bar{\Lambda}$ spin will always point in the z direction. This could, of course, have already been seen from simple spin-conservation arguments. Averaging over $\lambda = \pm 1$ gives equation (2.8) which, therefore, is uninteresting from the point of view of EPR-like correlations—the Bell inequalities are always satisfied. (The z direction may be thought of as a hidden variable.)

The interesting cases thus occur when we have one linear (transverse or longitudinal) polarization state. If spin-averaged quantities are considered then one may lose interesting EPR-like correlations, especially if, as in equation (2.8), one averages over only two polarizations both of equal weight.

3. Relativistic Effects in J/ψ Decay

3.1. General

We assume that the Dirac form factor, involving the $\bar{u}\gamma_\mu u$ coupling, dominates over the Pauli form factor. This is supported both by the correctly-predicted angular dependence of the $e^+ e^- \to \Lambda\bar{\Lambda}$ cross section and by

QCD.[15] The derivation of the relativistic correlation is simple if we remember that the amplitude of the longitudinal polarization component is suppressed by the factor m/E (where m and E are the fermion mass and energy respectively). Therefore the initial polarization of the J/ψ is entirely transverse to the e^+e^--axis [the longitudinal J/ψ being suppressed by a factor of $(2m_e/m_\psi)^2 \approx 10^{-7}$].

It is useful to choose for the two initial polarization components of the J/ψ: (1) the polarization component orthogonal to the scattering plane, N_x; and (2) the polarization component in the scattering plane, N_y (but orthogonal to the e^+e^--axis). By calculating the contributions from these two components separately and summing, the derivation is considerably simplified.

For case (1) the result is the same as in the nonrelativistic case, since N_x is orthogonal also to the $\Lambda\bar{\Lambda}$-axis:

$$R_x(\hat{\mathbf{a}}, \hat{\mathbf{b}}) \propto 1 - \alpha^2(\hat{\mathbf{a}} \cdot \hat{\mathbf{b}} - 2\hat{a}_x\hat{b}_x) \tag{3.1}$$

Case (2) is slightly more complicated since it has components both orthogonal and parallel to the $\Lambda\bar{\Lambda}$-axis $\mathbf{k}$. The parallel component $\sin\theta\,\hat{\mathbf{e}}_k$, where θ is the c.m.s. scattering angle, is reduced by the factor m_Λ/E_Λ. Therefore one gets a result equivalent to that from a nonrelativistic case with an initial polarization (cf. Figure 2)

$$\mathbf{N}_y' = \hat{\mathbf{e}}_y - \left(1 - \frac{m_\Lambda}{E_\Lambda}\right)\sin\theta\,\hat{\mathbf{e}}_k \tag{3.2}$$

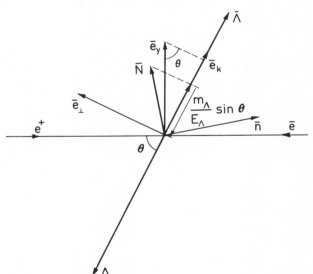

Figure 2. The definition of the vectors $\hat{\mathbf{e}}_y$, $\hat{\mathbf{k}}$, $\mathbf{N}$, and $\hat{\mathbf{n}}$ in the $e^+e^- \to \Lambda\bar{\Lambda}$ scattering plane.

whose norm is

$$N'_y = \left(1 - \frac{k^2}{E_\Lambda^2} \sin^2 \theta\right)^{1/2} \leq 1 \qquad (3.3)$$

We can now use our previous formula, equation (2.6a), and easily find

$$R_y(\hat{\mathbf{a}}, \hat{\mathbf{b}}) \propto N'^2_y(1 - \alpha^2\hat{\mathbf{a}} \cdot \hat{\mathbf{b}}) + 2\alpha^2\hat{\mathbf{a}} \cdot \mathbf{N}'_y\hat{\mathbf{b}} \cdot \mathbf{N}'_y$$

$$= N'^2_y(1 - \alpha^2\hat{\mathbf{a}} \cdot \hat{\mathbf{b}})$$

$$+ 2\alpha^2\left[\hat{a}_y\hat{b}_y - \left(1 - \frac{m_\Lambda}{E_\Lambda}\right) \sin \theta(\hat{a}_y\hat{b}_k + \hat{a}_k\hat{b}_y)\right.$$

$$\left. + \left(1 - \frac{m_\Lambda}{E_\Lambda}\right)^2 \sin^2 \theta \, \hat{a}_k\hat{b}_k\right] \qquad (3.4)$$

Averaging with the component orthogonal to the scattering plane [equation (3.1)] gives

$$R(\hat{\mathbf{a}}, \hat{\mathbf{b}}) \propto (1 + N'^2_y)(1 - \alpha^2\hat{\mathbf{a}} \cdot \hat{\mathbf{b}})$$

$$+ 2\alpha^2\left[\hat{a}_x\hat{b}_x + \hat{a}_y\hat{b}_y - \left(1 - \frac{m_\Lambda}{E_\Lambda}\right) \sin \theta(\hat{a}_y\hat{b}_k + \hat{a}_k\hat{b}_y)\right.$$

$$\left. + \left(1 - \frac{m_\Lambda}{E_\Lambda}\right)^2 \sin^2 \theta \, \hat{a}_k\hat{b}_k\right] \qquad (3.5)$$

A more useful and transparent form can be found if we use the fact, observed in equation (2.8), that the sum over two orthogonal polarizations of equal weight gives a simply-factorized form. We separate the x component of equation (3.1) into two parts, one proportional to N'^2_y and the other proportional to $1 - N'^2_y$. The first part can be combined with quantity R_y of equations (3.4) to give a factorized part, as in equation (2.8), while the second remains as a "net" polarization in the x direction. We let $\mathbf{n}$ be the unit vector orthogonal to $\mathbf{N}'_y$ and $\hat{\mathbf{e}}_x$ (i.e., it lies in the scattering plane, cf. Figure 2). Then

$$\mathbf{n} = \left(-\frac{m_\Lambda}{E_\Lambda} \sin \theta\hat{\mathbf{e}}_\perp + \cos \theta\hat{\mathbf{e}}_k\right) N'^{-1}_y \qquad (3.6)$$

where $\hat{\mathbf{e}}_\perp$ is perpendicular to $\hat{\mathbf{e}}_k$. Then equation (3.5) can be written

$$R(\hat{\mathbf{a}}, \hat{\mathbf{b}}) \propto 2\left(1 - \frac{k^2}{E_\Lambda^2}\sin^2\theta\right)(1 - \alpha^2\hat{a}_n\hat{b}_n) + \frac{k^2}{E_\Lambda^2}\sin^2\theta[1 - \alpha^2(\hat{\mathbf{a}}\cdot\hat{\mathbf{b}} - 2\hat{a}_x\hat{b}_x)]$$

$$(3.7)$$

In this final result, the first term is an average, as in equation (2.8), and is therefore uninteresting from the standpoint of EPR-like correlations. On the other hand, the second term results from the "net" linear polarization orthogonal to the scattering plane and thus is interesting from the standpoint of EPR-like correlations.

3.2. Special Cases in J/ψ Decay

Using equation (3.7) one can easily see the result in certain special cases:

1. For forward or backward scattering ($\theta = 0°$ or $180°$), or in the nonrelativistic limit ($k/E_\Lambda \to 0$), it reduces to the "uninteresting" form

$$R(\hat{\mathbf{a}}, \hat{\mathbf{b}}) \propto 1 - \alpha^2\hat{a}_z\hat{b}_z \qquad (3.8)$$

2. For 90° scattering the second "interesting" term is maximal:

$$R(\hat{\mathbf{a}}, \hat{\mathbf{b}}) \propto 2\frac{m_\Lambda^2}{E_\Lambda^2}(1 - \alpha^2\hat{a}_z\hat{b}_z) + \frac{k^2}{E_\Lambda^2}[1 - \alpha^2(\hat{\mathbf{a}}\cdot\hat{\mathbf{b}} - 2\hat{a}_x\hat{b}_x)] \qquad (3.9)$$

3. In the ultrarelativistic limit we have

$$R(\hat{\mathbf{a}}, \hat{\mathbf{b}}) \propto 2(1 - \alpha^2\hat{a}_k\hat{b}_k)\cos^2\theta + [1 - \alpha^2(\hat{\mathbf{a}}\cdot\hat{\mathbf{b}} - 2\hat{a}_x\hat{b}_x)]\sin^2\theta$$

$$(3.10)$$

Thus, without polarized beams, the interesting term is larger the closer the scattering angle is to 90° and the more relativistic the decay is. At the J/ψ, the factor k^2/E_Λ^2 is 0.48 and one is "half way" to the relativistic limit. At the Y, on the other hand, the same quantity is 0.94, i.e., one is essentially in the ultrarelativistic limit.

If sufficient events are observed to permit a more detailed study, it would be of interest to study both space-like and time-like separated $\Lambda\bar{\Lambda}$ decays, as discussed in Section 2.3 in the case of η_c decay.

4. The Domains of the Bell Inequalities and Quantum Mechanics

In this section we shall discuss Bell's inequalities[16,17] as well as bounds which follow from quantum mechanics in a way that makes their physical interpretation and comparison easy.[8] These comments are not restricted to the particular reactions discussed in previous sections.

The violation of Bell's inequalities by quantum mechanics has been historically of great importance in removing any remaining doubt that a local theory (in the EPR sense) is incompatible with quantum mechanics. These inequalities are usually written in terms of correlations, such that for the case of a spin-0 state decaying into two spin-$\frac{1}{2}$ particles the spin-correlation function $E(\hat{\mathbf{a}}, \hat{\mathbf{b}})$ obeys the inequality

$$|E(\hat{\mathbf{a}}, \hat{\mathbf{b}}) - E(\hat{\mathbf{a}}, \hat{\mathbf{c}})| \le 1 + E(\hat{\mathbf{b}}, \hat{\mathbf{c}}) \qquad (4.1a)$$

and similar inequalities with $\hat{\mathbf{a}}$, $\hat{\mathbf{b}}$, and $\hat{\mathbf{c}}$ permuted. Here $\hat{\mathbf{a}}$, $\hat{\mathbf{b}}$, and $\hat{\mathbf{c}}$ denote unit vectors along which the spin components are measured, and the two arguments refer to the two spin-$\frac{1}{2}$ particles. For the case of $\eta_c \to \Lambda\bar{\Lambda}$ discussed in Section 2.2, these directions can be identified with the pion directions in the respective c.m.s. of the Λ or $\bar{\Lambda}$, and $E = (1 - r)/\alpha^2$.

The correlations $E(\hat{\mathbf{a}}, \hat{\mathbf{b}})$ are, in practice, obtained through measurements of differences and ratios of differential cross sections (either in the form of multiple scattering in polarimeters, or in the form of multiparticle distributions as in the η_c or $J/\psi \to \Lambda\bar{\Lambda} \to \pi^- p\pi^+ \bar{p}$ decays discussed above). Denoting the cross section for observation of $(\pm\frac{1}{2}, \pm\frac{1}{2})$ spin components in the directions $(\hat{\mathbf{a}}, \hat{\mathbf{b}})$ by $\sigma_{\pm\pm}(\hat{\mathbf{a}}, \hat{\mathbf{b}})$, the relation is simply

$$r(\hat{\mathbf{a}}, \hat{\mathbf{b}}) = \frac{\sigma_{++}(\hat{\mathbf{a}}, \hat{\mathbf{b}})}{\sigma(\hat{\mathbf{a}}, \hat{\mathbf{b}})} = \tfrac{1}{2}[1 + E(\hat{\mathbf{a}}, \hat{\mathbf{b}})] \qquad (4.2)$$

where $\sigma(\hat{\mathbf{a}}, \hat{\mathbf{b}}) = \sigma_{++}(\hat{\mathbf{a}}, \hat{\mathbf{b}}) + \sigma_{+-}(\hat{\mathbf{a}}, \hat{\mathbf{b}})$ and where we have assumed $\sigma_{++} = \sigma_{--}$ and $\sigma_{+-} = \sigma_{-+}$ using general symmetry arguments.

Bell's inequalities then take the form of triangle inequalities between the ratios of cross sections r defined in equation (4.2) (see also Wigner[18]),

$$|r(\hat{\mathbf{a}}, \hat{\mathbf{b}}) - r(\hat{\mathbf{a}}, \hat{\mathbf{c}})| \le r(\hat{\mathbf{b}}, \hat{\mathbf{c}}) \qquad (4.1b)$$

and similar inequalities with $\hat{\mathbf{a}}$, $\hat{\mathbf{b}}$, and $\hat{\mathbf{c}}$ permuted. Or, equivalently, in a symmetric form:

$$\lambda[r^2(\hat{\mathbf{a}}, \hat{\mathbf{b}}), r^2(\hat{\mathbf{b}}, \hat{\mathbf{c}}), r^2(\hat{\mathbf{c}}, \hat{\mathbf{a}})] \le 0 \qquad (4.1c)$$

using the triangle function $\lambda[x, y, z] = (x + y - z)^2 - 4xy$.

In quantum mechanics the correlation $E(\hat{\mathbf{a}}, \hat{\mathbf{b}})$ is simply $-\hat{\mathbf{a}} \cdot \hat{\mathbf{b}}$. More specifically, one gets for r

$$r(\hat{\mathbf{a}}, \hat{\mathbf{b}}) = |\langle \chi_a^\dagger | \chi_b \rangle|^2 = \tfrac{1}{4} \text{Tr}[(1 - \boldsymbol{\sigma} \cdot \hat{\mathbf{a}})(1 + \boldsymbol{\sigma} \cdot \hat{\mathbf{b}})] = \tfrac{1}{2}(1 - \hat{\mathbf{a}} \cdot \hat{\mathbf{b}}) = \tfrac{1}{4}(\hat{\mathbf{a}} - \hat{\mathbf{b}})^2 \tag{4.3}$$

where $\hat{\mathbf{a}}^2 = 1$ and $|\chi_a\rangle$ is the spinor with spin in the direction $\hat{\mathbf{a}}$. Thus the norm of the amplitude $\langle \chi_a^\dagger | \chi_b \rangle$ is half the length of the vector $\hat{\mathbf{a}} - \hat{\mathbf{b}}$:

$$|\langle \chi_a^\dagger | \chi_b \rangle| = \tfrac{1}{2} |\hat{\mathbf{a}} - \hat{\mathbf{b}}| \tag{4.4}$$

The phase is, apart from phase conventions, a rather complicated expression in terms of $\hat{\mathbf{a}}$ and $\hat{\mathbf{b}}$. However, this phase is quite uninteresting from our point of view, since all measurable quantities can be formed from quantities like the one in equation (4.4).

For the three absolute values of the amplitudes $\langle \chi_a^\dagger | \chi_b \rangle$, $\langle \chi_b^\dagger | \chi_c \rangle$, and $\langle \chi_c^\dagger | \chi_a \rangle$ we can make a useful geometrical construction (see Figures 3 and 4):

The three unit vectors $\hat{\mathbf{a}}$, $\hat{\mathbf{b}}$, and $\hat{\mathbf{c}}$ form a tetrahedron (cf. Figure 3) and the end points of these vectors form a triangle whose sides are proportional to the norms of the amplitudes [equation (4.4)]. Thus quantum mechanics implies triangle inequalities for the norms of these amplitudes or, more physically, for the square roots of the cross sections:

$$|r^{1/2}(\hat{\mathbf{a}}, \hat{\mathbf{b}}) - r^{1/2}(\hat{\mathbf{a}}, \hat{\mathbf{c}})| \leq r^{1/2}(\hat{\mathbf{b}}, \hat{\mathbf{c}}) \tag{4.5a}$$

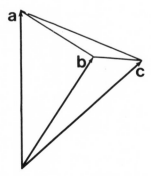

Figure 3. The tetrahedron formed by the three unit vectors **a**, **b**, and **c**. The end points of the three vectors form a triangle whose sides are proportional to the absolute values of the amplitudes [equation (4.4)]. The form of the triangle determines a point in Figure 4 which is inside the Bell bounds if the triangle is obtuse, but outside the Bell bounds if it is acute.

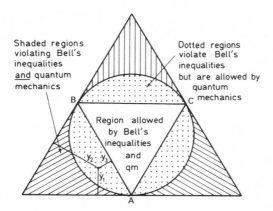

Figure 4. The triangle diagram with the normalized ratios y_i as the distances to the external triangle. The inscribed circle is the bound from quantum mechanics [equation (4.5)], while the inscribed triangle is the Bell bound [equation (4.1)]. The points A, B, and C correspond to the situation where two of the three directions $\hat{\mathbf{a}}$, $\hat{\mathbf{b}}$, and $\hat{\mathbf{c}}$ are equal.

and similar inequalities with $\hat{\mathbf{a}}$, $\hat{\mathbf{b}}$, and $\hat{\mathbf{c}}$ permuted, all of which can be written, using the triangle function, as a single inequality:

$$\lambda[r(\hat{\mathbf{a}}, \hat{\mathbf{b}}), r(\hat{\mathbf{b}}, \hat{\mathbf{c}}), r(\hat{\mathbf{c}}, \hat{\mathbf{a}})] \leq 0 \qquad (4.5b)$$

This inequality must always be satisfied for any choice of directions $\hat{\mathbf{a}}$, $\hat{\mathbf{b}}$, and $\hat{\mathbf{c}}$, although for given directions quantum mechanics is, of course, much more restrictive and provides a definite prediction for the amplitudes.

It is instructive to picture the domains separated by these inequalities in a barycentric coordinate system (Figure 4) in which one plots the normalized ratios

$$y_1 = \frac{r(\hat{\mathbf{a}}, \hat{\mathbf{b}})}{r_{\text{sum}}}, \qquad y_2 = \frac{r(\hat{\mathbf{b}}, \hat{\mathbf{c}})}{r_{\text{sum}}}, \qquad y_3 = \frac{r(\hat{\mathbf{c}}, \hat{\mathbf{a}})}{r_{\text{sum}}} \qquad (4.6)$$

where $r_{\text{sum}} = r(\hat{\mathbf{a}}, \hat{\mathbf{b}}) + r(\hat{\mathbf{b}}, \hat{\mathbf{c}}) + r(\hat{\mathbf{c}}, \hat{\mathbf{a}})$. The y_i are the distances to the sides of the external triangle.

The inequality (4.5) defines the domain of quantum mechanics to be the inside of the inscribed circle, while Bell's inequalities (4.1) define a region inside the inscribed equilateral triangle shown in Figure 4. The three regions between the inscribed circle and the incribed triangle thus violate Bell's inequalities but are consistent with quantum mechanics.

In practice, when testing experimentally these inequalities, a measure of the degree of violation would be useful. The diagram of Figure 4 suggests the distance B to the Bell bound:

$$B = \min_i(\tfrac{1}{2} - y_i) \qquad (4.7)$$

as such a measure. Negative values of B imply that the Bell bound is violated.

Given the directions $\hat{\mathbf{a}}$, $\hat{\mathbf{b}}$, and $\hat{\mathbf{c}}$, quantum mechanics defines a point inside the circle. The position of the point depends on the form (but not size) of the triangle formed by the end points of the three vectors $\hat{\mathbf{a}}$, $\hat{\mathbf{b}}$, and $\hat{\mathbf{c}}$. If this triangle is equilateral, the point lies at the center of Figure 4; if the triangle is right-angled, the point lies on the boundary of the inscribed triangle (the Bell bound); and if the triangle "collapses" to a line (a "flat triangle"), the point lies on the circle. Obviously, inside the Bell bound the triangle is acute (i.e., each of the angles of the triangle is less than 90°), while outside this bound the triangle is obtuse (i.e., one angle is larger than 90°).

The corners A, B, and C of the inscribed triangle correspond to situations where two of the directions are equal, e.g., point A corresponds to $\hat{\mathbf{a}} = \hat{\mathbf{b}}$. Here the situation is that of maximal anticorrelation: $E(\hat{\mathbf{a}}, \hat{\mathbf{b}}) = -1$, and there is only one remaining independent amplitude $(|\langle\chi_b^\dagger|\chi_c\rangle| = |\langle\chi_c^\dagger|\chi_a\rangle|)$. Then, with or without Bell's inequalities, no conflict arises— hidden-variable theories are here assumed to agree with quantum mechanics.

In quantum mechanics we superimpose probability amplitudes rather than probabilities. In choosing the different directions $\hat{\mathbf{a}}$, $\hat{\mathbf{b}}$, and $\hat{\mathbf{c}}$ we form new linear combinations of amplitudes which can then correspond to any point inside the circle. In particular, if $\hat{\mathbf{b}}$ and $\hat{\mathbf{c}}$ are kept fixed and $\hat{\mathbf{a}}$ is varied, the amplitudes are linear combinations of the components of $|\chi_b\rangle$ and $|\chi_c\rangle$. Equalities $\hat{\mathbf{a}} = \hat{\mathbf{b}}$ and $\hat{\mathbf{a}} = \hat{\mathbf{c}}$ correspond to the corners A and C respectively, while other directions give points inside the circle.

If, instead, one sums or averages over probabilities (or cross sections) as in EPR-local (hidden-variable) theories, one gets a point in Figure 4 on the line connecting two corners. Thus with the corners of the inscribed triangle being allowed by ansatz, such theories can only allow points inside the Bell bound.

It should be noted that since the Bell inequalities involve three directions $\hat{\mathbf{a}}$, $\hat{\mathbf{b}}$, and $\hat{\mathbf{c}}$, a single event such as $\eta_c \to \Lambda\bar{\Lambda} \to \pi^- p \pi^+ \bar{p}$ is not sufficient to give a point in the diagram of Figure 4. To obtain a point one must compare three differential cross sections in different regions of phase space. Thus one point in the diagram necessarily requires a large number of events. Therefore the diagram of Figure 4 is perhaps less useful in presenting data, than it is in demonstrating the Bell bounds in a simple and instructive way.

In the case of spin-1 decay it is easy to derive an inequality of the Bell type using equations (2.6b) or (2.6c):

$$|E(\hat{\mathbf{a}}, \hat{\mathbf{b}}_r) - E(\hat{\mathbf{a}}, \hat{\mathbf{c}}_r)| \leq 1 + E(\hat{\mathbf{b}}, \hat{\mathbf{c}}_r) \tag{4.8}$$

where the correlation E is related to r through $E = (1 - r)/\alpha^2$, and where the rotated vectors $\hat{\mathbf{a}}_r$, $\hat{\mathbf{b}}_r$, and $\hat{\mathbf{c}}_r$ are defined in equation (2.12).

5. The DM2 Experiment (Tixier et al.)

Preliminary results from an experiment by the DM2 collaboration designed to test the predictions of equation (3.5) have recently been published.[3,4] Another experiment by the SLAC Mark III group is in progress.[5] We shall briefly review the results published by the DM2 group to date.

The DM2 collaboration observed 7.7×10^6 J/ψ events of which 1284 were identified as $J/\psi \rightarrow \Lambda\bar{\Lambda} \rightarrow \pi^- p \pi^+ \bar{p}$. The pion momenta were measured by the DM2 detector (Figure 5), whose acceptance is maximal for particles nearly 90° off the e^+e^--axis. This is fortunate because, for large angles θ, the interesting second term of equation (3.7) is also the largest.

They studied the distribution in the angle Θ_{ab}, which we defined at the end of Section 2.3 and plotted in Figure 2 for the case of η_c decay. This is the angle between the two pions "corrected" for the boosts from their respective c.m.s. to the overall J/ψ c.m.s. (i.e., the π^- direction in the Λ c.m.s. and the π^+ direction in the $\bar{\Lambda}$ c.m.s. are calculated and then the two reference frames superimposed).

For the same reason, the group made a separate study of those events for which the scattering angle θ lay between 105° and 75°. This took account of 422 of the 1284 events. Since between these angles $\sin^2 \theta > 0.933$, the study was, in fact, very close to the second case enumerated in Section 3.2. They divided the variable $\cos \Theta_{ab}$ into 20 equally-spaced bins and obtained the experimental histogram shown in Figure 6.

In order to compare this result with the theoretical prediction of equation (3.7) they performed a Monte Carlo simulation, using as input the prediction of equation (3.7) and a standard description for the $J/\psi \rightarrow \Lambda\bar{\Lambda}$ differential cross section. The number of simulated events was normalized to the 422 events observed. Thus if equation (3.7) is a good description of the situation, the two histograms ought to agree.

As can be seen from Figure 6 the two histograms agree within statistical uncertainties. In particular, we note that the experimental histogram seems to show a small downward slope to the right. This is the correct direction, in accordance with the equations of Section 3 (note the term $-\alpha^2\hat{\mathbf{a}} \cdot \hat{\mathbf{b}}$), and the opposite direction compared to the one expected for the η_c case of

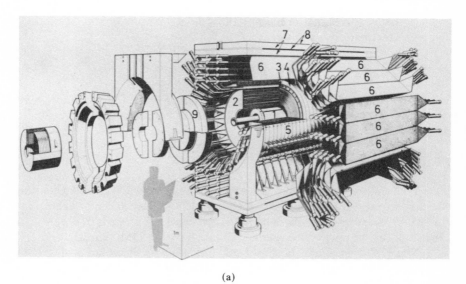

(a)

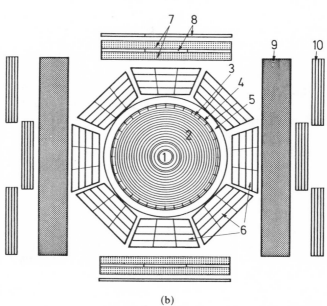

(b)

Figure 5. (a) A general view of the DM2 detector. (b) A schematic picture in a plane orthogonal to the beam line. Note the positions of (1) multiwire prportional counters (two layers), (2) drift chambers (13 layers), (3) Čerenkov counters, (4) time-of-flight counters, (5) coil, (6) photon detectors, (7) iron, (8) muon detectors, (9) end cap detector, and (10) concrete (shown in (b) only).

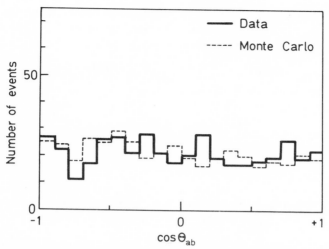

Figure 6. The data (solid line) of the DM2 group showing the distribution of the angle Θ_{ab} between pion momenta (cf. Figure 1). For this data sample, the scattering angle θ lies between 75° and 105°, for which both the detector efficiency and the interesting EPR-like correlation are maximal. A Monte Carlo simulation, using equation (3.7) (normalized to the observed 422 events) provided the quantum-mechanical prediction shown by the dashed histogram. As can be seen in despite the low statistics, the histograms are in quite close agreement. Note, in particular, the gradual downward slope to the right (in contrast to that of the η_c in Figure 1).

Figure 1. The test is, however, not very significant since there is insufficient statistics for so detailed a study of this variable.

The author believes that the test could be made more significant by finding a single number which measures the quantum-mechanical spin correlation. As suggested by Willutski,[14] the forward–backward asymmetry in the variable Θ_{ab} could be such a measure. As in the spin-singlet case, one would expect a smaller value for asymmetry to be predicted by theories which satisfy the Bell inequalities than what is predicted by quantum mechanics.

A possibly better number to measure the validity of equation (3.7) would be obtained if, by using the maximum-likelihood method, one determines the best experimental value of α by fitting equation (3.7) to the data. If, within experimental error, the number turns out to be 0.642, then the prediction of quantum mechanics can be said to be a good one. Alternatively, since the second term of equation (3.7) is the most interesting from the point of view of the EPR problem, one could try a fit in which α is kept at its physical value and, instead introduce another parameter as a multiplier for the second term. If the maximum-likelihood fit gives a value equal to one for this parameter, the quantum-mechanical prediction is, again, a good one. On the other hand, if one has reason to suspect that the Bell inequalities are satisfied (or that a hidden spin measurement has been done before the decay), both of the above-mentioned parameters would turn out to be smaller than the quantum-mechanical prediction.

ACKNOWLEDGMENT. I especially wish to thank A. Falvard (Clermont) for correspondence and many useful comments.

References

1. A. Einstein, B. Podolsky, and N. Rosen, *Phys. Rev.* **47**, 777 (1935).
2. M. Eaton *et al.*, *Phys. Rev. D* **29**, 804 (1984).
3. M. H. Tixier *et al.* (the DM2 collaboration, LAL Orsay, LPC, Clermont, Padova, Frascati), EPR Experiments Using the Reaction $J/\psi \to \Lambda\bar{\Lambda}$ with the DM2 Collaboration, presented at: *Conference on Microphysical Reality and Quantum Formalism*, Urbino, Italy (1985).
4. A. Falvard, private communication.
5. The SLAC Mark III collaboration (unpublished), private communication through J. Brown, M. Eaton, and H. Willutski.
6. N. A. Törnqvist, *Found. Phys.* **11**, 171 (1981).
7. N. A. Törnqvist, *Phys. Lett. A.* **117**, 1 (1986).
8. N. A. Törnqvist, *Europhys. Lett.* **1**, 377 (1986).
9. D. H. Perkins, *Introduction to High Energy Physics*, second edn., Addison–Wesley, Reading, Mass. (1982), p. 237.
10. M. Aguilar Benitez *et al.* (the Particle Data Group), *Phys. Lett. B* **170**, 1 (1986).
11. P. Chauvat *et al.* (the R608 collaboration), *Phys. Lett. B* **163**, 273 (1986).
12. D. Bohm, *Quantum Theory*, Prentice-Hall, Englewood Cliffs, N.J., pp. 614–622 (1951).
13. D. Bohm, *Phys. Rev.* **85**, 169 (1952).
14. H. Willutski, private communication.
15. S. J. Brodsky and G. P. Lepage, *Phys. Rev. D* **24**, 2848 (1981).
16. J. S. Bell, *Physics* **1**, 195 (1965).
17. J. S. Bell, in: *Foundations of Quantum Mechanics* (Proceedings of Enrico Fermi International Summer School), Academic Press, New York (19).
18. E. P. Wigner, *Am. J. Phys.* **38**, 1005 (1970).

5

Einstein-Podolsky-Rosen Paradox for the K^0-$\bar{K}^0$ and B^0-$\bar{B}^0$ Systems

DIPANKAR HOME

1. Introductory Remarks

The peculiarities of nonlocal quantum correlations between space-like separated systems were highlighted by Einstein, Podolsky, and Rosen in their seminal paper[1] which has stimulated lively debate over the past fifty years. A well-known illustration of the Einstein-Podolsky-Rosen (EPR) argument is Bohm's example[2] of a spin-0 system decaying into two spin-$\frac{1}{2}$ particles.

Apart from the epistemological interest that it has excited, the EPR debate has stimulated the study of viable experiments to test the nonseparability of the two-particle quantum wave function and its incompatibility with various local-realistic models.* The photon-polarization correlation measurements using radiative atomic cascade transitions, conducted by Aspect et al.,[3-5] are a significant mark forward in this direction. However, the interpretation of these experimental results has been the subject of vigorous controversy[6-11] owing to the low efficiency of the photomultiplier detectors used and uncertainties in the estimation of background counting rates. Specific examples of local-realistic models† have been proposed which reproduce the pertinent data of the atomic cascade experiments equally

---

* "Local realism" implies the following notion: individual measuremental results pertaining to the physical properties of a given system are independent of measurements performed on another spatially-separated system with which the given system may have interacted in the past but at present is not interacting.
† For an up-to-date review of such local-realistic models, see Marshall et al.[12] and Marshall.[13]

DIPANKAR HOME ● Department of Physics, Bose Institute, Calcutta 700009, India.

well as quantum mechanics. One therefore suspects that the final verdict on the question of local realism has yet to be passed.

A significant recent trend[14-18] has been the search for new examples of EPR-type situations, which may be amenable to experiment, in order to examine the incompatibility between quantum mechanics and local realism. An interesting example of such an experiment is provided, in the area of particle physics, by the decay of a $J^{PC} = 1^{--}$ vector meson into a pair of neutral pseudoscalar mesons. Lee and Yang[19,20]* were the first to point out the EPR-type features of this case (pertaining to the pair of kaons K^0-$\bar{K}^0$ resulting from a $p\bar{p}$ annihilation), followed by D'Espagnat,[22] Six,[23] and Selleri.[24]

In this chapter we seek to provide a critical analysis of the present status of this example and shall indicate possibilities for future investigations. We begin with a résumé of the important features of this example, considering specifically the decay of the spin-1 Φ (1020) resonance, by strong interaction, into a pair of neutral kaons K^0-$\bar{K}^0$.

2. EPR-Type Situation for the K^0-$\bar{K}^0$ System

If we invoke charge-conjugation invariance of the strong interaction, the wave function of the K^0-$\bar{K}^0$ pair at its time of production ($t = 0$) from the decay of the $J^{PC} = 1^{--}$ state is given by

$$|\Psi_0\rangle = (|K^0\rangle_L|\bar{K}^0\rangle_R - |\bar{K}^0\rangle_L|K^0\rangle_R)/\sqrt{2} \qquad (1)$$

where L (R) refers to the left (right) hemisphere.

The subsequent time-development of the K^0-$\bar{K}^0$ pair is described in terms of the eigenstates of the effective Hamiltonian (which includes weak interactions). In the situation under consideration, the weak interactions induce decays of both K^0 and $\bar{K}^0$ and also give rise to K^0-$\bar{K}^0$ transitions.† The effective Hamiltonian is written $H = M - i\Gamma/2$ where M and Γ are the Hermitian mass and decay matrices, respectively. The eigenstates of H are $|K_L\rangle$ and $|K_S\rangle$ with eigenvalues $\lambda_L = m_L - i\gamma_L/2$ and $\lambda_S = m_S - i\gamma_S/2$, respectively, where m_L (m_S) and γ_L (γ_S) are the mass and the decay width, respectively, of $|K_L\rangle$ ($|K_S\rangle$); $m_L - m_S = 0.53 \times 10^{10}\hbar s^{-1}$ and $\gamma_s = 582\gamma_L =$

---

* From the historical perspective it is interesting to note that, according to Jammer,[21] Lee first mentioned this example of the K^0-$\bar{K}^0$ system, during a talk at the Argonne National Laboratory in 1960, in which he anticipated the central idea of Bell's Theorem in the context of this example. Lee assigned further elaboration of this example to his assistant J. Schurtz who, unfortunately, began to work on another problem.

† A pure $|K^0\rangle$ ($|\bar{K}^0\rangle$) state, evolving in time, becomes a superposition of $|K^0\rangle$, $|\bar{K}^0\rangle$, and decay products. This gives rise to K^0-$\bar{K}^0$ oscillations.

$1.12 \times 10^{10} \hbar s^{-1}$. Throughout this section we assume CP invariance; the implications of CP violation for this type of example will be treated in Section 4.

The eigenstates $|K_L\rangle$ and $|K_S\rangle$ are given by

$$|K_L\rangle = (|K^0\rangle + |\bar{K}^0\rangle)/\sqrt{2}, \quad \text{and} \quad |K_S\rangle = (|K^0\rangle - |\bar{K}^0\rangle)/\sqrt{2}$$

and they time evolve as

$$U(t, 0)|K_L\rangle = |K_L\rangle \exp(-i\lambda_L t) + |\phi_L(t)\rangle \tag{2}$$

with a corresponding equation for $|K_S\rangle$. Here $|\phi_L(t)\rangle$ ($|\phi_S(t)\rangle$) represents the decay products from $|K_L\rangle$ ($|K_S\rangle$); $|\phi_L\rangle$ ($|\phi_s\rangle$) is taken orthogonal to the state $|K_L\rangle$ ($|K_S\rangle$). CP invariance requires that $\langle K_L|K_S\rangle = 0$.

The wave function $|\Psi_0\rangle$ given by equation (1) can be written in terms of the states $|K_L\rangle$ and $|K_S\rangle$ as

$$|\Psi_0\rangle = (|K_S\rangle_L |K_L\rangle_R - |K_L\rangle_L |K_S\rangle_R)/\sqrt{2} \tag{3}$$

The time evolution of the nonseparable form of the two-particle wave function $|\Psi_0\rangle$, given by equations (1) and (3), correlates the oscillations between the $|K^0\rangle$ and $|\bar{K}^0\rangle$ states such that it carries the essence of nonlocal correlation, reminiscent of the EPR-type situation. If the left (right) kaon is observed to be a K^0 (Strangeness, $S = +1$) at a particular instant, then the right (left) kaon can be predicted, with certainty, to be observable as a $\bar{K}^0$ ($S = -1$) at that same instant. Alternatively, if the left (right) kaon decays in the K_s mode (CP $= +1$) then the right (left) kaon is bound to decay as a K_L (CP $= -1$) at some future time. It is to be noted that there is a subtle distinction between the K^0-$\bar{K}^0$ and K_L-K_S correlations; while the former holds only for equal proper times, the latter is a time-independent consequence of the nonseparable form of the wave function. This aspect has been clearly discussed by Selleri.[24]

Six[23] suggested that experimental test of this EPR-type situation would be the measuring of the joint probability $P_{--}^{(t_1 t_2)}$ of a double $\bar{K}^0$ observation (i.e., on two sides), at times t_1 and t_2 on the left and right sides, respectively. The quantum-mechanical prediction for $P_{--}(t_1, t_2)$ is given by

$$P_{--}(t_1, t_2) = |\langle \bar{K}_L^0 \bar{K}_R^0 | \Psi(t_1, t_2)\rangle|^2$$

where $|\Psi(t_1, t_2)\rangle$ is the state evolved from $|\Psi_0\rangle$ at $t = 0$:

$$|\Psi(t_1, t_2)\rangle = (1/\sqrt{2})\{|K_S\rangle_L |K_L\rangle_R \exp[-i(\lambda_S t_1 + \lambda_L t_2)]$$

$$- |K_L\rangle_L |K_S\rangle_R \exp[-i(\lambda_L t_1 + \lambda_S t_2)]\} \tag{4}$$

whence one obtains

$$P_{--}(t_1, t_2) = \tfrac{1}{8}\{\exp[-(\gamma_S t_1 + \gamma_L t_2)] + \exp[-\gamma_L t_1 - \gamma_S t_2)]$$
$$- 2\exp[-\gamma(t_1 + t_2)]\cos \Delta m(t_1 - t_2)\} \qquad (5)$$

where $\gamma = (\gamma_L + \gamma_S)/2$ and $\Delta m = m_L - m_s$.

Selleri derived an upper bound on $P_{--}(t_1, t_2)$ for the $K^0 - \bar{K}^0$ system $[P^u_{--}(t_1, t_2)]$ using a general argument based on the notion of local realism:

$$P^u_{--}(t_1, t_2) = \tfrac{1}{8}\{\exp[-(\gamma_S t_1 + \gamma_L t_2)] + \exp[-(\gamma_L t_1 + \gamma_S t_2)]\} \qquad (6)$$

This local-realistic upper bound differs from the quantum-mechanical prediction (5) by the absence of the interference term. Therefore, quantum mechanics leads to a prediction that violates equation (6) whenever the interference term is positive, that is, whenever $\cos \Delta m(t_1 - t_2) < 0$. The maximum possible discrepancy is calculated to be about 12% for $\gamma_S(t_2 - t_1) \simeq 5$. Some experimentalists* have shown interest in checking this instance of incompatibility between quantum mechanics and local realism, but no significant findings have yet been reported.

It is important to note that the experimental study envisaged in connection with equations (5) and (6) has an intrinsic handicap: for meaningful results, t_1 and t_2 must be shorter than the lifetimes of K_L and K_S respectively; i.e., one requires $t_1, t_2 \lesssim 10^{-10}$ s. The uncertainties involved in ensuring that the observations are at the specified instants t_1 and t_2 would be quite appreciable within such a small time interval. This difficulty may be circumvented by considering the time-integrated joint probabilities. This aspect of the envisaged experiment has recently been examined by Datta and Home[26] for the case of the B^0-$\bar{B}^0$ system. This system is almost identical to the K^0-$\bar{K}^0$ system, the only difference being that $\gamma_L = \gamma_S(=\gamma)$.† The eigenstates of the B^0-$\bar{B}^0$ system are analogous to the $|K_L\rangle$ and $|K_S\rangle$ states, and are denoted by $|B_H\rangle$ and $|B_L\rangle$ with masses m_H and m_L, respectively ($m_H > m_L$).

3. Proposal for a New Test Using the B^0-$\bar{B}^0$ System

Recently, experiments on the decay of the spin-1 $\Upsilon(4s)$ vector meson into a pair of neutral pseudoscalar mesons B^0-$\bar{B}^0$ have attracted considerable

---

* Such as O. Piccioni (University of California, San Diego), D. Jovanovic (Fermi Laboratory, Batavia), and S. Zenone (Concordia University, Montreal). Some relevant experimental data on the two-kaon state given by the wave function (1) is available[25] but has never been analyzed in the context of the EPR problem.

† $|K_L\rangle$ has a longer lifetime than $|K_S\rangle$ because the phase space available in its principal decay mode $|K_L\rangle \to 3\pi$ is smaller than that available in the decay mode $|K_S\rangle \to 2\pi$. For the decay of the $|B_H\rangle$ and $|B_L\rangle$ states, the phase space available is roughly the same. Hence their decay widths are taken to be identical.

attention in the search for evidence of B^0-$\bar{B}^0$ mixing. (See, for example, Buras *et al.*[27]) Following the treatment by Datta and Home,[26] we shall in this section analyze the possibility of investigating experimentally the EPR-type quantum nonlocal correlations within the framework of current experiments on the decay $Y(4s) \rightarrow B^0\bar{B}^0$. We shall focus our attention on the time-integrated joint probabilities, remembering that B^0 and $\bar{B}^0$ can be identified by their characteristic semileptonic mode of decay: $B^0 \rightarrow l^-\bar{\nu}X$; $\bar{B}^0 \rightarrow l^+\nu X$, where l and X denote lepton and hadron, respectively.

The experimental arrangement currently in use to study $Y(4s) \rightarrow B^0\bar{B}^0$ was designed to measure the parameter R defined as follows:

$$R = (N_{++} + N_{--})/(N_{+-} + N_{-+}) \tag{7}$$

where N_{++} is the total number of double $\bar{B}^0$ decays (corresponding to the observation of double l^+ decay products on both sides); N_{--} is the total number of double B^0 decays (corresponding to the observation of double l^- decay products on both sides); N_{+-} is the total number of $\bar{B}^0$ decays on the left associated with B^0 decays on the right (corresponding to the observation of l^+ decay products on the left associated with l^- decay products on the right); and N_{-+} is the total number of B^0 decays on the left associated with $\bar{B}^0$ decays on the right (corresponding to the observation of l^- decay products on the left associated with l^+ decay products on the right).

The parameter R is calculated by evaluating the quantities N_{ij} $(i, j = \pm)$. The general expression for N_{ij} is given by

$$N_{ij} = 2N_0\lambda^2 \int_0^\infty dt_1 \int_{t_1}^\infty dt_2\, P_{ij}(t_1, t_2) \tag{8}$$

where P_{ij} (t_1, t_2) is the joint probability for observing the decay products l^i and l^j on both sides at times t_1 and t_2, respectively; N_0 is the total number of $Y(4s)$ decays, and λ is the semileptonic decay width of B^0 decaying into a l^-, (which is equal to the semileptonic decay width of $\bar{B}^0$ decaying into a l^+).

The quantum-mechanical expressions for $P_{ij}(t_1, t_2)$ [derived from the nonseparable form of the wave function (4)] are given by

$$P_{++}(t_1, t_2) = P_{--}(t_1, t_2) = \tfrac{1}{8}\{2 \exp[-\gamma(t_1 + t_2)]$$
$$- 2 \exp[-\gamma(t_1 + t_2)] \cos \Delta m(t_2 - t_1)\} \tag{9}$$

$$P_{+-}(t_1, t_2) = P_{-+}(t_1, t_2) = \tfrac{1}{8}\{2 \exp[-\gamma(t_1 + t_2)]$$
$$+ 2 \exp[-\gamma(t_1 + t_2)] \cos \Delta m(t_2 - t_1)\} \tag{10}$$

where $\Delta m = m_H - m_L$.

Using equations (9) and (10), we obtain from equation (8) the following quantum-mechanical values for N_{ij}:

$$N_{++} = N_{--} = N_0\lambda^2(\tfrac{1}{4}\gamma^2 - \tfrac{1}{4}\alpha^2) \tag{11}$$

$$N_{+-} = N_{-+} = N_0\lambda^2(\tfrac{1}{4}\gamma^2 + \tfrac{1}{4}\alpha^2) \tag{12}$$

where $\alpha^2 = \gamma^2 + (\Delta m)^2$. This leads to the following quantum-mechanical prediction* for the parameter R defined by equation (7):

$$R_{QM} = x^2/(2 + x^2) \tag{13}$$

where $x = \Delta m/\gamma$.

The result given by equation (13) hinges on the quantum nonseparability which is built into the wave function (1) and is assumed to be maintained even after the particles have become well separated in space. The experimental verification of equation (13) will, therefore, constitute a test for quantum nonseparability in this EPR-type situation.

In this connection, it would be instructive to compare the prediction (13) with the corresponding prediction derived from the notion of local realism. As an example, let us consider Furry's hypothesis[29] in the following form: the wave function has the nonseparable form (1) at the time of production of the B^0-$\bar{B}^0$ pair, but after spatial separation of the two particles the wave function becomes an equal mixture (not superposition) of the two independent states $|B_H\rangle_L|B_L\rangle_R$ and $|B_L\rangle_L|B_H\rangle_R$. It is tempting to accept this hypothesis because it would enable us to avoid the conceptual anomalies arising from the quantum nonseparability presented by the EPR paradox. Einstein[30,31] himself favored such a proposal. Bohm and Aharonov[31] analyzed the tenability of Furry's hypothesis and pointed out the significance of testing whether this hypothesis leads to any conflict with the available experimental results.†

If in our present EPR-type example of the B^0-$\bar{B}^0$ system, we apply Furry's hypothesis to evaluate the general formula given by equation (8) we obtain

$$P_{++}(t_1, t_2) = P_{--}(t_1, t_2) = P_{+-}(t_1, t_2) = P_{-+}(t_1, t_2)$$

$$= \tfrac{1}{4}\exp[-\Gamma(t_1 + t_2)]$$

---

* Okun *et al.*[28] have derived equation (13) in another context.

† The suggestion by Bohm and Aharonov for verifying Furry's hypothesis, by measuring the polarization correlation of the two photons produced in a positron–electron annihilation, has certain critical aspects which have generated some controversy (see Peres and Singer[32] and Horne[33]).

whence

$$N_{++} = N_{--} = N_{+-} = N_{-+} = N\lambda^2(\tfrac{1}{4}\Gamma^2) \qquad (14)$$

which leads to the following prediction for the parameter R, according to Furry's hypothesis:

$$R_F = 1 \qquad (15)$$

Comparing the prediction (15) with the prediction (13) we observe that, within the present experimental framework for the study of the decay $\Upsilon(4s) \to B^0\bar{B}^0$, it is possible to discriminate between the predicted values of R_{QM} and R_F, unless $x^2 \gg 1$ (the case of maximal B^0-$\bar{B}^0$ mixing, corresponding to $\Delta m \gg \gamma$). In this context, it is interesting to note that the CLEO group[34] have already provided an experimental upper bound on R, i.e., $R < 0.3$. This may appear to disprove Furry's hypothesis. However, one must be careful to remember that this upper bound involves certain theoretical model-dependent inputs (such as the use of the spectator model of mesonic decay). Hopefully, with better statistics, it will be possible to set the empirical upper bound in a model-independent way.*

As regards the quantum-mechanical prediction for R given by equation (13), we note that R_{QM} is model-dependent. Confining our attention to within the ambit of the Glashow–Weinberg–Salam standard model of electroweak interactions, we note the following observation. There are two types of B^0 mesons: B_d^0 (the $b\bar{d}$ quark–antiquark bound state) and B_s^0 (the $b\bar{s}$ quark–antiquark bound state). $\Upsilon(4s)$ decays into the B_d^0-$\bar{B}_d^0$ system only. (The B_s^0-$\bar{B}_s^0$ channel is forbidden by kinematic considerations.) In this case the standard model predicts $\Delta m/\Gamma \lesssim 0.2\beta$, where β is estimated to be within the range of 0.33 to 1.5.[38] It therefore follows that $R_{QM} \ll 1$ according to the standard model, suggesting that it would be quite feasible to distinguish experimentally between R_F and R_{QM} based on the standard model.

By way of suggesting further work along these lines we wish to point out that, apart from calculating the parameter R using the various local-realistic models (analogous to the types[12,13] used for analyzing the EPR atomic-cascade experiments), it seems important to derive general bounds on R, from local realism, which are independent of the details of any particular model. This would enable us to make decisive use of the current

---

* Recent experimental studies[35] set an upper bound on R of 0.12. It has also been observed that the pertinent empirical data can accommodate $R = 1$ only if a large departure (i.e., greater than 50%) from the spectator-model predictions for B-mesonic decay is allowed.[36] Other experimental results, such as the limits on B^+ and B^0 meson lifetimes, are consistent with the spectator-model predictions and theoretical arguments suggest that the deviations from the spectator model, if any, should be not more than 30%.[37]

experimental studies on R in order to test the notion of local realism against quantum mechanics. Such tests would constitute a valuable complement to the current EPR experiments. They would, incidentally, also be the first EPR tests to involve electroweak interactions.

It is appropriate to recall here that Selleri[24] derived an upper bound on P_{--} (t_1, t_2) for the K^0-$\bar{K}^0$ system using a general argument based on local realism; interestingly, this bound coincides with the value obtained from Furry's hypothesis. (This is, of course, mere coincidence!) It would be interesting to examine whether Selleri's treatment can be extended to the B^0-$\bar{B}^0$ system in order to obtain general local-realistic bounds on the parameter R. Then empirical investigations may be restricted to the domain of incompatibility between such bounds and the quantum-mechanical prediction for R. This possibility is currently under study.

4. Quantum Nonlocality and CP Violation: A Curious Gedanken Example

Recently Datta, Home, and Raychaudhuri[39] (DHR) have examined the effect of CP violation on the EPR-type gedanken example, taking it to be a generalized situation of the K^0-$\bar{K}^0$ or B^0-$\bar{B}^0$ type system. A crucial feature, introduced by CP noninvariance, is that the eigenstates of the effective weak interaction Hamiltonian, which exhibit exponential decay, are not mutually orthogonal. DHR argue that this property leads to an intriguing incompatibility of quantum mechanics with Einstein's locality condition* at the statistical level, at least in the gedanken formulation. In the present section, we shall look at the details of the DHR example.

Let us consider that a vector meson V with $J^{PC} = 1^{--}$ decays, by strong interaction, into a pair of spatially-separated neutral pseudoscalar mesons (one of which is the antiparticle of the other) denoted by M^0-$\bar{M}^0$—typical examples are the decays of the $\Phi(1020)$ resonance into K^0-$\bar{K}^0$ and $\Upsilon(4s)$ into B^0-$\bar{B}^0$. Our analysis will be within the framework of the formalism discussed at the beginning of Section 2.

Corresponding to equations (1) and (2) respectively we now have

$$|\Psi_0\rangle = (|M^0\rangle_L|\bar{M}^0\rangle_R - |\bar{M}^0\rangle_L|M^0\rangle_R)/\sqrt{2} \qquad (16)$$

and

$$U(t, 0)|M_L\rangle = |M_L\rangle \exp(-i\lambda_L t) + |\phi_L(t)\rangle \qquad (17)$$

---

* In his own words: "On one supposition we should, in my opinion, absolutely hold fast: the real factual situation of the system S_2 is independent of what is done with the system S_1, which is spatially separated from the former."[40]

and a corresponding equation for $|M_S\rangle$. Taking CP noninvariance into account, we have

$$|M_L\rangle = N[(1 + \varepsilon)|M^0\rangle + (1 - \varepsilon)|\bar{M}^0\rangle]$$

$$|M_S\rangle = N[(1 + \varepsilon)|M^0\rangle - (1 - \varepsilon)|\bar{M}^0\rangle]$$

where the normalization factor N is equal to $[2(1 + |\varepsilon|^2)]^{-1/2}$ and the parameter ε is a measure of CP violation. CP noninvariance requires $|M_L\rangle$ and $|M_S\rangle$ to be mutually nonorthogonal: $\langle M_L|M_S\rangle = 4N^2\mathrm{Re}\,\varepsilon$. Unitarity of $U(t, O)$ implies that

$$\langle \phi_L(t)|\phi_L(t)\rangle = 1 - \exp(-\gamma_L t)$$

$$\langle \phi_S(t)|\phi_S(t)\rangle = 1 - \exp(-\gamma_S t) \qquad (18)$$

and

$$\langle \phi_L(t)|\phi_S(t)\rangle = \langle M_L|M_S\rangle[1 - \exp(i\Delta m - \gamma)t]$$

where $\Delta m = m_L - m_S$ and $\gamma = \frac{1}{2}(\gamma_L + \gamma_S)$.

Considering now M^0-$\bar{M}^0$ oscillations, the probability $P_{M^0 \to \bar{M}^0}(t, 0)$ $[P_{\bar{M}^0 \to \bar{M}^0}(t, 0)]$ of finding $\bar{M}^0$ at time t in a beam which was pure M^0 $(\bar{M}^0)$ at $t = 0$ is given by

$$P_{M^0 \to \bar{M}^0}(t, 0) = \tfrac{1}{4}|(1 - \varepsilon)/(1 + \varepsilon)|^2[\exp(-\gamma_L t) + \exp(-\gamma_S t)$$

$$- 2\exp(-\gamma t)\cos(\Delta m t)] \qquad (19)$$

$$P_{\bar{M}^0 \to \bar{M}^0}(t, 0) = \tfrac{1}{4}[\exp(-\gamma_L t) + \exp(-\gamma_S t) + 2\exp(-\gamma t)\cos(\Delta m t)]$$

The wave function $|\Psi(t)\rangle$, which has evolved from the wave function (16) at $t = 0$, is given by

$$|\Psi(t)\rangle = [|\psi_1\rangle + |\psi_2\rangle - |\psi_3\rangle]/2\sqrt{2}N^2(1 - \varepsilon^2) \qquad (20)$$

where

$$|\psi_1\rangle = [U(t, 0)|M_S\rangle]_L|M_L\rangle_R \exp(-i\lambda_L t) - [U(t, 0)|M_L\rangle]_L|M_S\rangle_R \exp(-i\lambda_S t)$$

$$|\psi_2\rangle = [U(t, 0)|M_S\rangle]_L|\phi_L(t)\rangle_R$$

$$|\psi_3\rangle = [U(t, 0)|M_L\rangle]_L|\phi_S(t)\rangle_R$$

We now consider two possible situations:

1. At time $t = T$, one detects the total number n_0 of $\bar{M}^0$ in the left hemisphere. Using

$$\langle \bar{M}^0 | U(t, 0) | M_L \rangle = N(1 - \varepsilon) \exp(-i\lambda_L t)$$

$$\langle \bar{M}^0 | U(t, 0) | M_S \rangle = -N(1 - \varepsilon) \exp(-i\lambda_S t)$$

and equations (18) and (20) we obtain

$$n_0(T) = \tfrac{1}{8} N_0 \{ (1 + \eta^2)[\exp(-\gamma_S T) + \exp(-\gamma_L T)]$$
$$+ 2(1 - \eta^2) \exp(-\gamma T) \cos(\Delta m T) \} \tag{21}$$

where N_0 is the initial number of $V \to M^0 \bar{M}^0$ decays and $\eta = |(1 - \varepsilon)/(1 + \varepsilon)|$ is the CP violation parameter which is independent of phase convention. The second term on the right-hand side of equation (21) arises out of the nonorthogonality between $|M_L\rangle$ and $|M_S\rangle$ which implies that $\langle \phi_L | \phi_S \rangle \neq 0$.

2. At time $t = T'$ $(<T)$, measurement is performed in the right hemisphere with three possible results: (a) undecayed M^0-$\bar{M}^0$; (b) decay products $|\phi_L(T')\rangle$; or (c) decay products $|\phi_S(T')\rangle$. Since $\langle \phi_L | \phi_S \rangle \neq 0$, the collapse of the wave function $|\Psi(T')\rangle$ to a mixture of the states $|\psi_1\rangle$, $|\psi_2\rangle$, and $|\psi_3\rangle$ is only "partial", i.e., there remains an additional term involving a superposition of $|\psi_2\rangle$ and $|\psi_3\rangle$. However, the exact treatment required for this "partial collapse" is rather unclear. We therefore proceed by assuming the collapse to be "total" and afterward estimate the inaccuracy involved. Thus we consider the wave function given by equation (20) to collapse to a mixture (not superposition) of the states:

$$|\psi_1\rangle = [(|M^0\rangle_L |\bar{M}^0\rangle_R - |\bar{M}^0\rangle_L |M^0\rangle_R)/\sqrt{2}] \exp[-i(\lambda_L + \lambda_S)T']$$

$$|\psi_2\rangle = |M_S\rangle_L \exp(-i\lambda_S T') |\phi_L(T')\rangle_R / 2\sqrt{2} N^2 (1 - \varepsilon^2)$$

$$|\psi_3\rangle = |M_L\rangle_L \exp(-i\lambda_L T') |\phi_S(T')\rangle_R / 2\sqrt{2} N^2 (1 - \varepsilon^2) \tag{22}$$

where we have omitted the terms involving $|\phi_L(T')\rangle_L$ and $|\phi_S(T')\rangle_L$ since they do not contribute to the total number of $\bar{M}^0$ detected in the left hemisphere at the subsequent time T. This number $n(T, T')$ can easily be calculated using equations (18), (19), and (22):

$$n(T, T') = \tfrac{1}{8} N_0 \{ (1 + \eta^2)[\exp(-\gamma_S T) + \exp(-\gamma_L T)]$$
$$+ 2(1 - \eta^2) \exp[-\gamma(T + T')] \cos \Delta m(T - T') \} \tag{23}$$

The surprising feature is that the quantum-mechanically predicted value of $n(T, T')$ turns out to be dependent on the time T' at which measurement is performed in the right hemisphere, thereby violating Einstein's locality condition at the statistical level. A quantitative measure of this nonlocal effect can be defined as $\alpha = |(n - n_0)/n_0|$.

Putting $T = T' + \delta T$ and assuming δT to be sufficiently small so that $\Delta m \delta T \ll 1$ and $\gamma \delta T \ll 1$, we get from equations (21) and (23) the following lower bound on α:

$$\alpha \geq 2 \left| \langle M_L | M_S \rangle \frac{\exp(-2\gamma T) - \exp(-\gamma T) + \exp(-2\gamma T)\gamma\delta T}{\exp(-\gamma_S T) + \exp(-\gamma_L T)} \right| \quad (24)$$

In the presence of CP invariance ($\langle M_L | M_S \rangle = 0$), this nonlocal effect vanishes. In the presence of CP violation, the existence of this nonlocal effect ($\alpha \neq 0$) apparently contradicts a general theorem[41,42] in quantum mechanics which states that all statistical measurements on any observable of one of the systems in a correlated pair, are independent of measurements performed on the other system. However, the applicability of this theorem to the case of a correlated pair consisting of weakly decaying particles, in the presence of CP noninvariance, has not hitherto been analyzed. A key point to be noted is that the proof of this theorem relies on the condition that the measurement alluded to involves collapse of the pure-state wave function to a mixture of mutually-orthogonal states. This is, evidently, not satisfied in our gedanken example because $\langle \phi_L | \phi_S \rangle \neq 0$ in the presence of CP violation.

We now turn our attention to the identification procedure for the decay products associated with the states $|\phi_L(T')\rangle$ and $|\phi_S(T')\rangle$. It is well known that the probability distribution $P(E)$ of the invariant mass of the decay products follows the Breit–Wigner form:

$$P_{L,S}(E) \propto 1/(E - m_{L,S})^2 + \tfrac{1}{4}\gamma_{L,S}^2$$

where E is the invariant mass. For the decays of $|M_L\rangle$ and $|M_S\rangle$ the probability distributions $P_L(E)$ and $P_S(E)$ in general overlap. If there was no overlap, one could unambiguously distinguish the decay products corresponding to the states $|\phi_L\rangle$ and $|\phi_S\rangle$. Equation (23) has been derived assuming such unambiguous distinction. The estimate of the nonlocal effect given by equation (24), therefore, involves an error which can be quantified by specifying the ratio $r = $ Overlap area between the two curves/Total area under any one curve. For the nonlocal effect to be perceptible, one must have $r \ll \alpha$.

A conservative upper bound on r (r_u) is given by $r_u = $ Height of the probability distribution at the point of intersection/Peak height of any one of the distributions. This yields

$$r_u = 16(\Delta m)^2 \gamma_S^2 / [16(\Delta m)^4 + 8(\gamma_L^2 + \gamma_S^2)(\Delta m)^2 + (\gamma_L^2 - \gamma_S^2)^2] \quad (25)$$

With equations (24) and (25) the condition $r_u \ll \alpha$ becomes, for the special case $(\Delta m)^2 \gg \gamma_L^2, \gamma_S^2$,

$$\gamma_S^2/(\Delta m)^2 \ll 2 \left| \langle M_L | M_S \rangle \frac{\exp(-2\gamma T) - \exp(-\gamma T) + \exp(-2\gamma T)\gamma\delta T}{\exp(-\gamma_S T) + \exp(-\gamma_L T)} \right| \quad (26)$$

The requirement of unitarity sets an upper bound on $\langle M_L | M_S \rangle$ given by[43]

$$\langle M_L | M_S \rangle \leq [\gamma_L \gamma_S / (\Delta m)^2 + \gamma^2]^{1/2}$$

whence equation (26) reduces to

$$\gamma_S^2 \ll 2 \left| (\gamma_L \gamma_S)^{1/2} \frac{\exp(-2\gamma T) - \exp(-\gamma T) + \exp(-2\gamma T)\gamma\delta T}{\exp(-\gamma_S T) + \exp(-\gamma_L T)} \right| \quad (27)$$

[in units of $(\Delta m)^2 = 1$].

Let us further assume that, for our gedanken example, $\gamma_S = \gamma_L = \gamma$. Then the condition (27) becomes

$$\gamma^2 \ll \gamma |[\exp(-\gamma T) - 1]| \quad (28)$$

which can be satisfied for suitable hypothetical values of γ; for example, taking $\gamma = 0.01$, the left-hand side equals 10^{-4}, while the right-hand side is about 6×10^{-3}.

So far our discussion has been in the context of hypothetical CP-violating systems. In reality, the K^0-$\bar{K}^0$ system provides the only well-studied example of such a system. Using the actual values $\Delta m = 0.53 \times 10^{10}\hbar s^{-1}$ and $\gamma_S = 582\gamma_L = 1.12 \times 10^{10}\hbar s^{-1}$, we see that the condition $r \ll \alpha$ cannot be satisfied for any choice of T and T'. The B^0-$\bar{B}^0$ system is another candidate for such a system. Theoretically, $\gamma_L = \gamma_S$ for this system and some theoretical models predict that $(\Delta m)^2 \gg \gamma_L^2, \gamma_S^2$. It is not yet clear whether the condition (28) can be satisfied for the B^0-$\bar{B}^0$ system.

The T'-dependence exhibited by equation (23), albeit hypothetical, is quite puzzling because it seems to permit, at least in principle, faster-than-light communication by Morse signaling, i.e., the performing of a measurement on the right at a distance vT' from the source (v is the average speed of M^0 and $\bar{M}^0$) and the subsequent counting of the $\bar{M}^0$ particles at a distance vT to the left, with T as close to T' as one chooses. In this connection, the following observation, due to D'Espagnat,[44] is worth considering: equation (23) really expresses the mean number of the observed $\bar{M}^0$ at a given instant.

In fact, this number fluctuates, and for superluminal signaling one requires this fluctuation to be smaller than any variation due to the measurement on the right. Whether this condition can actually be satisfied in the present example calls for careful analysis.*

It may interest the reader to refer, in the context of this example, to Bohm's quantum-potential approach[47,48] as an aid toward a deeper understanding of the nature of quantum nonlocality in the presence of CP violation.

To sum up our discussion of the DHR example, we may say that it suggests that if there is an EPR-type gedanken situation involving basis states which are mutually nonorthogonal but nevertheless at least partially distinguishable through some relevant physical observables then, in principle, there exists the possibility of nonlocal effects manifesting themselves at the statistical level. It appears possible that a situation of this type may occur in the unique case of CP nonconservation which leads to nonorthogonality between the physically-observable states. (These states can, at least in principle, be partially distinguished from each other through differences in the nonoverlap areas between the probability distributions of the invariant masses of the decay products corresponding to these states.)

ACKNOWLEDGMENTS. I am grateful to F. Selleri for encouragement and valuable suggestions. It is my pleasure to thank A. Datta and A. Raychaudhuri for stimulating collaboration which introduced me to the fascinating physics of K^0-$\bar{K}^0$ and B^0-$\bar{B}^0$ systems. A part of this chapter is based on an invited talk given at the Satellite Meeting of the Second International Symposium on Foundations of Quantum Mechanics held at Tokyo (September, 1986) under the auspices of Hitachi Ltd. and the Physical Society of Japan. I gratefully acknowledge the warm hospitality of the sponsors of this conference and the pleasure of many inspiring discussions with its participants.

---

* As pointed out by A. Peres, one could raise the following objection to the DHR example: in view of the so-called quantum doctrine, which holds that it is not possible to simultaneously measure quantities that correspond to noncommuting observables, it appears that one cannot simultaneously determine precisely whether there are decay products in each of the two states $|\phi_L\rangle$ and $|\phi_S\rangle$ because these states are nonorthogonal in the presence of CP nonconservation.

However, it must be stressed that the justification for the above quantum doctrine is rather dubious. Lande[45] raised serious doubts about its validity. Later, Park and Margenau[46] provided a comprehensive critical examination of this question and justified Lande's doubts. They showed the precise points of vulnerability in the standard arguments put forward to defend this dogma. They also furnished counterexamples of simultaneous-measurement schemes for noncommuting observables. It is, therefore, felt that such a controversial proposition should not be invoked to assess the DHR example which admittedly involves a variety of intricacies that require further clarification, particularly since "imprecise" measurements are considered here.

References

1. A. Einstein, B. Podolsky, and N. Rosen, *Phys. Rev.* **47**, 777 (1935).
2. D. Bohm, *Quantum Theory*, pp. 614–619, Prentice-Hall, Englewood Cliffs, N.J. (1951).
3. A. Aspect, P. Grangier, and G. Roger, *Phys. Rev. Lett.* **47**, 460 (1981); **49**, 91 (1982).
4. A. Aspect, J. Dalibard, and G. Roger, *Phys. Rev. Lett.* **49**, 1804 (1982).
5. A. Aspect and P. Grangier, *Lett. Nuovo Cim.* **43**, 345 (1985).
6. T. W. Marshall, E. Santos and F. Selleri, *Phys. Lett. A* **98**, 5 (1983).
7. T. W. Marshall, *Phys. Lett. A* **99**, 163 (1983).
8. A. Garuccio and F. Selleri, *Phys. Lett. A* **103**, 99 (1984).
9. T. W. Marshall and E. Santos, *Phys. Lett. A* **108**, 373 (1985).
10. D. Home and T. W. Marshall, *Phys. Lett. A* **113**, 183 (1985).
11. M. Ferrero and E. Santos, *Phys. Lett. A* **116**, 356 (1986).
12. T. W. Marshall, E. Santos, and F. Selleri, in: *Open Questions in Quantum Physics* (G. Tarozzi and A. Van der Merwe, eds.), pp. 87–101, Reidel, Dordrecht (1985).
13. T. W. Marshall, in: *Microphysical Reality and Quantum Formalism* (G. Tarozzi and A. Van der Merwe, eds.), Reidel, Dordrecht (1987).
14. T. K. Lo and A. Shimony, *Phys. Rev. A* **23**, 3003 (1981).
15. E. Santos, *Phys. Rev. A* **30**, 2128 (1984).
16. A. Shimony, *Phys. Rev. A* **30**, 2130 (1984).
17. F. Selleri, *Phys. Lett. A* **108**, 197 (1985).
18. C. Dewdney, M. A. Dubois, P. R. Holland, A. Kyprianidis, L. Laurent, M. Pain, and J. P. Vigier, *Phys. Lett. A* **113**, 135 (1985).
19. T. D. Lee and C. N. Yang, unpublished.
20. D. R. Inglis, *Rev. Mod. Phys.* **33**, 1 (1961).
21. M. Jammer, in: *The Philosophy of Quantum Mechanics*, p. 308, Wiley, New York (1974).
22. B. D'Espagnat, *Conceptual Foundations of Quantum Mechanics*, pp. 85–86, W. A. Benjamin, London (1976).
23. J. Six, *Phys. Lett. B* **114**, 200 (1982).
24. F. Selleri, *Lett. Nuovo Cim.* **36**, 521 (1983).
25. R. Armenteros *et al.*, at: The International Conference on High Energy Physics, CERN, Geneva (1962).
26. A. Datta and D. Home, *Phys. Lett. A* **119**, 3 (1986).
27. A. J. Buras, W. Slominski, and H. Steger, *Nucl. Phys. B* **245**, 369 (1984).
28. L. B. Okun, V. I. Zakharov, and B. M. Pontecorvo, *Lett. Nuovo Cim.* **13**, 218 (1975).
29. W. H. Furry, *Phys. Rev.* **49**, 393 (1936).
30. A. Einstein, private communication to D. Bohm; cf. Ref. 31, p. 1071.
31. D. Bohm and Y. Aharonov, *Phys. Rev.* **108**, 1070 (1957); *Nuovo Cim.*, **17**, 964 (1960).
32. A. Peres and P. Singer, *Nuovo Cim.* **15**, 964 (1960).
33. M. Horne, Ph.D. Thesis, pp. 82–85, Boston University (1970).
34. P. Avery, *Phys. Rev. Lett.* **53**, 1309 (1984).
35. Reported at: The XXIII International Conference on High Energy Physics, Berkeley, USA (July, 1986).
36. A. Bean *et al.*, *CLEO-86-11* (1986).
37. I. I. Bigi, *Phys. Lett. B* **169**, 101 (1986).
38. A. J. Buras, W. Slominski, and H. Steger, *Nucl. Phys. B* **238**, 529 (1984).
39. A. Datta, D. Home, and A. Raychaudhuri, *Phys. Lett. A* **123**, 4 (1987).
40. In: *Albert Einstein: Philosopher–Scientist* (P. A. Schilpp, ed.) p. 85, The Library of Living Philosophers, Inc., Evanston, Illinois (1949).
41. G. C. Ghirardi, A. Rimini, and T. Weber, *Lett. Nuovo Cim.* **27**, 293 (1980).
42. B. D'Espagnat, *Phys. Rep.* **110**, 201 (1984).

43. T. D. Lee and L. Wolfenstein, *Phys. Rev. B* **138**, 1490 (1965).
44. B. D'Espagnat, private communication.
45. A. Lande, *New Foundations of Quantum Mechanics*, p. 124, Cambridge University Press (1965).
46. J. L. Park and H. Margenau, in: *Perspectives in Quantum Theory* (W. Yourgran and A. Van der Merwe, eds.), pp. 37–70, MIT Press (1971).
47. D. Bohm, C. Dewdney, and B. J. Hiley, *Nature* **315**, 294 (1985).
48. D. Bohm and B. J. Hiley, *Phys. Rev. Lett.* **55**, 2511 (1985).

Even Local Probabilities Lead to the Paradox

FRANCO SELLERI

1. Introduction

The essence of the Einstein–Podolsky–Rosen (EPR) paradox[1] is the incompatibility at the experimental level between some empirical predictions of quantum theory and the consequences of local realism. This incompatibility has become fully evident after the 1965 paper by Bell[2] in which a class of local hidden-variable models was shown to lead to the validity of an inequality ("Bell's inequality") that is sometimes grossly violated by quantum mechanics. That pioneering work led slowly to the awareness that, more generally, it was the philosophy of local realism itself (in any traditional definition of terms *locality* and *realism*) that disagreed at the empirical level with the existing quantum theory. This striking diagreement led Stapp[3] to the conclusion that "Bell's theorem is the most profound discovery of science."

In spite of this there are several authors who are not convinced that the incompatibility between local realism and quantum theory really exists and many papers have been written in which the validity of "Bell's theorem" is questioned. It has even been observed that all the authors of such papers, collected together, could give rise to an international conference entirely devoted to "refutations" of the EPR paradox!

The *motivation* of such points of view is very respectable as it arises from a double conviction: The idea that local realism is an extremely natural

FRANCO SELLERI • Department of Physics, University of Bari, 70126 Bari, Italy.

point of view in physics, and the idea that the great successes of the existing quantum theory prove its power and its strength. The arguments usually made for the restoration of a locally realistic quantum physics are basically of two types:

1. Those accepting the existence of the EPR paradox as a theoretical statement, but questioning the meaning of the experimental investigations performed up to now, especially as far as the validity of the ergodic hypothesis is concerned.
2. Those questioning the very existence of the EPR paradox and attributing its demonstrations to wrong and/or insufficiently general reasoning.

Ideas of the first type are well represented in the present book. It is rather against arguments of type (2) that the present chapter is addressed. Two rigorous proofs of the EPR paradox are reviewed and developed.

The first part of the chapter (up to Section 5) consists of a demonstration based on the relatively strong assumptions of local realism[4] which are those of the original *deterministic* approach of Einstein, Podolsky, and Rosen and of Bell.

The second part of the chapter is based on some far weaker assumptions of local realism, so weak that most people would probably believe them to be natural consequences of the existing quantum theory. In fact, only *local probabilities* are assumed.[5] The fact that the EPR paradox still results shows how strikingly strong the quantum-mechanical rejection of local realism is. Our results imply that local realism is *a priori* incompatible with the "additional assumptions" made in the analysis of Bell-type experiments.

2. The Singlet State

We assume that a spin-0 object ε is given and that it decays into two spin-$\frac{1}{2}$ objects α and β. Let us suppose that, in the final state, the spatial part of the total wave function is separated, meaning that it can be written as a product of two terms appreciably different from zero in two different regions of space, R_α and R_β, separated by a very large distance. We denote the Pauli matrices representing the spin angular momentum for α [β] by $\sigma_1(\alpha)$, $\sigma_2(\alpha)$, and $\sigma_3(\alpha)$ [$\sigma_1(\beta)$, $\sigma_2(\beta)$, and $\sigma_3(\beta)$]. Furthermore, we let $u_+(\alpha)$ and $u_-(\alpha)$ [$u_+(\beta)$ and $u_-(\beta)$] be the eigenvectors corresponding to the eigenvalues $+1$ and -1 of the Pauli matrix $\sigma_3(\alpha)$ [$\sigma_3(\beta)$] representing the third component of the spin angular momentum for α [β]:

$$\sigma_3(\alpha)u_\pm(\alpha) = \pm u_\pm(\alpha)$$
$$\sigma_3(\beta)u_\pm(\beta) = \pm u_\pm(\beta)$$

$$(1)$$

There exist actual physical situations in which the spin state vector for the system (α, β) has to be the "singlet" state vector given by

$$\eta_0 = [u_+(\alpha)u_-(\beta) - u_-(\alpha)u_+(\beta)]/\sqrt{2} \qquad (2)$$

The singlet state is invariant under rotation, so that it can be written

$$\eta_0 = [u_+^n(\alpha)u_-^n(\beta) - u_-^n(\alpha)u_+^n(\beta)]/\sqrt{2} \qquad (3)$$

where $u_\pm^n(\alpha)$ and $u_\pm^n(\beta)$ denote eigenstates of $\sigma_n(\alpha) \equiv \boldsymbol{\sigma}(\alpha) \cdot \hat{\mathbf{n}}$ and of $\sigma_n(\beta) \equiv \boldsymbol{\sigma}(\beta) \cdot \hat{\mathbf{n}}$, respectively, $\hat{\mathbf{n}}$ being an *arbitrary* unit vector. Therefore

$$\sigma_n(\alpha)u_\pm^n(\alpha) = \pm u_\pm^n(\alpha)$$

$$\sigma_n(\beta)u_\pm^n(\beta) = \pm u_\pm^n(\beta) \qquad (4)$$

From these well-known properties of the singlet state vector three important empirical consequences follow:

(C1) As already mentioned, η_0 describes some empirically well-known pairs of quantum objects. An example is that of the (rare) decay of the neutral pion into an electron–positron pair.

(C2) Let us suppose that an observer O_α $[O_\beta]$ performs a measurement of $\sigma_n(\alpha)$ $[\sigma_n(\beta)]$ in the region R_α $[R_\beta]$. Then, even if the two observers perform their measurements at different times, they will find opposite results for all (α, β) pairs. This holds for arbitrary $\hat{\mathbf{n}}$.

(C3) η_0 predicts large (approximately 41%) violations of Bell's inequality.

In the following discussion quantum theory will be assumed *only* in the sense that (C1), (C2), and (C3) are being used. These three empirical consequences of the theory will therefore suffice for establishing the EPR paradox.

3. EPR Paradox for a Complete Quantum Theory

Consider a large ensemble E composed to N decays $\varepsilon \to \alpha + \beta$. The singlet state η_0, which by assumption describes all these (α, β) pairs, implies, by consequence (C2), that if the observer O_α measures $\sigma_n(\alpha)$ and finds +1 (−1), then a subsequent measurement of $\sigma_n(\beta)$, performed by the second observer O_β, will give −1 (+1) with certainty.

The latter observation is the basis of the EPR paradox and we shall formulate it by assuming that O_α and O_β use *perfect instruments* which never fail in detecting the measured systems and in recording the right values of the measured observables.

Let us assume that O_α measures $\sigma_n(\alpha)$ on the α-objects of an ensemble E_1, where $E_1 \subset E$ and contains l pairs. He finds the results

$$\sigma_n(\alpha) = A_1, A_2, \ldots, A_l \tag{5}$$

each of which equals ± 1.

The results of subsequent measurements of $\sigma_n(\beta)$, performed by O_β, on the l β-objects of E_1 can then be predicted *with certainty*, by consequence (C2), to be

$$\sigma_n(\beta) = -A_1, -A_2, \ldots, -A_l \tag{6}$$

This prediction can be checked, e.g., on an esemble $E_2 \subset E_1$. We are of course assuming that quantum theory is correct, so that every conceivable check of its predictions will give a positive answer.

The next three steps in our reasoning use Einstein locality in an essential way, and allow us to introduce elements of reality into all the β-objects of E_2, and to deduce their necessary existence in E_1 and E as well (see Figure 1 for the logical structure of the proof):

1. *Reality.* The prediction with certainty expressed by equation (6) allows us to use the EPR reality criterion and to attribute to each

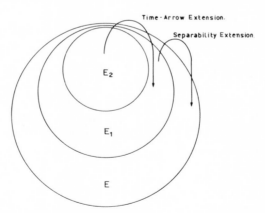

Figure 1. Structure of the proof of the EPR paradox: E_2 is the set of (α, β) pairs for which both $\sigma_n(\alpha)$ and $\sigma_n(\beta)$ are measured. The EPR reality criterion attributes an element of reality to every α of this set; E_1 is the set of (α, β) pairs for which only $\sigma_n(\beta)$ is measured. The "time arrow" assumption attributes an element of reality to every α of this set; and E is the set of all (α, β) pairs. In general, no measurements on either the α or the β are made. The assumption of separability extends the attribution of an element of reality to every α of this set.

of the β-objects of E_2 an element of reality r which determinates the result of the future measurement of $\sigma_n(\beta)$.*

2. *Time arrow.* Assuming that the element r is not generated retroactively in time by the future measurement on β, we may conclude that r belongs to all $\beta \in E_1$ and not only to those β (that is, $\beta \in E_2$) for which a measurement of $\sigma_n(\beta)$ will actually be performed.†

3. *Separability.* If the objects α and β are separated by a very large distance, we assume that the element of reality r of β could not have been created by the measurement performed on α. Consequently, r exists even if no measurement on α is carried out and, therefore, it can be attributed to all $\beta \in E$ and not only to those β (that is, $\beta \in E_1$) for which a measurement of $\sigma_n(\alpha)$ has actually been carried out.‡

One should note that the elements of reality introduced must actually be of two types, since different effects presuppose different causes. We will denote by r_+ that value (or those values) of r which leads to the prediction "+1" in equation (6), and by r_- that value (or those values) of r which leads to the prediction "-1" in equation (6). Therefore the element of reality introduced can assume a plurality of values and is, in fact, a "variable."

We continue our reasoning by assuming that quantum mechanics is complete. If this is true then, by definition of completeness, the elements of reality r_+ and r_- must have counterparts in the quantum theoretical description of the β-objects. But the only state vectors which describe the outcome of a measurement of $\sigma_n(\beta)$ as certain and equal to +1 or -1, are $u_+^n(\beta)$ and $u_-^n(\beta)$, respectively. One of these two descriptions must therefore apply to each β of E.

Actually, $u_+^n(\beta)$ and $u_-^n(\beta)$ must be found in E with an equal frequency (of $\frac{1}{2}$ each), since the two results +1 and -1 are present in equation (6) with equal probability. This last point is, however, inessential as far as the EPR paradox is concerned.

By consequence (C2), it is clear that the $\hat{n}$-component of the *total* spin of (α, β) must be zero in all cases. The only state vectors of (α, β) describing β by either $u_+^n(\beta)$ or $u_-^n(\beta)$ and giving zero for the $\hat{n}$-component of the

---

* The negation of this assumption was, essentially, the basis of Bohr's refutation of the EPR paper (see Bohr[6]).

† The rejection of this assumption leads to solutions of the EPR paradox in which retroactions in time are explicitly introduced (see, for example, Chapter 10).

‡ The rejection of this assumption leads to solutions of the EPR paradox in which action-at-a-distance is explicitly admitted (see, for example, Chapter 9).

total spin of the two particles are

$$u_+^n(\alpha)u_-^n(\beta) \qquad \text{and} \qquad u_-^n(\alpha)u_+^n(\beta) \qquad (7)$$

We thus reach the conclusion that the total ensemble E is actually a mixture of the previous two (factorizable) state vectors. But this conclusion quite clearly contradicts the consequence (C3) since it implies the validity of Bell's inequality.*

We thus arrive at an absurd conclusion [i.e., a contradiction with the consequence (C3)].

This implies that the set of hypotheses made during our earlier reasoning is self-contradictory and that at least one of these assumptions must be discarded. It was assumed that:

1. The empirical consequences of quantum theory (C1), (C2), and (C3) are correct.
2. Einstein locality (i.e., reality + time arrow + separability) holds.
3. Quantum theory is complete.

The obvious way out of the paradox, as proposed by Einstein, Podolsky, and Rosen in 1935, is to declare quantum mechanics incomplete: after all, it is only natural that a statistical theory should give an incomplete description of reality. What is far more important is that this description should be correct!

It will be shown in the following section that, following Bell's discovery of his inequality in 1966, this obvious solution to the paradox no longer exists.

4. The Case of Incompleteness

Let us accept the foregoing EPR reasoning up to and including the point at which it was recognized that the element of reality r can assume the different values r_+ and r_-.

Instead of assuming that quantum mechanics is complete, we shall now consider the contrary hypothesis, i.e., that quantum theory is *not* complete, whence it follows that the discovery of an exactly predictable result does not imply that one should adopt an eigenvector of the corresponding observable.

The first part of the earlier EPR reasoning allowed us to conclude that r belongs to β even if no measurement on either α or β is carried out. In this way, r could be attributed to every β of E.

---

* All mixtures of factorizable state vectors lead to the validity of Bell's inequality. This was first shown by Capasso *et al.*[7]

This means that *the assignment of an element of reality has been completely disentangled from all acts of measurement.*

Given that consequence (C2) holds for arbitrary $\hat{n}$, one may introduce an arbitrary number of elements of reality for every β of E: It does not matter whether they arise from incompatible observables (described by noncommuting operators) since all that is assumed is Einstein locality and the correctness of the consequences (C1), (C2), and (C3).

Given the symmetry between α and β, one can also introduce an arbitrary number of elements of reality for every α of E.

In particular, given four unit vectors $\hat{a}$, $\hat{a}'$, $\hat{b}$, and $\hat{b}'$ one can assign two elements of reality, s and s', to α and two elements of reality, t and t', to β, as follows:

$$s, \text{corresponding to } \sigma(\alpha) \cdot \hat{a}$$

$$s', \text{corresponding to } \sigma(\alpha) \cdot \hat{a}'$$

$$t, \text{corresponding to } \sigma(\beta) \cdot \hat{b} \qquad (8)$$

$$t', \text{corresponding to } \sigma(\beta) \cdot \hat{b}'$$

This new notation could be redundant, since a new symbol has been introduced every time a new observable had led to the discovery of an element of reality. There is, of course, no reason why these elements of reality should all be independent of one another. They could be related in various ways. It is even conceivable that nature is so simple that even when considering different observables we invariably uncover the same one element of reality which fixes all observables. This would only mean that our notation is too rich, but not that it is wrong! In all cases we could deduce the existence of the 2^4 subensembles which are the key to the proof of Bell's inequality given below.

Given the dichotomic nature of all the observables of our problem, every one of the elements of reality (8) will assume two (types of) values, depending on the predicted value ($+1$ or -1) of the corresponding observable:

$$s = s_+ \text{ or } s_-$$

$$s' = s'_+ \text{ or } s'_-$$

$$t = t_+ \text{ or } t_- \qquad (9)$$

$$t' = t'_+ \text{ or } t'_-$$

Let us introduce the following new symbols:

$A(a, s)$ is the predicted value of the observable $\sigma(\alpha) \cdot \hat{\mathbf{a}}$

$A(a', s')$ is the predicted value of the observable $\sigma(\alpha) \cdot \hat{\mathbf{a}}'$

$$(10)$$

$B(b, t)$ is the predicted value of the observable $\sigma(\beta) \cdot \hat{\mathbf{b}}$

$B(b', t')$ is the predicted value of the observable $\sigma(\beta) \cdot \hat{\mathbf{b}}'$

Depending on the alternatives (9), each of these predicted values can assume either the value +1 or the value −1. The statistical ensemble E of N pairs (α, β) thus splits into 2^4 subensembles, each of which has a fixed value ("plus" or "minus") of every one of these four elements of reality. Denoting by $n(s, s', t, t')$ the number of pairs with fixed values of s, s', t, and t', one clearly has

$$\sum n(s, s', t, t') = N \tag{11}$$

where the sum is taken over the 2^4 possible sets of values of the four variables, that is, over the choices (9) of s, s', t, and t'.

By the very definition of correlation function (i.e., average product of results of correlated measurements on α and β) it follows that it is given by

$$P(a, b) = \frac{1}{N}\sum n(s, s', t, t')A(a, s)B(b, t) \tag{12}$$

where the sum is again taken over the 2^4 possible sets of values of the four variables. Similar expressions to (12) are easily formed for the remaining three correlation functions $P(a, b')$, $P(a', b)$, and $P(a', b')$. Remembering that all the functions (10) can assume only the values +1 and −1, one easily obtains

$$|P(a, b) - P(a, b')| \le \frac{1}{N}\sum n(s, s', t, t')|B(b, t) - B(b', t')|$$

$$(13)$$

$$|P(a', b) + P(a', b')| \le \frac{1}{N}\sum n(s, s', t, t')|B(b, t) + B(b', t')|$$

By summing these inequalities and using the obvious relation

$$|B(b, t) - B(b', t')| + |B(b, t) + B(b', t')| = 2$$

one obtains Bell's inequality

$$\Delta \equiv |P(a, b) - P(a, b')| + |P(a', b) + P(a', b')| \leqslant 2 \qquad (14)$$

In this way, we have again obtained the same contradiction with consequence (C3) as we did in the case of a complete quantum theory.*

5. The Nature of the Paradox

The same result (i.e., the validity of Bell's inequality) has now been deduced under the opposite assumptions of completeness and lack of completeness of the existing quantum theory. This shows that *the question of completeness is totally irrelevant, as far as the EPR paradox is concerned.* We may therefore conclude that Einstein locality is incompatible with quantum theory at the empirical level, and that this incompatibility is expressed by, for instance, Bell's inequality, which was shown in the previous section to be an inevitable result of the paradoxical reasoning, but which is violated by quantum mechanics.

Einstein locality has been defined as the set of three assumptions: *reality, time arrow,* and *separability.* We must thus conclude that either these assumptions, or the empirical consequences of quantum theory introduced before [(C1), (C2), and (C3)], are not correct.

It should be noted that the above formulation of the paradox, like all the considerations developed in this chapter, is based on the existence of perfect detectors. This is so because predictions with certainty were assumed. possible.

Recent papers that discuss the EPR paradox in the case of low-efficiency detectors[9-14] have shown that the earlier atomic-cascade experiments[15-21] were far from providing conclusive evidence against Einstein locality. The reason is that the quantum-mechanical predictions *in the case of low-efficiency detectors* do not violate Bell's inequality.

The introduction of additional assumptions[22] permits the deduction, from Einstein locality, of other inequalities, *much stronger than Bell's inequality,* that are violated by the quantum-mechanical predictions also in the case of low-efficiency detectors. The experimental evidence indeed shows that these new inequalities (which can be called strong inequalities) are violated, but this is probably only because the additional assumptions are not true in nature! In fact Einstein locality, by itself, reduces the numerical value of Δ, defined by equation (14), to 2, from its value of $2\sqrt{2}$ as predicted by quantum theory. The introduction of the additional assumptions (for

---

* This proof of Bell's inequality is similar to that given by Wigner.[8]

example, "no enhancement") allows one to prove, typically, that $\Delta \leq 0.02$. Now, this inequality is violated experimentally, but it appears far more reasonable to question the correctness of an arbitrary *ad hoc* assumption introduced in the reasoning, than to question a basic scientific principle such as Einstein locality.

The fundamental idea behind the above proof of the paradox is that of *determinism*, with the variables s, s', t, and t' fixing in each case, *a priori* and completely, the result of a future measurement [refer, for example, to equation (10)]. The limited scope of determinism was realized following the discovery of Bell's inequality and, in consequence, probabilistic proofs were looked for. The most popular approach was that of Clauser–Horne (CH) factorizability,[23] but it has since been criticized in various ways. Its weak point is understood to lie in its conception of probability for a *single* quantum object and for a *single* pair of correlated objects which is alien to our current understanding of probability theory.*

In the following discussion, CH factorizability will be avoided altogether and probabilities will only be introduced for statistical ensembles.

6. Probabilistic Reality

Let us consider a set S of physical objects $\alpha_1, \alpha_2, \ldots, \alpha_N$ all of the same type (e.g., photons):

$$S = \{\alpha_1, \alpha_2, \ldots, \alpha_N\} \tag{15}$$

and let a dichotomic physical quantity $A(a)$ be given which can be measured on the α-objects composing S, and which can only assume the values ± 1. We assume that the following reality criterion holds:

Probabilistic Reality Criterion. If it is possible to predict the existence of a subset S' of S:

$$S' = \{\alpha_{i_1}, \alpha_{i_2}, \ldots, \alpha_{i_l}\} \tag{16}$$

and if it is possible to correctly predict that future measurements of $A(a)$ on S' will give the results $+1$ and -1 with respective probabilities p_+ and p_-, and if the previous predictions can be made without in any way disturbing the α-objects of S and of S', then it will be said that a physical property Λ' belongs to S' that fixes the probabilities

$$p_+ = p(a_+, \Lambda') \quad \text{and} \quad p_- = p(a_-, \Lambda') \tag{17}$$

---

* For a critical discussion of CH factorizability see Selleri.[24]

Of course, $p_+ + p_- = 1$.

The previous statement is called the "probabilistic reality criterion" (PRC) because it provides a natural generalization (i.e., to sets of objects) of the criterion for physical reality (of individual objects) put forward by Einstein, Podolsky, and Rosen in 1935. It can be applied to EPR experiments, where each of the α-objects composing S is physically correlated with a (separated) object β. Let the set of these new objects be

$$T = \{\beta_1, \beta_2, \ldots, \beta_N\} \tag{18}$$

For instance, the objects α_i and β_i might have been produced together in the decay of an unstable system ε_i ($i = 1, 2, \ldots, N$).

Thus, in a typical EPR experiment there are two observers: O_α who performs measurements of $A(a)$ on the set S of α-objects, and O_β who measures a second dichotomic observable $B(b)$ on the set T of β-objects. Assuming that O_β is the first to perform his measurements (in the laboratory frame) and that T' is the subset of T for which $B(b) = +1$ has been found, then O_β himself can predict the existence of the subset S' (composed of the α-objects which are individually correlated with the β-objects composing T') for which at some later time O_α will find the results $A(a) = +1$ and $A(a) = -1$, with respective probabilities p_+ and p_-.

Note that p_+ and p_- are, in general, different from the corresponding probabilities for the whole ensemble S. Therefore, if O_α does in fact find the predicted probabilities, it must be concluded that there is something in the physical reality of S' that somehow generates p_+ and p_-.

It is in this sense that the physical property Λ' is attributed to S'. This property is a part of the physical reality of S' that is not detected *directly* in the usual experiments, in which only eigenvalues and probabilities are measured. It is therefore of the same nature as the so-called "hidden variables," even though it obviously is more general than these since it is attributed to statistical ensembles rather than to individual systems. At present the existence of Λ' is conjectural, but the conjecture is based on causality and it should one day become possible to detect Λ' by means of suitable experiments.

We assumed that Λ' fixes the probabilities [see equation (17)]. In general, however, the probabilities, which become actual when the observable $A(a)$ is concretely measured, may also depend on the instrument that is used.

This is, in fact, the case in many real experiments where the precision of the analyzers, the efficiency of the counters, and so on, indeed affect the values of the probabilities observed. These probabilities thus have a dual nature, since they reflect actual properties of both the ensemble S' *and* the instrument used. The notation (17) therefore represents the true nature of probabilities by showing them to be dependent on both Λ' and $A(a)$ (the latter being denoted by the symbol a).

The foregoing discussion shows us one reason why the probabilities were not attributed *directly* to S' as real properties and, instead, the physical property Λ' was introduced: Had we defined the probabilities, themselves as real then, although none of the results to be derived below would be any different, we would have exposed our approach to the criticism of attributing reality to something that can only be made concrete by the intervention of a measuring apparatus. This criticism is known as "counterfactuality."

The exact nature of Λ' is not of interest here. This physical property results, in general, from the existence of a large number of "elements of reality" of the individual α-objects composing S' all of which cooperate to generate a physical situation for S' in which the probabilities are precisely p_+ and p_-.

7. Probabilistic Separability

Let us suppose that, at some time, the space parts of the quantum-mechanical wave packets describing an α-object and its corresponding β-object are separated by a very large distance. If this is the case for all (α, β) pairs, we shall then consider it a sufficient condition for the physical separability of the sets S and T. This leads to the following assumption:

Probabilistic Separability. Measurements performed on T (the set of β-objects) cannot generate physical properties (such as Λ', introduced above) which belong to S (the set of α-objects) or to any subset S' of S, and *vice versa*.

We shall use this assumption of separability in the following way. We consider a measurement of $B(b)$ made by O_β. The results $B(b) = +1$ and $B(b) = -1$ split T into two subsets: T' [all objects with $B(b) = +1$] and T'' [all objects with $B(b) = -1$]. Now, the conditions of the PRC are satisfied by virtue of our assuming probabilistic separability (PS), and a physical property Λ' can therefore be attributed to S'. Naturally, a symmetrical reasoning can be carried out for T'' by starting from the set S'' of α-objects related to it and verifying that the probabilities are correctly predicted in this case as well and, in consequence, introducing a new physical property Λ'' for S''.

The splitting of T into T' and T'' takes place when $B(b)$ is measured. It is this splitting that allows one to *identify* the sets S' and S'' to which the physical properties Λ' and Λ'' are attributed. But these physical properties cannot be created at-a-distance in S by the measurements on T, this being excluded by PS.

Therefore an unknown, but nevertheless actual splitting of S into S' (with property Λ') and S'' (with property Λ'') exists even if no measurement is made on the β-objects.

The mere conclusion of existence of S' and S'' (with their respective properties, Λ' and Λ'') suffices for establishing the EPR paradox, as we shall see, even when it is not possible to identify the α-objects composing S' and S''.

8. Time Arrow for Probabilities

Our next question is whether the splitting of S, into S' and S'', is generated by the instrument used for measuring $A(a)$ in S. It is not possible to believe that the subsets exhibit probabilities which may be very different from those of the set S as a whole, for example $p(a_+, \Lambda')$ and $p(a_-, \Lambda')$ in S', and $p(a_+, \Lambda'')$ and $p(a_-, \Lambda'')$ in S'', merely because of the intervention of the experimental apparatus. It is instead clear that the above probabilities are (at least partly) determined by physical properties of the subsets themselves. It is for this reason that a physical property Λ' (Λ'') has been assigned to S' (S'').

It is, of course, natural to assume that the physical property Λ' (Λ'') belongs to S' (S'') even if no actual measurement is carried out on these subsets, provided that extensive experience with S and T has shown that probabilities for subsets are correctly predicted.

By "extensive experience" we imply nothing more than the standard scientific procedure: If a large body of empirical evidence, gathered by repeated observations of the result $B(b) = +1$, proves that p_+ and p_- invariably show up in the subset S'—and, therefore, by our PRC, we can say that the property Λ' invariably belongs to S'—then we can conclude that Λ' is real even if no measurement on the set S' is made.

We therefore make the following assumption:

Time Arrow for Probabilities. Physical properties of sets (such as Λ' of S') are the cause of the observed probabilities and not vice versa. In other words, there exists an arrow of time from the past to the future that cannot be reversed.

With the assumption of a time arrow for probabilities (TAP) we exclude any possibility of future measurements creating the physical properties of the statistical ensembles on which they will be carried out.

9. Probabilistic Einstein Locality

The assumptions made in the three previous sections, taken *together*, constitute the idea of probabilistic Einstein locality:

$$probabilistic\ Einstein\ locality = \begin{cases} probabilistic\ reality\ criterion\ (PRC) \\ probabilistic\ separability\ (PS) \\ time\ arrow\ for\ probabilities\ (TAP) \end{cases}$$

This generalization of Einstein locality has two very important advantages: First, it is not limited conceptually to those rare (or even nonexistent) cases in which one is able to make *totally certain* predictions. Second, it can be applied to all types of quantum-mechanical descriptions (i.e., eigenstates, superpositions, and mixtures) and not only to eigenstates of the predicted observable. *Thus whenever one can predict values of probabilities our probabilistic approach can be applied, no matter what the source of information leading to the prediction* (e.g., trusted theory, previous experiment).

This present form of Einstein locality is so general, and rests on such a weak type of realism, that one wonders if even the physicists of the Copenhagen and Göttingen schools could not have accepted it. It will however be shown that it nevertheless suffices for the derivation of Bell-type inequalities. Since these latter inequalities are violated by quantum-mechanical predictions, it follows that not even the present weak form of local realism is compatible with existing quantum theory.

10. Locality for Conditional Probabilities

It will now be shown that Einstein locality, in its probabilistic formulation, leads to the conclusion that quantum probabilities must have a more detailed structure than is implied by the existing quantum theory.

Let us consider a large set E of correlated (α, β) pairs:

$$E = \{(\alpha_1, \beta_1), (\alpha_2, \beta_2), \ldots, (\alpha_N, \beta_N)\}$$

and we suppose that a dichotomic observable $A(a) = \pm 1$ is measured on the set S of α-objects:

$$S = \{\alpha_1, \alpha_2, \ldots, \alpha_N\}$$

and that either the observable $B(b) = \pm 1$, or the observable $B(b') = \pm 1$, is measured on the set T of β-objects:

$$T = \{\beta_1, \beta_2, \ldots, \beta_N\}$$

Thus E is the *physical union* of S and T (see Figure 2). The measurement of $B(b)$ divides T, and therefore also E, into two parts. Let $E'(b)$ $[E''(b)]$ be the subset of E for which $B(b) = +1$ $[B(b) = -1]$ has been obtained. Naturally

$$E = E'(b) \cup E''(b)$$

We introduce four (conditional) probabilities:

$\omega(a_+|b_+)$ is the probability of finding $A(a) = +1$ in $E'(b)$

$\omega(a_-|b_+)$ is the probability of finding $A(a) = -1$ in $E'(b)$

$\omega(a_+|b_-)$ is the probability of finding $A(a) = +1$ in $E''(b)$

$\omega(a_-|b_-)$ is the probability of finding $A(a) = -1$ in $E''(b)$

If these probabilities can be predicted correctly, as we suppose, then they can be considered to be determined by some real physical properties of the subsets $E'(b)$ and $E''(b)$ to which they belong, by virtue of our PRC. We will assume that *these physical properties belong locally to the sets of α-objects and are not generated at-a-distance by the measurements performed on the β-objects.* This is, of course, a consequence of our PS assumption, discussed in Section 7.

We stress that a qualitative formulation of separability is sufficient for our present purposes. A quantitative formulation of the same idea will be given in the next section [see equation (24) below].

From the separability condition, it follows that the probabilities $\omega(a_\pm|b_\pm)$ are necessary consequences of real properties of some *unknown* subsets even if no measurement of $B(b)$ has been performed. In fact, if this were not the case, we would have to say that it is precisely the measurement of $B(b)$ on the β-objects that creates at-a-distance the physical properties of the α-objects, in violation of PS.

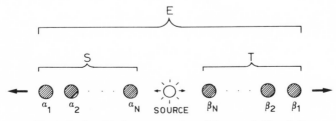

Figure 2. Graphical representation of the sets S, T, and E.

It should be noted that separability here does *not* mean that $\omega(a_\pm|b_\pm)$ is independent of b. This point can perhaps best be illustrated with an example. Let us consider the set E of married European men currently in Japan who have a wife in Europe, and define the dichotomic properties:

a = is French   (true = +1, false = −1)

b = has a French wife in Europe (true = +1, false = −1)

In addition, we consider the splitting of E into $E'(b)$ and $E''(b)$, generated by the b-property above.

Obviously $\omega(a_+|b_+)$ will be very close to one, thus showing that there is a dependence on b. It could, however, take a very different value if we considered a different condition, as for example:

c = weighs more than 75 kg (true = +1, false = −1)

In this case one would expect a smaller value for $\omega(a_+|c_+)$, close to the fraction of French men in E. Thus $\omega(a_+|b_+)$ depends on b, but this does not imply a violation of separability since the physical correlation between a and b was established during the common past of the pairs considered.

Returning to our general argument, we can then say that separability only means that the b-dependence of $\omega(a_\pm|b_\pm)$ does not arise through an action-at-a-distance caused by some event concerning the β-objects. Therefore, we deduce from Einstein locality that *even if no measurement of $B(b)$ is made* there exists a subensemble* $E'(b)$ of E with two properties:

1. Its population $N'(b)$, divided by the total population N of E, equals the *a priori* probability $Q(b_+)$ of measuring $B(b)$ on the β-objects and finding +1:

$$Q(b_+) = N'(b)/N$$

2. It has a real physical property which gives rise to the probabilities $\omega(a_+|b_+)$ and $\omega(a_-|b_+)$ of measuring $A(a)$ on the α-objects and finding ±1, respectively.

---

* Actually, there must be several equivalent subsets of this type since, in a probabilistic approach, it cannot be fixed *a priori* which β-systems will produce the result $B(b) = 1$ upon measurement. This complication is, however, irrelevant to our purposes.

Probabilities can also be introduced for a different splitting of E into $E'(b')$ and $E''(b')$ arising from a (possible, but not necessarily actual) measurement of $B(b')$ on the β-systems. For example, we could introduce $\omega(a_+|b'_+)$ as the probability of finding $A(a) = +1$ in $E'(b')$. Once more, we apply the PRC and declare $\omega(a_+|b'_+)$ to result from a real property of $E'(b')$. We have thus introduced three probabilities:

$$\omega(a_+|b_+) \text{ which applies to } E'(b)$$

$$\omega(a_+|b_-) \text{ which applies to } E''(b)$$

$$\omega(a_+|b'_+) \text{ which applies to } E'(b')$$

They all refer to the result $A(a) = +1$. As conditional probabilities they must depend on b or b', that is, on observables defined for the β-systems, but we stress that this only true in the sense that the ensembles $E'(b)$, $E''(b)$, and $E'(b')$, to which they apply, depend on b or b'.

Since $E = E'(b) \cup E''(b)$, it follows that $E'(b')$, which is part of E, is necessarily composed of a fraction γ of pairs of $E'(b)$ and a fraction $1 - \gamma$ of pairs of $E''(b)$, where $0 \leq \gamma \leq 1$. If the ensembles $E'(b)$ and $E''(b)$ are homogeneous, in the sense that every part of each of them gives a probability for the $A(a) = +1$ which is exactly equal to that for the whole ensemble, then one necessarily has

$$\omega(a_+|b'_+) = \gamma\omega(a_+|b_+) + (1 - \gamma)\omega(a_+|b_-) \tag{19}$$

Equation (19) gives $\omega(a_+|b'_+)$ as a weighted average of $\omega(a_+|b_+)$ and $\omega(a_+|b_-)$. This means that $\omega(a_+|b'_+)$ must lie in the interval between the other two probabilities. But the latter condition is in general not satisfied. Take, for instance, the quantum-mechanical predictions for the singlet state which are given by

$$\omega(a_+|b_+) = \sin^2[(a - b)/2]$$

$$\omega(a_+|b_-) = \cos^2[(a - b)/2] \tag{20}$$

$$\omega(a_+|b'_+) = \sin^2[(a - b')/2]$$

If we substitute the functions (20) into equation (19) and take $a - b = \pi/2$, then the right-hand side equals $\frac{1}{2}$, while the left-hand side varies according to the value of $a - b'$.

Therefore, if our formulation of Einstein locality is to be accepted, the quantum-mechanical probabilities, such as given by the functions (20), cannot arise from homogeneous ensembles but must result from averages of variously sized probabilities which arise from subensembles of which $E'(b)$, $E''(b)$, and $E'(b')$ are themselves the unions. In short, we can say that if quantum probabilities arise from something real, then they must be endowed with a structure that is not given by the existing quantum theory.

11. New Proof of Bell's Inequality

The most general way of giving structure to a probability defined for an ensemble of similar objects, is to introduce *individual probabilities*. By so doing one is only in danger of overgeneralizing. It could happen, for instance, that individual probabilities can actually assume only a few different values, so that only probabilities for subensembles would have to be introduced. This obviously less-general situation can be recovered as a particular case of an approach which utilizes individual probabilities. Their introduction in this paper is therefore only for the sake of simplicity.

We can then say that the physical properties, required by the PRC, for ensembles such as $E'(b)$ and $E''(b)$, *arise from physical averages of individual physical properties which in general vary between different objects.*

For every object α of E it is therefore true that *the object α possesses the physical property λ which gives rise to the probabilities $p(a_+, \lambda)$ and $p(a_-, \lambda)$ for the two results $A(a) = \pm 1$, respectively, if and when the observable $A(a)$ is measured on this object.* Of course,

$$p(a_+, \lambda) + p(a_-, \lambda) = 1 \tag{21}$$

The above conclusion holds for all objects α of the set E. This is true because similar conclusions of inhomogeneity hold for $E'(b)$ and $E''(b)$, and because the union of these two ensembles is E.

As was stressed before, the situation is symmetrical in α and β so that similar individual probabilities, $q(b_+, \lambda)$ and $q(b_-, \lambda)$, can be introduced for the observation of $B(b) = \pm 1$, respectively, with

$$q(b_+, \lambda) + q(b_-, \lambda) = 1 \tag{22}$$

By means of the individual probabilities, it is now possible to express all the interesting *ensemble* probabilities. Let $P(a_+)$ and $P(a_-)$ [$Q(b_+)$ and $Q(b_-)$] be the probabilities of measuring $A(a)$ [$B(b)$] on the α-objects

[β-objects] of E and of finding the results ± 1, respectively. One can obviously write

$$P(a_+) = \langle p(a_+, \lambda) \rangle_E, \qquad P(a_-) = \langle p(a_-, \lambda) \rangle_E$$

$$Q(b_+) = \langle q(b_+, \lambda) \rangle_E, \qquad Q(b_-) = \langle q(b_-, \lambda) \rangle_E \tag{23}$$

where the symbols $\langle \cdot \cdot \cdot \rangle$ denote, as usual, an average.

The conditional probability of finding $A(a) = \pm 1$ on the α-objects *if* $B(b) = \pm 1$ on the β-objects has previously been found is given by

$$\omega(a_\pm | b_+) = \langle p(a_\pm, \lambda) \rangle_{E'(b)}$$

$$\omega(a_\pm | b_-) = \langle p(a_\pm, \lambda) \rangle_{E''(b)} \tag{24}$$

According to the general rules of probability calculus, the joint probabilities of finding $A(a) = \pm 1$ on the α-objects *and* $B(b) = \pm 1$ on the correlated β-objects are given by

$$\Omega(a_\pm, b_+) = \omega(a_\pm | b_+) Q(b_+)$$

$$\Omega(a_\pm, b_-) = \omega(a_\pm | b_-) Q(b_-) \tag{25}$$

With the help of the joint probabilities we can now calculate the correlation function, which is given by

$$P(a, b) = \Omega(a_+, b_+) - \Omega(a_+, b_-) - \Omega(a_-, b_+) + \Omega(a_-, b_-) \tag{26}$$

By using equations (23), (24), and (25) it is easy to show that correlation function (26) becomes

$$P(a, b) = Q(b_+) \langle \pi(a, \lambda) \rangle_{E'(b)} - Q(b_-) \langle \pi(a, \lambda) \rangle_{E''(b)} \tag{27}$$

where

$$\pi(a, \lambda) = p(a_+, \lambda) - p(a_-, \lambda) \tag{28}$$

is a difference of two probabilities, so that

$$|\pi(a, \lambda)| \leq 1 \tag{29}$$

Considering next a new observable $A(a')$ of the α-objects and the previous one $B(b)$ for the β-objects, we see that all the previous considerations can be repeated to obtain

$$P(a', b) = Q(b_+) \langle \pi(a', \lambda) \rangle_{E'(b)} - Q(b_-) \langle \pi(a', \lambda) \rangle_{E''(b)} \tag{30}$$

From equations (30) and (27) we easily obtain

$$P(a, b) - P(a', b)$$
$$= Q(b_+)\langle[\pi(a, \lambda) - \pi(a', \lambda)]\rangle_{E'(b)} - Q(b_-)\langle[\pi(a, \lambda) - \pi(a', \lambda)]\rangle_{E''(b)}$$

whence

$$|P(a, b) - P(a', b)| \leq Q(b_+)\langle|\pi(a, \lambda) - \pi(a', \lambda)|\rangle_{E'(b)}$$
$$+ Q(b_-)\langle|\pi(a, \lambda) - \pi(a', \lambda)|\rangle_{E''(b)} \tag{31}$$

But the right-hand side of equation (31) is just the weighted average of $|\pi(a, \lambda) - \pi(a', \lambda)|$ over the whole ensemble $E = E'(b) \cup E''(b)$. In fact, $Q(b_+)$ $[Q(b_-)]$ is the *a priori* probability of the set $E'(b)$ $[E''(b)]$ and, of course,

$$Q(b_+) + Q(b_-) = 1$$

holds as a consequence of equations (22) and (23). Therefore

$$|P(a, b) - P(a', b)| \leq \langle|\pi(a, \lambda) - \pi(a', \lambda)|\rangle_E \tag{32}$$

A completely analogous reasoning can be carried out for the same two observables $A(a)$ and $A(a')$ of the α-objects, but for a different dichotomic observable $B(b') = \pm 1$ of the β-objects. Taking now the sum of equations (27) and (30), with b' in place of b, we obtain

$$|P(a, b') + P(a', b')| \leq \langle|\pi(a, \lambda) + \pi(a', \lambda)|\rangle_E \tag{33}$$

Since, besides (29), one also has

$$|\pi(a', \lambda)| \leq 1 \tag{34}$$

it is easy to prove that the inequality

$$|\pi(a, \lambda) - \pi(a', \lambda)| + |\pi(a, \lambda) + \pi(a', \lambda)| \leq 2 \tag{35}$$

holds for arbitrary λ. By summing equations (32) and (33) and using equation (35), one easily obtains

$$|P(a, b) - P(a', b)| + |P(a, b') + P(a', b')| \leq 2 \tag{36}$$

which is Bell's inequality.

The previous proof is based on a completely general probabilistic formulation of Einstein locality and the usual definition of probability, considered as a frequency in a statistical set of similar objects. In particular, no use was made of the CH factorizability condition. *It can thus be said that the previous proof of the inequality is the first satisfactory formulation of the EPR paradox.* Note the difference between the deterministic proof of the paradox, given in the first part of the chapter, and the probabilistic reasoning: Here, no discussion of completeness of quantum theory was needed, since a kind of incompleteness (the structure of conditional probabilities) was *deduced* directly from probabilistic Einstein locality.

12. All the Inequalities of Einstein Locality

Given an arbitrary linear combination of correlation functions, it has been shown[25] that a corresponding inequality must be satisfied if the CH formulation of Einstein locality is accepted. Therefore, the physical content of Einstein locality is not exhausted by Bell's inequality, which appears to be only one example of an infinite set of inequalities of the same nature. It has, furthermore, been shown that there are correlation functions which satisfy Bell's inequality for all possible choices of their arguments while violating *other* inequalities, thus showing their incompatibility with Einstein locality (see Chapter 3).

In the present section, all these inequalities will be placed on a firmer footing by deducing them again from the probabilistic approach of the previous sections.

Given the dichotomic observables $A(a_\mu)$ ($\mu = 1, 2, \ldots, m$) for the α-objects and $B(b_\nu)$ ($\nu = 1, 2, \ldots, n$) for the β-objects, let us consider the linear combination of correlation functions

$$\sum_{\mu\nu} c_{\mu\nu} P(a_\mu, b_\nu) \tag{37}$$

where the $c_{\mu\nu}$ are $m \times n$ real coefficients. We will prove that Einstein locality implies

$$\left| \sum_{\mu\nu} c_{\mu\nu} P(a_\mu, b_\nu) \right| \leq M_0 \tag{38}$$

where

$$M_0 = \text{Max} \left\{ \sum_{\mu\nu} c_{\mu\nu} \xi_\mu \eta_\nu \right\} \tag{39}$$

the right-hand side being calculated for that choice of the sign factors $\xi_\mu(=\pm 1)$ and $\eta_\nu(=\pm 1)$ which maximizes it.

In order to proceed with the proof, we need to enrich our notation slightly. It has already been specified that N is the number of pairs contained in the (total) set E. Now we also introduce $N'(b)$, the number of pairs in $E'(b)$, and $N''(b)$ the number of pairs in $E''(b)$. Naturally

$$N'(b) + N''(b) = N \tag{40}$$

Furthermore we specify

$$I = \{1, 2, \ldots, N\}$$

the set of integers from 1 to N, corresponding to the (α, β) pairs in E, and similarly $I'(b)$, the set of integers corresponding to the pairs in $E'(b)$, and $I''(b)$, the set of integers corresponding to the pairs in $E''(b)$. Since

$$E'(b) \cup E''(b) = E$$

one must have

$$I'(b) \cup I''(b) = I \tag{41}$$

We will use an index i, running from 1 to N, to indicate a particular (α, β) pair. The same index can be used in the probabilities introduced in the previous sections, which can now be written

$$\omega(a_\pm | b_+) = \frac{1}{N'(b)} \sum_{i \in I'(b)} p(a_\pm, \lambda_i)$$

$$\omega(a_\pm | b_-) = \frac{1}{N''(b)} \sum_{i \in I''(b)} p(a_\pm, \lambda_i) \tag{42}$$

Note that the probabilities of measuring $B(b)$ and finding $+1$ and -1 [denoted by $Q(b_+)$ and $Q(b_-)$, respectively] are measured by the relative populations of the ensembles [$E'(b)$ and $E''(b)$, respectively] in which the two results have been registered:

$$Q(b_+) = N'(b)/N, \qquad Q(b_-) = N''(b)/N \tag{43}$$

It follows from equations (25), (42), and (43) that

$$\Omega(a_\pm, b_+) = \frac{1}{N} \sum_{i \in I'(b)} p(a_\pm, \lambda_i)$$

$$\Omega(a_\pm, b_-) = \frac{1}{N} \sum_{i \in I''(b)} p(a_\pm, \lambda_i) \tag{44}$$

so that the correlation function becomes

$$P(a, b) = \frac{1}{N} \sum_{i \in I'(b)} \pi(a, \lambda_i) - \frac{1}{N} \sum_{i \in I''(b)} \pi(a, \lambda_i) \tag{45}$$

where

$$\pi(a, \lambda_i) = p(a_+, \lambda_i) - p(a_-, \lambda_i) \tag{46}$$

Obviously equation (21) implies that

$$|\pi(a, \lambda_i)| \leq 1 \tag{47}$$

By introducing the sign function

$$\eta_i(b) = \begin{cases} +1 & \text{if } i \in I'(b) \\ -1 & \text{if } i \in I''(b) \end{cases} \tag{48}$$

we can write the correlation function as

$$P(a, b) = \frac{1}{N} \sum_{i \in I} \eta_i(b) \pi(a, \lambda_i) \tag{49}$$

Note the different natures of $\eta_i(b)$ and $\pi(a, \lambda_i)$ (i.e., dichotomic and continuous, respectively).

Since the result (49) holds for arbitrary values of a and b, we can write

$$\sum_{\mu\nu} c_{\mu\nu} P(a_\mu, b_\nu) = \frac{1}{N} \left\{ \sum_{i \in I} \sum_{\mu\nu} c_{\mu\nu} \pi(a_\mu, \lambda_i) \eta_i(b_\nu) \right\} \leq \frac{1}{N} \sum_{i \in I} M_0^{(i)} \tag{50}$$

where

$$M_0^{(i)} = \text{Max} \left\{ \sum_{\mu\nu} c_{\mu\nu} \pi(a_\mu, \lambda_i) \eta_i(b_\nu) \right\} \tag{51}$$

The maximum on the right-hand side of equation (51) is to be calculated for fixed values of the (given) coefficients $c_{\mu\nu}$, but for all conceivable values of the sign factors $\eta_i(b_\nu)$ and of the functions $\pi(a_\mu, \lambda_i)$. The right-hand side of equation (51) is linear in each one of the quantities $\pi(a_\mu, \lambda_i)$ and its maximum can therefore be found on the boundary, that is, remembering equation (47), for

$$\pi(a_\mu, \lambda_i) = \pm 1 = \xi_i(a_\mu) \tag{52}$$

where the new terms $\xi_i(a_\mu)$ are sign factors $(=\pm 1)$, just like $\eta_i(b_\nu)$. We can then write

$$M_0^{(i)} = \text{Max}\left\{ \sum_{\mu\nu} c_{\mu\nu} \xi_i(a_\mu) \eta_i(b_\nu) \right\} \tag{53}$$

We recall that there are 2^{m+n} different choices of the sign factors. From among these we have to choose that which maximizes the right-hand side of equation (53). But this choice depends only on the coefficients $c_{\mu\nu}$ and has nothing to do with either the index i, or the parameters a_μ and b_ν. The result is therefore

$$M_0^{(i)} = M_0 \leqslant \text{Max}\left\{ \sum_{\mu\nu} c_{\mu\nu} \xi_\mu \eta_\nu \right\} \tag{54}$$

implying that $M_0^{(i)}$ is independent of i, and that the sign factors must always be chosen in the same way, independently of a_μ and b_ν.

Substitution of equation (54) into (50) then gives

$$\sum_{\mu\nu} c_{\mu\nu} P(a_\mu, b_\nu) \leqslant M_0 \tag{55}$$

Since it is straightforward to show that the minimum value of the left-hand side of equation (55) is simply $-M_0$, equation (55) itself becomes equivalent to equations (38) and (39), which is what we set out to prove.

A discussion of the physical meaning of these new inequalities is presented in Chapter 3 of this book.

13. The Need for New Experiments

In Section 12, Bell-type inequalities were deduced from a very general formulation of Einstein locality resting on:

1. The probabilistic reality criterion

2. Probabilistic separability
3. The time-arrow assumption

These three ideas give rise to such a weak form of realism that one wonders if the physicists of the Copenhagen and Göttingen schools could not perhaps have accepted it. It was nevertheless possible to deduce Bell-type inequalities, and since these are violated by quantum-mechanical predictions it follows that not even the present form of Einstein locality is compatible with existing quantum theory.

The results obtained are important also from another point of view. All the conditions in the present chapter were made under the assumption of perfect detectors, this being evident from the condition

$$E'(b) \cup E''(b) = E$$

introduced in Section 10, which is equivalent to the idea that every time a measurement of $B(b)$ is performed, either $+1$ or -1 is obtained. (What could instead happen is that the arrival of some β-objects goes undetected.) Nevertheless the necessity of introducing variable probabilities was proved, in full agreement with the conclusions deduced from experiments that have been performed on EPR pairs. It has in fact been shown elsewhere[9-14] that published experiments can be explained by means of local realistic models *precisely because the quantum efficiency of the detectors that were employed is small.* This means that the usual (arbitrary) additional hypotheses that are made in connection with these experiments are false and therefore that *detection* probabilities, which differ between the separate subsets of the ensemble of quantum pairs that were studied, must exist.

This is, clearly, a very similar conclusion to the one which we reached in this chapter, although we could not deal with variable detection probabilities, given our assumption of certain detection of the α- and β-objects by their respective instruments. In this chapter we obtained evidence, so to say, of variable "transmission" probabilities, while elsewhere[9-14] variable detection probabilities were introduced. All this points toward considering quantum objects as endowed of variable properties, even within a general probabilistic scheme.

But this simply means that the usual additional assumptions are themselves in direct contradiction to Einstein locality. It cannot therefore be correct to deduce that Einstein locality does not hold in nature, from experimental results (like violations of the strong inequalities), which are obtained using assumptions that already contain the impossibility of Einstein locality!

A new set of experiments is therefore clearly needed. Fortunately, there are several interesting proposals which can, in principle, lead to a solution of the EPR paradox (see Chapters 4, 5, 8, and 16).

References

1. A. Einstein, B. Podolsky, and N. Rosen, *Phys. Rev.* **47**, 777 (1935).
2. J. S. Bell, *Physics* **1**, 195 (1965).
3. H. P. Stapp, *Nuovo Cim. B* **40**, 191 (1977).
4. F. Selleri and G. Tarozzi, *Phys. Lett. A* **119**, 101 (1986).
5. F. Selleri, in: *Microphysical Reality and Quantum Formalism* (A. van der Merwe *et al.*, eds.), D. Reidel, Dordrecht (1988).
6. N. Bohr, *Phys. Rev.* **48**, 696 (1935).
7. V. Capasso, D. Fortunato, and F. Selleri, *Int. J. Theor. Phys.* **7**, 319 (1973).
8. E. P. Wigner, *Am. J. Phys.* **38**, 1005 (1970).
9. T. W. Marshall, E. Santos, and F. Selleri, *Phys. Lett. A* **98**, 5 (1983).
10. T. W. Marshall, *Phys. Lett. A* **99**, 163 (1983).
11. S. Caser, *Phys. Lett. A* **102**, 152 (1984).
12. A. Garuccio and F. Selleri, *Phys. Lett. A* **108**, 197 (1985).
13. D. Home and T. W. Marshall, *Phys. Lett. A* **113**, 183 (1985).
14. E. Ferrero and E. Santos, *Phys. Lett. A* **116**, 356 (1986).
15. S. J. Freedman and J. F. Clauser, *Phys. Rev. Lett.* **28**, 938 (1972).
16. R. A. Holt and F. M. Pipkin, University of Harvard, preprint (1974).
17. E. S. Fry and R. C. Thompson, *Phys. Rev. Lett.* **37**, 465 (1976).
18. A. Aspect, P. Grangier, and G. Roger, *Phys. Rev. Lett.* **47**, 460 (1981).
19. A. Aspect, P. Grangier, and G. Roger, *Phys. Rev. Lett.* **49**, 91 (1982).
20. A. Aspect, J. Dalibard, and G. Roger, *Phys. Rev. Lett.* **49**, 1804 (1982).
21. W. Perrie, A. J. Duncan, H. J. Beyer, and H. Kleinpoppen, *Phys. Rev. Lett.* **54**, 1790 (1985).
22. J. F. Clauser, M. A. Horne, A. Shimony, and R. A. Holt; *Phys. Rev. Lett.* **23**, 880 (1969).
23. J. F. Clauser and M. A. Horne, *Phys. Rev. D* **10**, 526 (1974).
24. F. Selleri, in: *Determinism in Physics* (E. Bitsakis and N. Tambakis, eds.), Gutenberg, Athens (1985).
25. A. Garuccio and F. Selleri, *Found. Phys.* **10**, 209 (1980).

7

The Experimental Investigation of the Einstein–Podolsky–Rosen Question and Bell's Inequality

A. J. DUNCAN AND H. KLEINPOPPEN

1. Introduction

The argument of Einstein, Podolsky, and Rosen, in 1935, concerning the completeness of quantum mechanics and the possible existence of hidden variables,[1] was originally couched in terms of the position and momentum coordinates of a pair of particles which could assume a continuous range of values. Subsequently, in 1951, Bohm[2] put the argument in terms of an initially spin-0 system which dissociates into two spin-$\frac{1}{2}$ systems, the components of the spin of which could only take on discrete values. Later, in 1957, Bohm and Aharonov[3] discussed the problem with reference to the polarization properties of the γ-ray photons resulting from the annihilation of para-positronium. Then, in 1964, Bell[4] derived his inequality, which allowed a quantitative distinction to be made between the predictions of quantum mechanics and local realism, and, in 1969, Clauser et al.[5] showed how this inequality might be tested experimentally by examining the polarization properties of the two photons emitted in certain atomic cascade processes.

A. J. DUNCAN AND H. KLEINPOPPEN • Atomic Physics Laboratory, University of Stirling, Stirling FK9 4LA, Scotland, United Kingdom.

1.1. The EPR Argument for Photons

The EPR argument itself can, in fact, be discussed most conveniently in terms of photons. Consider, for example, an isotropic source emitting two photons, one in the $+z$ direction and one in the $-z$ direction. Then it can be shown from simple consideration of angular momentum and parity that the state vector must be in one of the forms $|\psi\rangle = (|R\rangle_1|R\rangle_2 \pm |L\rangle_1|L\rangle_2)/\sqrt{2}$, where $|R\rangle_1$ and $|L\rangle_1$ represent photons of right-handed and left-handed helicity, respectively, propagating in the $-z$ direction, with a corresponding definition for $|R\rangle_2$ and $|L\rangle_2$. In terms of linear polarization basis states the state vector takes one of the following forms:

$$|\psi\rangle_+ = \frac{1}{\sqrt{2}} (|x\rangle_1|x\rangle_2 + |y\rangle_1|y\rangle_2) \tag{1.1}$$

$$|\psi\rangle_- = \frac{1}{\sqrt{2}} (|x\rangle_1|y\rangle_2 - |y\rangle_1|x\rangle_2) \tag{1.2}$$

where $|x\rangle_1$ and $|y\rangle_1$ represent photons propagating in the $-z$ direction with linear polarization, respectively, in the x direction and y direction, with a corresponding definition for $|x\rangle_2$ and $|y\rangle_2$. Most experiments have made use of a 0-1-0 cascade process in the source, which gives rise to the $|\psi\rangle_+$ even-parity form for the two-photon state vector. A few, however, have made use of a 1-1-0 cascade, in which case, and in the case of positronium annihilation, $|\psi\rangle_-$, the odd-parity form, is the appropriate state vector.

The EPR argument can now be stated very simply: Assuming, for example, that the two photons are described by the state vector given by equation (1.1), then, if we make a measurement and obtain the result that, say, the photon propagating in the $-z$ direction is polarized in the x direction, we know the state vector has been reduced to the form $|x\rangle_1|x\rangle_2$. We can then say, with certainty, that the photon propagating in the $+z$ direction will be found, upon measurement, to be polarized also in the x direction. Thus, the polarization of the second photon in the x direction corresponds to an "element of reality." But, because of the rotational symmetry which exists about the z axis, the choice of x direction is arbitrary, so we are forced to conclude that the polarization of the second photon in any direction at right angles to the z axis is specified and corresponds to an "element of reality." A similar conclusion may be reached for the other photon of the pair. The specification of the polarization of a photon in more than one direction at a time is more than is allowed by quantum mechanics and, therefore, quantum mechanics, according to Einstein, Podolsky, and Rosen, cannot be a complete theory.

The correlation between the polarization properties of the photons and the EPR argument are both a consequence of the nonlocal predictions which follow from the nonfactorizable form of the state vector given in equations

(1.1) and (1.2). Clearly the EPR argument would not hold if the two-photon state were to change spontaneously from one which is described by state vectors of the kind shown in equations (1.1) and (1.2) to one described by a mixture of states of the form $|x\rangle_1|x\rangle_2$ [equation (1.1)] or $|x\rangle_1|y\rangle_2$ [equation (1.2)], where x and y take on all possible orientations at right angles to the z axis. Such a suggestion was considered by both Schrödinger[6] and Furry,[7] and is sometimes referred to as the Schrödinger–Furry hypothesis.

Almost all the experimental work in this field has centered on tests of Bell's inequality, making use of polarization measurements on photon pairs. For the sake of completeness Bell's inequality will now be derived in its various forms before a review of the relevant experimental work is undertaken. For further information the reader is referred to the review papers of Clauser and Shimony[8] and of Pipkin.[9] First, it is necessary to give quantitative meaning to the concept of polarization correlation.

1.2. Polarization Correlation

Consider first the ideal situation where pairs of photons, frequencies ν_1 and ν_2, emitted in the $-z$ direction and $+z$ direction, respectively, from an atomic source, are analyzed by two-channel polarizers π_1 and π_2 as shown in Figure 1. The detectors D_{ij} ($i, j = 1, 2$) are assumed to be 100% efficient. The transmission axes of the polarizers are set in the directions $\hat{a}$ and $\hat{b}$, where $\hat{a}$ and $\hat{b}$ are unit vectors parallel to the x-y plane. The use of two-channel polarizers allows the polarization components of radiation both parallel to and perpendicular to the transmission axis of each polarizer to be monitored simultaneously. For simplicity of notation we shall, from now on, omit the unit vector sign from $\hat{a}$ and $\hat{b}$.

It is normal in this type of experiment, where we are looking for correlations between the polarizations of the two photons, to assign the

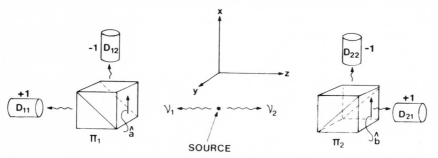

Figure 1. Diagram to illustrate the ideal measurement of polarization correlation. π_1 and π_2 are ideal two-channel polarizers set with their transmission axes in the directions $\hat{a}$ and $\hat{b}$, respectively; D_{ij} ($i, j = 1, 2$) are 100% efficient detectors. The source emits pairs of photons, frequencies ν_1 and ν_2, in the $-z$ and $+z$ directions respectively.

value +1 to detection in D_{11} or D_{21} and −1 to detection in D_{12} or D_{22}. Detection to the left of the source can thus be represented by a variable A, say, which can take on the values ±1, and detection to the right by a variable B which can also take on the values ±1. It follows that a measure of the extent of the correlation between A and B for given settings a and b of the polarizers is the correlation coefficient $E(a, b)$ defined as

$$E(a, b) = \overline{AB} \qquad (1.3)$$

where the bar denotes an average over an ensemble of emitted pairs. If we denote by $P_{++}(a, b)$ the probability of a photon pair giving a result +1 to the left and +1 to the right with similar definitions for $P_{+-}(a, b)$, $P_{-+}(a, b)$, and $P_{--}(a, b)$, then

$$E(a, b) = P_{++}(a, b) + P_{--}(a, b) - P_{+-}(a, b) - P_{-+}(a, b) \qquad (1.4)$$

Alternatively, if, in a given time, N photon pairs are emitted resulting in $N_{++}(a, b)$ detection events in which +1 is registered to the left and +1 to the right, then, provided N is sufficiently large, $P_{++}(a, b) = N_{++}(a, b)/N$, with similar expressions for $P_{+-}(a, b)$, $P_{-+}(a, b)$, and $P_{--}(a, b)$ so that we can write

$$E(a, b) = \frac{N_{++}(a, b) + N_{--}(a, b) - N_{+-}(a, b) - N_{-+}(a, b)}{N} \qquad (1.5)$$

In quantum-mechanical terms, we can say that there is an observable, represented by the operator $A^*(a)$ with eigenvectors $|a^{\pm}\rangle$ and eigenvalues $A = \pm 1$, respectively, describing the results of measurements of photon ν_1 parallel and perpendicular to a, and an observable, represented by the operator $B^*(b)$ with eigenvectors $|b^{\pm}\rangle$ and eigenvalues $B = \pm 1$, describing the results of measurements of photon ν_2 parallel and perpendicular to b. It is then easy to see that, in terms of the linear polarization basis vectors $|x\rangle$ and $|y\rangle$,

$$|a^+\rangle = \cos\theta_1|x\rangle_1 + \sin\theta_1|y\rangle_1, |a^-\rangle = -\sin\theta_1|x\rangle_1 + \cos\theta_1|y\rangle_1$$
$$|b^+\rangle = \cos\theta_2|x\rangle_2 + \sin\theta_2|y\rangle_2, |b^-\rangle = -\sin\theta_2|x\rangle_2 + \cos\theta_2|y\rangle_2 \qquad (1.6)$$

where $|x\rangle_1$ and $|y\rangle_1$ denote polarization states on the left, $|x\rangle_2$ and $|y\rangle_2$ on the right, while θ_1 and θ_2 are the angles between the x axis and a and b, respectively. It follows that

$$A^*(a) = (+1)|a^+\rangle\langle a^+| + (-1)|a^-\rangle\langle a^-| \qquad (1.7)$$

and

$$B^*(b) = (+1)|b^+\rangle\langle b^+| + (-1)|b^-\rangle\langle b^-| \qquad (1.8)$$

since then

$$A^*(a)|a^+\rangle = (+1)|a^+\rangle \qquad \text{etc.} \qquad (1.9)$$

If the two-photon state vector is represented by $|\psi\rangle$, then we can calculate the expectation value for the product $A^* \otimes B^*$ according to

$$\langle\psi|A^* \otimes B^*|\psi\rangle = |\langle b^+|\langle a^+|\psi\rangle|^2 + |\langle b^-|\langle a^-|\psi\rangle|^2 - |\langle b^+|\langle a^-|\psi\rangle|^2 - |\langle b^-|\langle a^+|\psi\rangle|^2$$

$$= P_{++}(a, b) + P_{--}(a, b) - P_{+-}(a, b) - P_{-+}(a, b)$$

$$= E(a, b) \qquad (1.10)$$

The correlation coefficient $E(a, b)$ can thus be identified as the expectation value of the direct product operator $A^* \otimes B^*$.

If equation (1.1) for $|\psi\rangle$ is inserted into equation (1.10) for $E(a, b)$ and equation (1.6) used for $|a^+\rangle$, $|b^+\rangle$, $|a^-\rangle$ and $|b^-\rangle$, we find

$$P_{++}(a, b) = P_{--}(a, b) = \tfrac{1}{2}\cos^2(\theta_1 - \theta_2)$$

$$P_{+-}(a, b) = P_{-+}(a, b) = \tfrac{1}{2}\sin^2(\theta_1 - \theta_2) \qquad (1.11)$$

and hence

$$E_{QM}(a, b) = \cos 2(\theta_1 - \theta_2) = \cos 2(a, b) \qquad (1.12)$$

where $E_{QM}(a, b)$ represents the quantum-mechanical prediction for $E(a, b)$ and $(a, b) = \theta_1 - \theta_2$ denotes the relative angle between a and b. Similarly, using equation (1.2) for $|\psi\rangle$, we find $E_{QM}(a, b) = -\cos 2(a, b)$. In both cases $E(a, b)$ ranges from -1 to $+1$, both extremes corresponding to complete correlation.

1.3. Bell's Inequality for the Ideal Case

The nonlocality and lack of realism inherent in quantum mechanics has inspired many attempts through the years to explain the results in terms of a theory which is both local and realistic. Without a specific local realistic theory it is, of course, not possible to predict a value for $E(a, b)$ to compare with the quantum-mechanical value $E_{QM}(a, b)$ in equation (1.12). However, in 1964 J. S. Bell[4] showed for the first time that such theories place constraints on $E(a, b)$, or rather combinations of $E(a, b)$, for different values of a and b.

To understand Bell's approach consider the experimental situation represented in Figure 1. We assume that the initial state of the two photons can be described in terms of hidden variables λ with a probability density $\rho(\lambda)$. The variable λ may denote a single variable or a set of variables which may be discrete or continuous. However, for simplicity we write as if λ is a single continuous variable. We also assume that

$$\int_{\Lambda} \rho(\lambda) \, d\lambda = 1 \qquad (1.13)$$

where Λ is the space of the states λ. The result A to the left then depends on λ and on a, the result B to the right on λ and b, but if we wish our theory to be local, A cannot depend on b nor B on a. If λ determines uniquely the measurement outcome for each photon pair, we can define

$$E(a, b) = \int_{\Lambda} A(\lambda, a) B(\lambda, b) \rho(\lambda) \, d\lambda \qquad (1.14)$$

in accord with our discussion regarding the correlation function (1.3). More generally, for a given λ describing an emitted pair of photons, the quantities A and B may take on values $+1$ or -1 with a probability depending on λ. In this case, λ does not determine uniquely the outcome of each measurement of A and B but only their average values $\bar{A}$ and $\bar{B}$ over an ensemble of emissions.[10] We can then define

$$E(a, b) = \int_{\Lambda} \bar{A}(\lambda, a) \bar{B}(\lambda, b) \rho(\lambda) \, d\lambda \qquad (1.15)$$

where now the averages $\bar{A}$ and $\bar{B}$ will be, owing to locality, independent of b and a respectively. It follows that instead of $A = \pm 1$ and $B = \pm 1$ we now require only that $|\bar{A}| \leq 1$ and $|\bar{B}| \leq 1$ and these latter conditions are sufficient to derive an interesting restriction on these so-called local stochastic realistic theories.

Let a' and b' be alternative settings of the polarizers and consider the expression

$$S(a, b, a', b') = E(a, b) - E(a, b') + E(a', b) + E(a', b') \qquad (1.16)$$

By substituting for $E(a, b)$ etc, in the form of equation (1.15) and noting the above restriction on $|\bar{A}|$ and $|\bar{B}|$, it then follows easily that

$$-2 \leq S(a, b, a', b') \leq 2 \qquad (1.17)$$

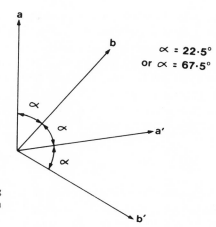

Figure 2. Orientations of the polarizers leading to extremum values of the function $S(a, b, a', b')$.

Essentially the same inequality was derived by Clauser, Horne, Shimony, and Holt[5] and it is sometimes referred to as the Bell, Clauser, Horne, Shimony, and Holt (BCHSH) inequality. Its importance lies in the fact that it represents a general restriction on the predictions of theories based on local realism.

For example, if we take the quantum-mechanical form $E_{QM}(a, b)$ given by equation (1.12) and evaluate $S(a, b, a', b')$ for various orientations of the coplanar vectors a, b, a', and b', it is easy to show that $S(a, b, a', b')$ takes on extremum values for the situations, shown in Figure 2, where $(a, b) = (b, a') = (a', b') = 22.5°$, $(a, b') = 67.5°$ when $S = +2\sqrt{2}$, and where $(a, b) = (b, a') = (a', b') = 67.5°$, $(a, b') = 22.5°$ when $S = -2\sqrt{2}$. Both extreme values for $S(a, b, a', b')$ clearly violate Bell's inequality showing that local realistic theories must necessarily disagree with quantum mechanics over at least some of the range of relative polarizer orientations. This situation may not, at first, seem surprising given the general success of quantum mechanics but, in fact, before the development of Bell's inequality no necessary conflict between local realism and quantum mechanics had been demonstrated. Bell's inequality showed that conflict did indeed exist and brought the question concerning the possibility of a local realistic description of the physical world into the experimental domain. It also pointed out an area where there might conceivably have been a breakdown in conventional quantum mechanics.

1.4. Bell's Inequality in Experimental Situations

The various experiments that have been carried out differ mainly in their choice of source and type of polarizer used. Because of the angular correlation of the photon pairs emitted in two-photon decay processes from an atom, the finite solid angle of detection, and the low detection efficiency of the photodetectors, in practice only a very small proportion of the photon pairs emitted by the source is actually detected. In this situation, in order to have any hope of violating Bell's inequality it is necessary to take N in equation (1.5) for $E(a, b)$ to be the total number of photon pairs that would have been detected in the absence of the polarizers, rather than the total number actually emitted by the source. In evaluating $E(a, b)$ we therefore take

$$N = N_{++}(a, b) + N_{+-}(a, b) + N_{-+}(a, b) + N_{--}(a, b) \qquad (1.18)$$

Adoption of the above procedure, however, implicitly requires the additional assumption that, for each setting of the polarizers, the ensemble of detected pairs is a true representative sample of the ensemble of pairs emitted by the source.

The above form (1.17) of Bell's inequality applies to the situation where both orthogonal polarization states are detected on each side of the source. However, most experiments that have been carried out only detect the signals that are transmitted directly through the polarizers. One can imagine that in Figure 1 only the detectors D_{11} and D_{12} are in place. In these circumstances, a new form for Bell's inequality is required since the quantities $N_{+-}(a, b)$, $N_{-+}(a, b)$, and $N_{--}(a, b)$ cannot now be directly measured. These quantities, however, can be deduced from measurements with either one or both polarizers removed. If we denote by ∞ the absence of a polarizer, then we expect that

$$N_{++}(a, \infty) = N_{++}(a, b) + N_{+-}(a, b)$$

$$N_{++}(\infty, b) = N_{++}(a, b) + N_{-+}(a, b) \qquad (1.19)$$

and

$$N_{++}(\infty, \infty) = N_{++}(a, b) + N_{+-}(a, b) + N_{-+}(a, b) + N_{--}(a, b) \qquad (1.20)$$

From equations (1.19) and (1.20) it then follows that the inequality (1.17) can be rewritten in the form

$$-1 \leqslant S' \leqslant 0 \qquad (1.21)$$

with

$$S' = [N(a, b) - N(a, b') + N(a', b) + N(a', b')$$

$$- N(a', \infty) - N(\infty, b)]/N(\infty, \infty) \qquad (1.22)$$

omitting, for clarity, the subscripts on the N_{++}. Finally, if we assume that the results of measurement depend only on the relative angle between the axes of the polarizers and write the inequality (1.21) for the two sets of angles which give extreme values for S', i.e., $(a, b) = (b, a') = (a', b') = 22.5°$, $(a, b') = 67.5°$, and $(a, b) = (b, a') = (a', b') = 67.5°$, $(a, b') = 22.5°$, then it is easy to derive the inequality

$$\eta \equiv \left| \frac{N(22.5°) - N(67.5°)}{N(\infty, \infty)} \right| \leq 0.25 \qquad (1.23)$$

originally put forward by Freedman.[11] This form of inequality has proved to be one of the most useful experimentally, since it only requires three measurements to be made.

2. Experiments Utilizing an Atomic Source

Since 1972 a series of experiments, of increasing precision and making use of various atomic sources and detection arrangements, have been carried out to verify the form of the two-photon state vector given in equations (1.1) and (1.2), and to test one version or another of Bell's inequality. In the circumstances considered in Section 1.4, Bell's inequality allows a clear distinction to be made between the predictions of local realism and quantum mechanics. However, in a real experiment, several factors act to reduce the strength of the quantum mechanically predicted correlations, in certain circumstances to the extent that a violation of Bell's inequality can no longer be expected to occur. For example, in all the experiments to be described in this section, the radiation emitted by the atomic source is collected and collimated by a pair of lenses with a finite aperture before being analyzed by imperfect polarizers. If ϕ is the half-angle subtended at the source by the lenses, and ε_{M1} and ε_{m1} are, respectively, the transmission efficiencies for light polarized parallel to and perpendicular to the axis a of polarizer π_1, and ε_{M2} and ε_{m2} the corresponding quantities for polarizer π_2 with its axis orientated in the direction b, then it can be shown[5] that, according to quantum mechanics,

$$\frac{N(a, b)}{N(\infty, \infty)} = \tfrac{1}{4}[(\varepsilon_{M1} + \varepsilon_{m1})(\varepsilon_{M2} + \varepsilon_{m2})$$

$$\pm F(\phi)(\varepsilon_{M1} - \varepsilon_{m1})(\varepsilon_{M2} - \varepsilon_{m2}) \cos 2(a, b)] \qquad (2.1)$$

$$\frac{N(a, \infty)}{N(\infty, \infty)} = \tfrac{1}{2}(\varepsilon_{M1} + \varepsilon_{m1}), \qquad \frac{N(\infty, b)}{N(\infty, \infty)} = \tfrac{1}{2}(\varepsilon_{M2} + \varepsilon_{m2}) \qquad (2.2)$$

and hence, assuming symmetrical two-channel polarizers, that

$$E_{QM}(a, b) = \pm F(\phi) \frac{(\varepsilon_{M1} - \varepsilon_{m1})(\varepsilon_{M2} - \varepsilon_{m2})}{(\varepsilon_{M1} + \varepsilon_{m1})(\varepsilon_{M2} + \varepsilon_{m2})} \cos 2(a, b) \qquad (2.3)$$

The $\pm$ sign applies, respectively, to the situation where the photon pairs result from a 0-1-0 or 1-1-0 type of cascade. The quantity $F(\phi)$, which has a different mathematical form in the two cases and is equal to unity when $\phi = 0$, takes into account the depolarizing effect of the noncollinear emission of photon pairs.

In order to carry out a successful test of Bell's inequality, $F(\phi)$ must be greater than some minimum value which depends on the transmission efficiencies of the polarizers. Let us assume for simplicity that the transmission efficiency ε_M is the same for both polarizers, then, since $F(\phi)$ is a monotonically decreasing function of ϕ, there is an upper limit (which depends on ε_M) on the detector half-angle necessary for a test of Bell's inequality, as shown in Figure 3. Clearly, the use of a 0-1-0 cascade places a less stringent requirement on the apparatus parameters than does a 1-1-0

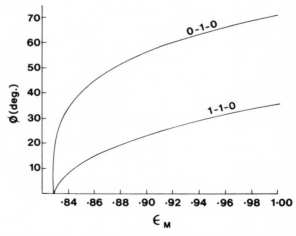

Figure 3. Upper limits on detector half-angle ϕ as a function of polarizer efficiency ε_M. To test Bell's inequality, the experiment must be performed with apparatus parameters chosen in the region below the appropriate curve—the upper curve for a 0-1-0 cascade, the lower for a 1-1-0 cascade (after Clauser et al.[5]).

cascade. It should be noted that it is not necessary to know the results (2.1), (2.2), and (2.3) in order to test Bell's inequality. They must be used, of course, if it is required to compare the experimental results with the quantum-mechanical predictions.

The original experiment of the type being considered here was, in fact, carried out by Kocher and Commins[12] in 1967 and, therefore, predated most of the theoretical work on Bell's inequality, in particular the important paper by Clauser, Horne, Shimony, and Holt[5] which showed how the work of Bell could be applied in a real experiment. Unfortunately, Kocher and Commins only made measurements for relative angles of 0° and 90° between the transmission axes of the polarizers and, in addition, the transmission characteristics of their polarizers did not satisfy the criterion illustrated in Figure 3 for a satisfactory test of Bell's inequality. However, their results were, at least, consistent with quantum mechanics and Clauser[13] was able to use them to demonstrate the inadequacy of semiclassical radiation theory in this situation and to show that the Schrödinger–Furry hypothesis was not tenable in this case. The first proper test of Bell's inequality was carried out by Freedman and Clauser[14] in 1972.

2.1. Freedman and Clauser (1972)

The arrangement of the apparatus used by Freedman and Clauser,[14] shown in Figure 4, is typical of all subsequent experiments which differ, mainly, only in the nature of the source, its method of excitation, and the type of polarizers used to analyze the emitted photons. As indicated in Figure 5, in this experiment the $3d4p^1P_1$ state of calcium in a beam was

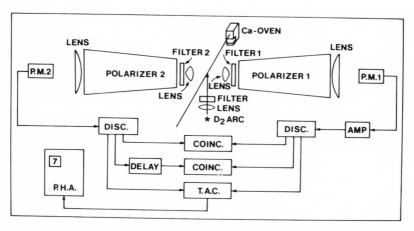

Figure 4. Schematic diagram of the apparatus of Freedman and Clauser.[14]

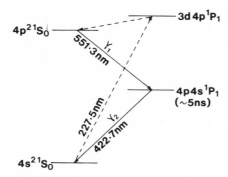

Figure 5. Level scheme for calcium. Dashed lines show the route for excitation to the initial state $4p^2\,{}^1S_0$ used in the experiment of Freedman and Clauser.[14]

excited by radiation of wavelength 227.5 nm from a deuterium arc lamp. About 10% of the atoms which do not return directly to the ground state go to the $4p^2\,{}^1S_0$ state which is the initial state of the $4p^2\,{}^1S_0$–$4s4p^1P_1$–$4s^2\,{}^1S_0$ cascade emitting photons of wavelengths 551.3 nm and 422.7 nm. Since the naturally occurring calcium used in this experiment contained 99.855% of the isotope with zero nuclear spin, there was no significant reduction to be expected in the polarization correlation due to the presence of hyperfine structure. It should also be noted that the calcium beam, whose density is about $3 \times 10^{10}\ \text{cm}^{-3}$, was orientated at an acute angle to the observation axis thus ensuring that Doppler broadening of the resonance line reduced any effects due to resonance trapping of the 422.7 nm photon. On each side of the source, as shown in Figure 4, the photons were collected and collimated by a lens, then passed through a filter and linear polarizer to a photomultiplier. In order to satisfy the requirement for large efficient linear polarizers, they used pile-of-plates polarizers each of which was about 1 m in length and consisted of ten 0.3-mm-thick glass sheets inclined nearly at Brewster's angle. The sheets were mounted on hinged frames so that they could be folded out of the optical path. The transmission efficiencies were measured to be $\varepsilon_{M1} = 0.97 \pm 0.01$, $\varepsilon_{m1} = 0.038 \pm 0.004$, $\varepsilon_{M2} = 0.96 \pm 0.01$, $\varepsilon_{m2} = 0.037 \pm 0.004$ and the half-angle subtended by the lenses at the source was $30°$ giving $F(\phi) = 0.99$. The photomultiplier pulses were fed to a coincidence circuit and coincidence measurements were made for 100 s periods, the periods during which all the plates were removed alternating with periods in which the plates were inserted.

The results obtained, as the relative orientation of the transmission axes of the polarizers was varied from $0°$ to $90°$, were found to be in agreement with the quantum-mechanical prediction as expressed in equation (2.1). Also, the results at $(a, b) = 22.5°$ and $(a, b) = 67.5°$ combined with those with both sets of polarizer plates removed gave $\eta = 0.300 \pm 0.008$, in clear violation of the Freedman form of Bell's inequality expressed in equation (1.23), and in agreement with the quantum-mechanical prediction $\eta_{QM} = 0.301 \pm 0.007$ obtained from equation (2.1).

2.2. Holt and Pipkin (1973)

Holt and Pipkin[15] observed the 567.6 nm and 404.7 nm photons emitted in the $9^1P_1-7^3S_1-6^3P_0$ cascade of the zero nuclear spin isotope ^{198}Hg of mercury. The relevant transitions are shown in Figure 6, from which it can be seen that the final cascade level is not the ground state of the atom. Thus, in this experiment no precautions had to be taken to avoid the effects of resonance trapping. To produce the required radiation, mercury vapor was excited to the 9^1P_1 state by a 100 eV electron beam, both the beam and the vapor being contained in an encapsulated source made from Pyrex glass. Since the source was of the 1-1-0 variety, the requirements on polarizer efficiency and collection solid angle were more stringent than in the 0-1-0 case. In addition, any lack of isotropy among the excited atoms in the 9^1P_1 state could have a significant effect on the results. A third photomultiplier viewed the 435.8 nm photons from the $7^3S_1-6^3P_1$ transition to monitor the lamp intensity and to produce a correction signal for the lamp stabilization circuitry. In contrast to the two previous experiments, calcite-type polarizers were used with transmission efficiencies $\varepsilon_{M1} = 0.910 \pm 0.001$, $\varepsilon_{M2} = 0.880 \pm 0.001$, ε_{m1}, $\varepsilon_{m2} < 10^{-4}$. This type of polarizer has a much better extinction ratio than pile-of-plates polarizers, but the values of ε_{M1} and ε_{M2} are not particularly high. What is more, since a 1-1-0 cascade was being used, the factor $F(\phi)$ took on the relatively low value 0.951 even with an acceptance half-angle of only 13°.

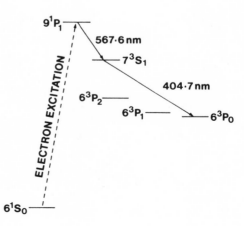

Figure 6. Level scheme for mercury showing the states used in the Holt and Pipkin experiment.[15]

Experimentally, it was found that $\eta = 0.216 \pm 0.013$, a result which disagrees with the quantum-mechanical prediction $\eta_{QM} = 0.266$ and clearly does not violate Bell's inequality. Although this discrepancy has never been completely explained, it is thought that the low value of η may have arisen as a result of stress-induced optical activity in the walls of the Pyrex glass envelope.[16] However, since this result is, in fact, the only one of its kind consistent with a local realistic interpretation of these long-range two-photon polarization correlations, proponents of such theories have suggested that there may be some significance to be attached to the use of calcite polarizers or to the fact that the final state of the cascade is not the ground state of the atom.

2.3. Clauser (1976)

In view of the unexpected result obtained by Holt and Pipkin,[15] Clauser[16] repeated their experiment using the same cascade but in the even isotope ^{202}Hg of mercury. Also, instead of calcite polarizers, he used pile-of-plates polarizers of the type used previously in the experiment described in Section 2.1 but with 15 rather than 10 plates to give transmission efficiencies $\varepsilon_{M1} = 0.965$, $\varepsilon_{m1} = 0.011$, $\varepsilon_{M2} = 0.972$, $\varepsilon_{m2} = 0.0084$. With a collection half-angle of 18.6°, it is expected from quantum mechanics that $\eta_{QM} = 0.2841$ while the experiment gave $\eta = 0.2885 \pm 0.0093$, violating Bell's inequality and in close agreement with the quantum-mechanical result. In addition, a calculation of the variation of $N(a, b)/N(\infty, \infty)$, from equation (2.1), with angle (a, b), taking into account the measured polarizer efficiencies, the collection half-angle, depolarization due to the presence of ^{199}Hg and ^{201}Hg isotopes, and alignment of the 9^1P_1 state, showed close agreement between the experimental results and quantum mechanics.

In an extension to the above experiment Clauser[17] measured the circular polarization correlation by inserting quarter-wave plates between each linear polarizer and the source. The quarter-wave plates were constructed by applying pressure to bars of commercial grade quartz. Assuming ideal quarter-wave plates, quantum mechanics predicts that equation (2.1) still holds and thus equation (1.23) remains a valid form of Bell's inequality. From the experimental results Clauser found $\eta = 0.235 \pm 0.025$ while, taking into account the transmission efficiencies of the linear polarizers, the collection half-angle, and the lack of stability of the quarter-wave plates, he predicted $\eta_{QM} = 0.252$. Thus, although within the limits of experimental error these circular-polarization results were in agreement with quantum mechanics, they failed to provide a conclusive test of Bell's inequality. However, Clauser did show that the results were not consistent with the Schrödinger–Furry hypothesis.

2.4. Fry and Thompson (1976)

Fry and Thompson[18] used the 435.8 nm and 253.7 nm photons emitted in the 7^3S_1–6^3P_1–6^1S_0 cascade in the zero nuclear spin isotope ^{200}Hg of mercury. The relevant transitions are shown in Figure 7. The 7^3S_1 state in a mercury beam was populated in a two-step process with electron bombardment excitation of the 6^3P_2 metastable state being followed downstream, where all short-lived states had decayed, by absorption of resonant 546.1 nm radiation from a tunable dye laser the output of which was polarized with its electric field vector in the direction of the observation axis. The magnetic field in the interaction volume was reduced to less than 5 mG in all directions. Although mercury of natural isotopic abundance was used, the laser bandwidth was narrow enough (15 MHz) that the ^{200}Hg isotope could be selectively excited. In addition, since there was a one-to-one correspondence between all 435.8 nm and 257.3 nm photons, data accumulation rates were obtained which were high compared to those achieved in previous experiments, a typical run lasting about 80 min. The polarizers used in this experiment were of some interest, being of the pile-of-plates variety with each polarizer consisting of two sets of 7 plates symmetrically arranged so as to cancel out transverse ray displacements.

Since the initial state of the cascade had $J = 1$, it was necessary to take into account possible effects resulting from unequal population of, and coherence between, the initial Zeeman sublevels, which Fry and Thompson did by measuring the polarization of the 435.8 nm fluorescence at appropriate angles. Allowing for these effects along with the transmission efficiencies of the polarizers $\varepsilon_{M1} = 0.98 \pm 0.01$, $\varepsilon_{M2} = 0.97 \pm 0.01$, $\varepsilon_{m1} = \varepsilon_{m2} = 0.02 \pm 0.005$, and the half-angle $19.9° \pm 0.3°$ of the collection optics, it was predicted on the basis of quantum mechanics that $\eta_{QM} = 0.294 \pm 0.007$ while from the experiment the value $\eta = 0.296 \pm 0.014$ was found, in agreement with the quantum-mechanical result but clearly violating Bell's inequality. Fry and Thompson also made a least-squares fit of the form

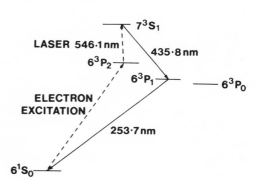

Figure 7. Level scheme for mercury showing the states used in the Fry and Thompson experiment.[18]

$A + B \cos 2\phi + C \sin 2\phi$ to their data and obtained $A = 0.242 \pm 0.003$, $B = -0.212 \pm 0.004$, $C = -0.003 \pm 0.004$ in good agreement with the values expected quantum mechanically.

2.5. Aspect, Grangier, and Roger (1981)

In common with Freedman and Clauser,[14] Aspect, Grangier, and Roger[19] made use of 551.3 nm and 422.7 nm photons from the $4p^2{}^1S_0$–$4s4p^1P_1$–$4s^2{}^1S_0$ cascade of calcium. However, in their case, the calcium atoms were excited to the $4p^2{}^1S_0$ state by a nonresonant two-photon absorption process using a krypton-ion laser beam of wavelength 406 nm and a dye laser beam tuned to 581 nm, both laser beams being at right angles to the calcium atomic beam emitted from a tantalum oven. The laser beams had parallel polarizations and were focused at the interaction region to provide a source about 60 μm in diameter by 1 mm long. The density was about 3×10^{10} cm^{-3}, which resulted in a typical cascade rate of 4×10^7 s^{-1}. The narrow resonance of the excitation process (less than 50 MHz) allowed selective excitation of the even ^{40}Ca isotope of calcium, thus preventing the strength of polarization correlation from being reduced by the effects of hyperfine structure. Feedback loops were used to control the wavelength of the tunable dye laser and the krypton-ion laser power output. After collection and collimation by a lens system which subtended a half-angle of about 32° at the source, the photons from the cascade were analyzed by polarizers and filters in much the same way as in previous experiments. The polarizers were of the pile-of-plates type each consisting of 10 optically flat plates set nearly at Brewster's angle, with efficiencies measured to be $\varepsilon_{M1} = 0.971 \pm 0.005$, $\varepsilon_{m1} = 0.029 \pm 0.005$, and $\varepsilon_{m2} = 0.028 \pm 0.005$. The pulses from two photomultipliers were fed in the usual way to a coincidence circuit, and the time correlation spectrum displayed on a multichannel analyzer. A typical spectrum obtained in this way[20] is shown in Figure 8. The asymmetry of the coincidence peak results mainly from the finite lifetime (5 ns) of the intermediate $4s4p^1P_1$ state of the cascade.

This experiment was noteworthy with regard to the strength of the source, with coincidence rates of up to 100 s^{-1} allowing measurements of 1% statistical accuracy to be obtained in only 100 s counting time. The results of these measurements in the form of a graph of $N(a, b)/N(\infty, \infty)$ against the angle (a, b) between the transmission axes of the two polarizers is shown in Figure 9, which also shows the predicted quantum-mechanical curve. The agreement between the theory and experiment is clearly excellent. Using the experimental results at 22.5° and 67.5° gave $\eta = 0.3072 \pm 0.0043$, in agreement with the quantum-mechanical prediction $\eta_{QM} = 0.308 \pm 0.002$, calculated using equation (2.1), and violating Bell's inequality by more than 13 standard deviations.

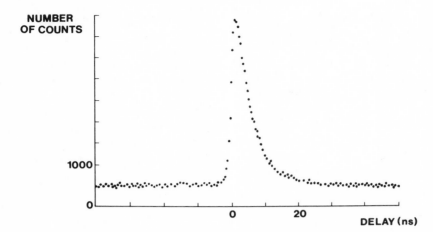

Figure 8. A typical time correlation spectrum obtained in the experiment of Aspect *et al.*[20] The asymmetry of the coincidence peak can be accounted for by an exponential decay with a lifetime $\tau = 4.7$ ns.

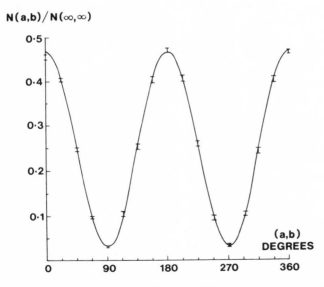

Figure 9. Normalized coincidence rate as a function of the relative orientation of the transmission axes of the polarizers in the experiment of Aspect, Grangier, and Roger.[19] The solid curve represents the quantum-mechanical prediction.

Aspect, Grangier, and Roger also used this experiment to test Bell's inequality in the form of equation (1.21), $-1 \leq S' \leq 0$, which does not assume the rotational invariance required for the Freedman form of the inequality. They found $S' = 0.126 \pm 0.014$, violating inequality (1.21) by 9 standard deviations, and in good agreement with the quantum-mechanical prediction $S'_{QM} = 0.118 \pm 0.005$.

Finally, it was observed that moving each polarizer up to 6.5 m from the source, i.e., to four coherence lengths of the wave packet associated with the lifetime of the intermediate state of the cascade (5 ns), produced no change in the results, thus providing further strong evidence against the Schrödinger–Furry hypothesis.

2.6. Aspect, Grangier, and Roger (1982)

As discussed in Section 1.4, the experiments described until now, in which only the N_{++} signals are detected, depart from the ideal arrangement discussed in Section 1.3. To remedy this situation, in 1982 Aspect, Grangier, and Roger[21] performed an experiment using the same source as described above but with two-channel polarizers instead of the previous one-channel pile-of-plates type. Their apparatus was essentially that of Figure 1 with the addition of collecting and collimating lenses. Each polarizer, in the form of a polarizing cube, constructed using the properties of dielectric thin films and antireflection coated, was rotatable about the observation axis. This arrangement allowed the quantity $E(a, b)$ defined in equations (1.5) and (1.18) to be measured directly in a single run, using a fourfold coincidence technique for each of the four relative orientations of the polarizers $(a, b) = (b, a') = (a', b') = 22.5°, (a, b') = 67.5°$. In this way Bell's inequality could be tested in the form of equation (1.17), $-2 \leq S \leq 2$, derived in Section 1.3.

From the experimental results the value $S = 2.697 \pm 0.015$ was found, while the quantum-mechanical prediction obtained using equation (2.3) for $E(a, b)$ with $\varepsilon_{M1} = 0.950 \pm 0.005$, $\varepsilon_{M2} = 0.930 \pm 0.005$, $\varepsilon_{m1} = \varepsilon_{m2} = 0.007 \pm 0.005$ gave $S_{QM} = 2.70 \pm 0.05$, in good agreement with the experimental result. In addition, as shown in Figure 10 a measurement of $E(a, b)$ as a function of the angle (a, b) between the transmission axes of the polarizers gave close agreement with the quantum-mechanical prediction based on equation (2.3).

As already pointed out in Section 1.4, for the experiment to provide a satisfactory test of Bell's inequality it is necessary, of course, to assume that the ensemble of pairs actually detected is a true representative sample of the ensemble of pairs which are emitted. Although it is not possible to prove that the detected sample is unbiased in this way, nevertheless added confidence is given to the result of the experiment by the observation that

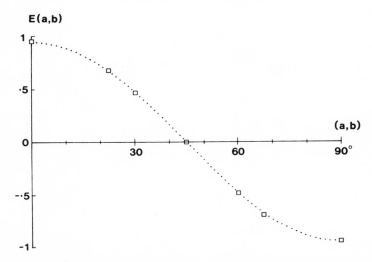

Figure 10. Correlation function $E(a, b)$ as a function of the relative angle (a, b) for the experiment of Aspect, Grangier, and Roger.[21] The indicated errors are ±2 standard deviations. The dashed curve is the quantum-mechanical prediction.

the quantity $N_{++} + N_{--} + N_{+-} + N_{-+}$ was indeed constant as the polarizers were rotated about the observation axis.

It should be noted here that an experiment similar to the one described in this section is being carried out by a group at the University of Catania who have published[22] descriptions of their apparatus.

2.7. Aspect, Dalibard, and Roger (1982)

In all the experiments described so far the orientations of the transmission axes of the polarizers have been fixed at various angles during the measurements. Thus, it could be argued that, in some way, the polarizers and the process of emission of photon pairs could reach some mutual rapport by exchange of signals with speed less than or equal to the speed of light. Such a possibility could be ruled out if the settings of the polarizers were changed in a time which was short compared to the time of flight of photons from the source to each polarizer. A possible scheme to achieve this ideal was suggested by Aspect[23] in 1976 and the experiment was realized in 1982 by Aspect, Dalibard, and Roger.[24]

In their experiment, which used the same source as described in Section 2.5, an optical switch rapidly redirected the light incident from the source to one of two polarizing cubes on each side of the source, as shown in Figure 11. In contrast to the previous experiment described in Section 2.6, however, only the transmitting channels of the polarizing cubes were used. The switching of the light was effected by what is essentially a Bragg

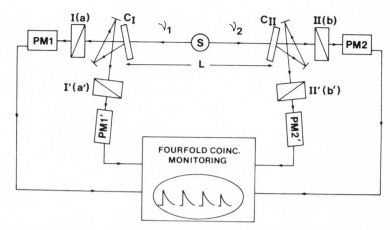

Figure 11. The experiment of Aspect, Dalibard, and Roger[24] with optical switches. Each switching device (C_I, C_{II}) is followed by two polarizers in two different orientations. The arrangement is equivalent to one in which a single polarizer on each side is switched quickly between two orientations. $L = 12$ m.

reflection from an ultrasonic standing wave in water. The light was completely transmitted without deflection when the amplitude of the standing wave was zero, and was almost fully deflected through 10 mrad when the amplitude was a maximum. Switching between the two channels occurred about once every 10 ns and since this time, as well as the lifetime of the intermediate state of the cascade (5 ns), was small compared to L/c (40 ns), where L was the separation between the switches (12 m) and c the speed of light, a detection event on one side and the corresponding change of orientation on the other side were separated by a space-like interval. The results were registered in a fourfold coincidence monitoring system as shown in Figure 11 but, because of the necessity to reduce the beam divergence in the optical system to achieve good switching, the coincidence rates were reduced to only a few per second, with an accidental background of about one per second.

If the two switches work at random, it is possible to write Bell's inequality in the slightly modified form of equation (1.21),

$$-1 \leq S'' \leq 0 \qquad (2.4)$$

where

$$S'' = \frac{N(a, b)}{N(\infty, \infty)} - \frac{N(a, b')}{N(\infty, \infty')} + \frac{N(a', b)}{N(\infty', \infty)}$$
$$+ \frac{N(a', b')}{N(\infty', \infty')} - \frac{N(a', \infty)}{N(\infty', \infty)} - \frac{N(\infty, b)}{N(\infty, \infty)}$$

Although the switching was, in practice, periodic rather than random, the switches on the two sides were driven by different generators at different frequencies and it was assumed that they functioned in an uncorrelated way. Experimentally, it was found that $S'' = 0.101 \pm 0.020$ in violation of Bell's inequality, while taking into account the solid angle of detection and the efficiencies of the polarizers gave the quantum-mechanical prediction $S''_{QM} = 0.112$. A measurement of the normalized coincidence rate as a function of the relative orientation of the polarizers also showed agreement with quantum mechanics, though with a statistical accuracy somewhat less than that achieved in the experiments described in Sections 2.5 and 2.6.

Finally, it should be noted that some criticisms[25-27] have been made of the experiments making use of a high-density calcium source, on the grounds that there may have been significant effects due to resonance trapping. A reply to these criticisms has been given by Aspect and Grangier.[20]

2.8. Perrie, Duncan, Beyer, and Kleinpoppen (1985)

Perrie, Duncan, Beyer, and Kleinpoppen[28] measured for the first time the polarization correlation of the two photons emitted simultaneously by metastable atomic deuterium in a true second-order decay process and used the results to test Bell's inequality. Single photon decay from the $2S_{1/2}$ state of deuterium is forbidden and, as illustrated in Figure 12, the main channel for the spontaneous de-excitation of this state is by the simultaneous emission of two photons which can have any wavelength consistent with conservation of energy for the pair, the most probable occurrence being the emission of two photons each of wavelength 243 nm. Since the decay proceeds through virtual intermediate states the effects of hyperfine structure can be neglected[29,30] and, hence, the angular and polarization correlations

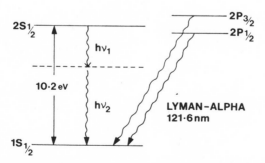

Figure 12. Level diagram for atomic deuterium, neglecting hyperfine structure (not to scale). The two photons, frequencies ν_1 and ν_2, can have any energy provided $h\nu_1 + h\nu_2 = 10.2$ eV (h is Planck's constant).

are predicted to be identical to those resulting from a 0–1–0 cascade in an atom with zero nuclear spin.

In the experiment illustrated in Figure 13, a 1 keV metastable atomic deuterium beam of density about $10^4\,\mathrm{cm}^{-3}$ was produced by charge exchange, in cesium vapor, of deuterons extracted from a radiofrequency ion source. Electric field pre-quench plates upstream from the observation region allowed the $2S_{1/2}$ component of the beam to be switched on and off by Stark mixing the $2S_{1/2}$ and $2P_{1/2}$ states and, at the end of the apparatus, the beam was fully quenched so that the resulting Lyman-α signal could be used to normalize the two-photon coincidence signal. As in previous experiments, the two-photon radiation was collected and collimated by a pair of lenses, each lens subtending a half-angle of 23° at the source. The polarizers were of the pile-of-plates type with 12 plates set nearly at Brewster's angle. In order to have a high transmission in the neighborhood of 243 nm, both the lenses and the plates of the polarizers were made from high-quality fused silica with a short-wavelength cutoff at 160 nm. However,

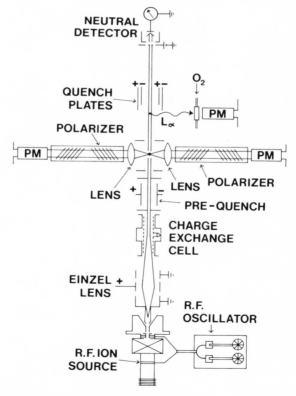

Figure 13. Schematic diagram of the apparatus of Perrie, Duncan, Beyer, and Kleinpoppen.[28]

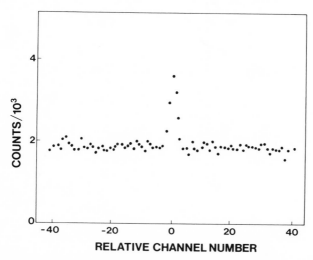

Figure 14. A typical time correlation spectrum for the experiment of Perrie, Duncan, Beyer, and Kleinpoppen[28] after subtraction of the spectrum obtained with the metastable component of the beam quenched. Polarizer plates removed. Time delay per channel 0.8 ns. Total collection time 21.5 h. Singles rate with metastables present (quenched) about $1.15 \times 10^4 \, \text{s}^{-1}$ ($0.85 \times 10^4 \, \text{s}^{-1}$). True two-photon coincidence rate $490 \, \text{h}^{-1}$.

in practice, because of the absorption in oxygen, the short-wavelength cutoff occurred at 185 nm which, in turn, implied a long-wavelength cutoff at 355 nm and hence an observation window between 185 nm and 355 nm. The transmission efficiencies of the polarizers were measured to be $\varepsilon_M = 0.908 \pm 0.013$ and $\varepsilon_m = 0.0299 \pm 0.0020$. On each side of the source the pulses from the photomultipliers were fed to a standard coincidence circuit with the time correlation spectra obtained with the metastable atoms present and quenched, being stored in separate segments of a multichannel analyzer memory and then subtracted at the end of a run. A typical spectrum obtained in this way is shown in Figure 14, from which it can be seen that the coincidence peak is symmetrical, as expected for a simultaneous emission process, in contrast to the situation illustrated in Figure 8 for a cascade process.

The results of measurement are shown in Figure 15, and clearly agree with the quantum-mechanical prediction calculated in the usual way taking into account the half-angle subtended by the lenses at the source and the efficiencies of the polarizers. In addition, using the results at 22.5° and 67.5° gave $\eta = 0.268 \pm 0.010$, in violation of Bell's inequality but in agreement with the quantum-mechanical result $\eta_{QM} = 0.272 \pm 0.008$.

In an extension to the above experiment[31,32] the circular polarization correlation was measured by placing achromatic quarter-wave plates in

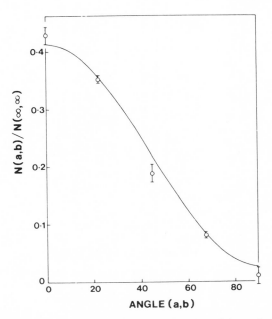

Figure 15. Coincidence signal $N(a, b)/N(\infty, \infty)$ as a function of the angle (a, b) between the transmission axes of the polarizers for the experiment of Perrie, Duncan, Beyer, and Kleinpoppen.[28] The solid curve represents the quantum-mechanical prediction.

each detection arm between linear polarizer and source. The results obtained were consistent with conservation of angular momentum for the photon pair along the observation axis but did not violate Bell's inequality. It seems most likely, however, that this failure to violate Bell's inequality was due to the fact that the retardation of the plates can vary by ±10% over the wavelength range 185–355 nm even for a parallel light beam while, in the experiment, the light rays from the source, after collimation, could be at an angle of up to ±2° to the observation axis.

2.9. Hassan, Duncan, Perrie, Beyer, and Kleinpoppen (1986a)

As we have seen in Sections 2.1 to 2.8, the vast majority of the various experiments have agreed with quantum mechanics and have violated Bell's inequality. However, as pointed out in Section 1.4, to reach this latter conclusion it is necessary to make some additional assumption regarding lack of bias or enhancement in the detection process. The first assumption of this kind was proposed by Clauser, Horne, Shimony, and Holt[5] in the form: If a pair of photons emerges from the a polarizers the probability of their joint detection is independent of the orientations a and b of the

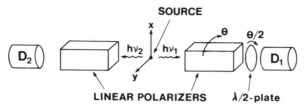

Figure 16. Arrangement of apparatus for the experiment of Hassan *et al.*[(34)] with $\lambda/2$-plate in place. The transmission axis of the right-hand linear polarizer is rotated through an angle θ, the fast axis of the $\lambda/2$ plate through $\theta/2$, relative to the x axis.

polarizers. Subsequently, Clauser and Horne[(33)] made the assumption, called by them the "no-enhancement" assumption, in the form: For every atomic emission, the probability of a count with a polarizer in place is less than or equal to the probability with the polarizer removed.

An attempt to test the assumption in the form given by Clauser, Horne, Shimony, and Holt was made by Hassan, Duncan, Perrie, Beyer, and Kleinpoppen[(32,34,35)] in an extension to the experiment described in Section 2.8, in which a half-wave plate was inserted in one detection arm of the apparatus between polarizer and photomultiplier as shown in Figure 16. By rotating the fast axis of the half-wave plate through half the angle of rotation of the transmission axis of the linear polarizer, it was possible to ensure that, insofar as it is acceptable to think in terms of individual photons emerging from the polarizers, the planes of polarization of the two photons were always parallel just prior to detection in the photomultipliers.

The results of the experiment are given in Figure 17 from which it is clear that, within the limits of experimental error, the results are in agreement with quantum mechanics. In addition, it was found that $\eta = 0.271 \pm 0.021$, violating Bell's inequality and agreeing with the quantum-mechanical prediction $\eta = 0.272 \pm 0.008$ found previously.

This experiment supports the assumption that there is no enhancement in the detection process but, because it is based on the idea that the Clauser, Horne, Shimony, and Holt statement can be put in terms of the polarization state of emerging photons, it does not rule out completely the possibility that enhancement may occur as a result of the settings of the polarizers themselves. It is quite easy, for example, to imagine that the linear polarizers may endow the photon pairs with properties, leading to enhanced detection, which are unaffected by passage through a half-wave plate.

2.10. Hassan, Duncan, Perrie, Beyer, and Kleinpoppen (1986b)

In a further experiment[(32,34,35)] making use of the same apparatus as in Section 2.8, an additional linear polarizer was inserted in one arm of the

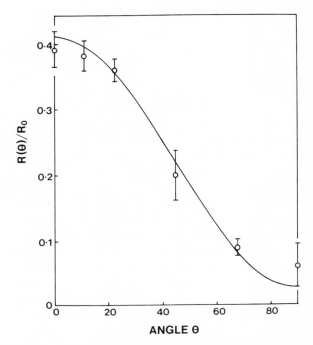

Figure 17. Coincidence signal $R(\theta)$ as a function of the angle θ between the transmission axes of the polarizers in the experiment of Hassan *et al.*,[33] relative to R_0, the coincidence signal with the polarizer plates removed, and the $\lambda/2$-plate in place. The solid curve represents the QM prediction using the median values for ε_M, ε_m.

detection system, as shown in Figure 18. The orientation of polarizer a was held fixed, while polarizer b was rotated through an angle β in a clock-wise sense and polarizer a' through an angle α' in the opposite sense. The ratio $R(\beta, \alpha')/R(\beta, \infty)$ was then measured as a function of α', for various angles β, where $R(\beta, \alpha')$ was the coincidence rate with both polarizers in place

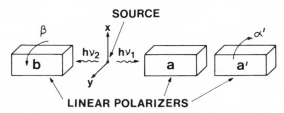

Figure 18. Arrangement of apparatus for the three-polarizer experiment of Hassan *et al.*[31,33] The orientation of polarizer a is fixed with its transmission axis parallel to the x axis, while the transmission axes of polarizers b and a' are rotated, respectively, through angles β and α' relative to the x axis.

and $R(\beta, \infty)$ the coincidence rate with the plates of polarizer a' removed. The results for $\beta = 0°$ are shown in Figure 19, and are in good agreement with the corresponding quantum-mechanical prediction. The experiment for $\beta = 0°$ can, in fact, be considered to be a test of Malus' law for the transmission of polarized light from a very weak source through polarizer a'.

This three-polarizer experiment was, originally, suggested by Garuccio and Selleri[36] in order to test a local realistic model which included the possibility of enhanced photon detection. In their model a photon has, in addition to a polarization vector **l**, a detection vector **λ** which is unaffected by passage through a linear polarizer, the angle between **l** and **λ** determining the detection probability for a photon. Their model is consistent with single-photon physics and all two-photon polarization correlation measurements so far described, but predicts a discrepancy with quantum mechanics in the case of the three-polarizer experiment. In particular, for various angles β, the model sets the upper limit on the ratio $R(\beta, a')/R(\beta, \infty)$ shown in Figure 19 by the broken line. The measurements at $\beta = 0°$ do not

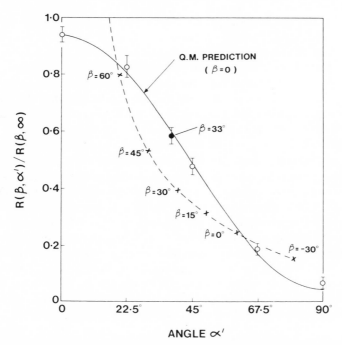

Figure 19. Variation of the ratio $R(\beta, a')/R(\beta, \infty)$ as a function of a' in the experiment of Hassan *et al.*[31,33] The points marked ○ correspond to the results from $\beta = 0°$. The solid curve represents the QM prediction for $\beta = 0$. The broken curve shows the upper limit for the ratio set by the Garuccio–Selleri model for various angles β. The single point, marked as ●, is the result of measurement at $\beta = 33°$.

suffice to provide a conclusive test between quantum mechanics and the Garuccio–Selleri model, but the single result at $\beta = 33°$ is clearly at variance with such a model.

3. Experiments Utilizing Electron–Positron Annihilation Radiation and Proton Spin

When a positron annihilates at rest, conservation of linear momentum requires that at least two photons with equal and opposite momenta are created. Wheeler[37] first showed that the polarization of the two photons should be correlated. Yang[38] showed that this two-photon polarization correlation is the consequence of invariance under rotation and parity transformation, and the appropriate form of the state vector is that given in equation (1.2). In contrast, however, to the situation in the visible and ultraviolet part of the spectrum, there are no standard optical tools such as polaroids, birefringent crystals, or quarter-wave plates available for a polarization-correlation measurement of the high-energy γ rays emitted in positron annihilation. Instead, the anisotropy of Compton scattering has been successfully applied to measure the linear polarization correlation of the annihilation radiation. Compton scattering acts like a linear-polarization analyzer for which Thomson scattering is the classical analogue. When a linearly polarized electromagnetic wave interacts with an electron, the electron vibrates in the direction of the electric vector and radiates like a dipole, so that the scattered γ rays have maximum intensity in a direction perpendicular to the electric vector of the incoming radiation.

In the case of the two-photon annihilation radiation, which is polarized at right angles, the Compton radiation from two Compton scatterers would tend to scatter in perpendicular directions. Various measurements of polarization correlations of annihilation radiation have been reported in the 1940s and 1950s. These measurements were mainly concerned with determining asymmetry parameters for Compton scattering angles. However, Bohm and Aharonov[3] made use of the results of Wu and Shaknov to show that equation (1.2) did indeed describe the two-photon system correctly and to provide evidence against the idea discussed by Schrödinger[6] and Furry[7] that the state vector might spontaneously change its form. Measurements of polarization correlation as tests of quantum mechanics versus local realistic theories were not reported until the 1970s.

3.1. Kasday, Ullman, and Wu (1975)

As an example of such an experimental arrangement for the linear polarization correlation of annihilation radiation, we refer to Figures 20–22 taken from Kasday et al.[39] Positrons from a ^{64}Cu source annihilate between

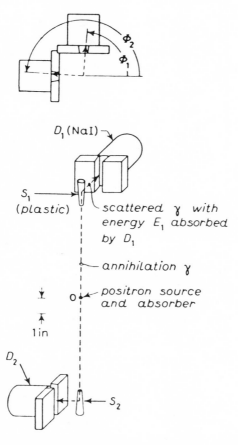

Figure 20. Schematic view of the apparatus of Kasday *et al.*[39] for measuring the polarization of the e^+e^- annihilation photons. The positron source and the absorber are located in the center. The annihilation photons travel in opposite directions and strike the Compton scatterers S_1 and S_2. The scintillation detectors D_1 and D_2 detect the Compton scattered γ rays. ϕ_1 and ϕ_2 are the azimuthal angles of the scattered photons.

two Compton scatterers. The main characteristic of the ^{64}Cu decay scheme is illustrated in Figure 21. The β^+ activity consists of a 19% branching ratio and about 1 MeV γ rays accompanying 0.5% of the positron emission. The positrons of the radioactive source were stopped and annihilated in the source itself and in a thin layer of the surrounding holder material (Figure 22). The annihilation γ rays were emitted in all directions; the vertical two opposite directions for the coincident two-photon detection were selected by a lead collimator which is shown in Figure 22. The Compton scatterer was of plastic material of conical shape, surrounded by a slightly larger

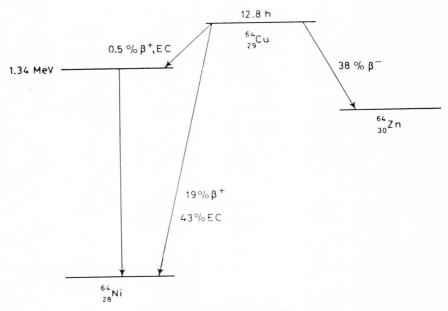

Figure 21. Decay scheme of ^{64}Cu.

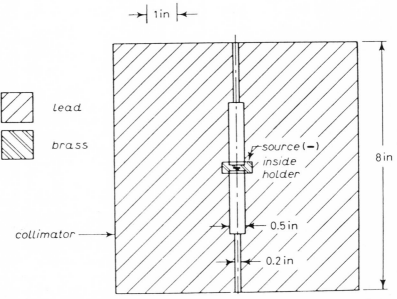

Figure 22. Collimator, source holder, and source of the two-photon annihilation experiment of Kasday et al.[39]

conical light reflector coated on the inside with MgO for more efficient diffuse reflection. The Compton scattered photons were detected by energy-sensitive scintillation detectors (2 in. long NaI crystals) with bi-alkali 12-stage photomultipliers. Since the kinematics of the Compton scattering give a definite relation between the electron scattering angle θ and the photon energy, one can write the probability of finding the two Compton scattered photons as a function of E_1, E_2, ϕ_1, and ϕ_2 in the form

$$P(E_1, E_2, \phi_1, \phi_2) = \frac{1}{4\pi^2} F(E_1)F(E_2)[1 - m(E_1)m(E_2) \cos 2(\phi_2 - \phi_1)]$$

(3.1)

where E_1 and E_2 are the two photon energies, while ϕ_1 and ϕ_2 are the azimuthal angles as shown in Figure 20. The quantity $F(E)$ is the usual Klein–Nishina cross section for Compton scattering, and $m(E) = -\sin^2 \theta / \chi(E_0, E)$, $\chi(E_0, E) = E_0/E + E/(E_0 - \sin^2 \theta)$, with E_0 the energy of incident photon. The ϕ-dependence of the coincident counting rate according to quantum mechanics is therefore of the form $A - B \cos 2(\phi_2 - \phi_1)$.

Instead of measuring the coincidence rate as a function of the azimuthal angle $(\phi_2 - \phi_1)$, Kasday et al.[39] measured the quantity R defined by

$$R(\phi_1, \phi_2) = (N/N_{ss})/(n_1/N_{ss})(n_2/N_{ss})$$

(3.2)

where N_{ss} is the number of times the two photons Compton-scatter, N the number of times the two photons Compton-scatter and both photons are detected, n_1 the number of times the two photons Compton-scatter and only photon 1 is detected, n_2 the number of times the two photons Compton-scatter and only photon 2 is detected, while ϕ_1 and ϕ_2 are the true azimuthal angles at which the slits are positioned.

Quantum mechanics, local hidden-variable theory, and the Schrödinger–Furry (or Bohm–Aharonov) hypothesis predict the following simple relations for the quantities A and B:

$A = 1$, $\quad B = m_1 m_2$ $\qquad$ if quantum mechanics is valid

$A = 1$, $\quad B = m_1 m_2 / \sqrt{2}$ $\qquad$ if local hidden-variable theory is valid

$A = 1$, $\quad B = m_1 m_2 / 2$ $\qquad$ if the Schrödinger–Furry hypothesis is valid $\qquad$ (3.3)

Kasday et al.[39] chose four energy regions of the two Compton-scattered photons in order to measure the coefficient B; Figure 23 illustrates the energy regions expressed in units of the electron mass. Figures 24 and 25

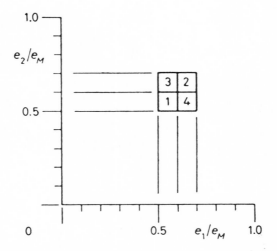

Figure 23. The four energy regions chosen by Kasday *et al.*[39] to study the cosine dependence of the quantity R in their two-photon annihilation experiment.

show results for the quantity R and B. The measured angular correlation function R (Figure 24) appears to be fitted very well to the predicted form $A - B \cos 2\phi$, where $\phi = \phi_2 - \phi_1$, using the Klein–Nishina formula and the known geometry of the apparatus. Taking into account limitations with regard to an ideal geometry of the experiment, A is extracted from the fit to the experimental data as $A = 1.01 \pm 0.05$, which is consistent with the value unity. The B values extracted from experimental data at the energies of Figure 23 are shown in Figure 25 and compared to the predictions of

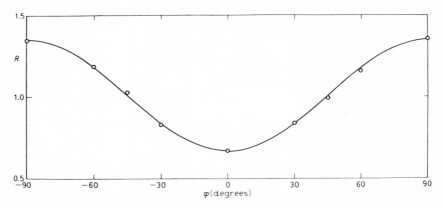

Figure 24. Plot of experimental data for R versus the relative azimuthal angle $\phi = \phi_1 - \phi_2$ in the two-photon annihilation experiment of Kasday *et al.*[39] The data is fitted to the function $A - B \cos 2\phi$. The size of the data points ○ represents a typical $\pm 1\sigma$ uncertainty.

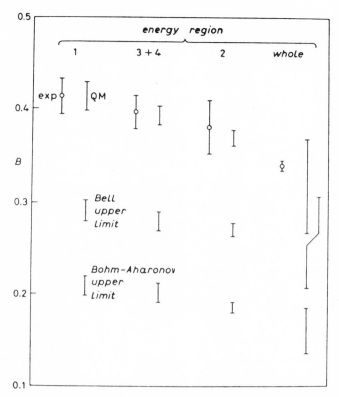

Figure 25. Experimental data for the quantity B from the two-photon annihilation experiment of Kasday *et al.*[(39)] for the four energy regions of the two photons indicated in Figure 23. For comparison, quantum predictions (QM) and the upper limits on B for Bell's inequality and the Schrödinger–Furry (or Bohm–Aharonov) hypothesis are included for the energy regions outlined in Figure 23. Note that according to the Schrödinger–Furry hypothesis, quantum mechanics should be valid for particles (or photons) which are close together. However, after the photons are some distance apart from each other their state vector would change into a product of states for individual photons. A measurement on photon 1 would affect the state vector of photon 1 but not the state vector of photon 2.

equation (3.3). For each energy, the experimental value of B agreed with quantum mechanics and exceeded the upper limits from Bell's inequality and from the Schrödinger–Furry (or Bohm–Aharonov) hypothesis.

3.2. Faraci, Gutkowski, Notarrigo, and Pennisi (1974)

While the experiment of Kasday *et al.*[(39)] gives most clear evidence for a quantum-mechanical description of the two-photon annihilation of electron–positron pairs, there is disagreement with the experiment of Faraci *et*

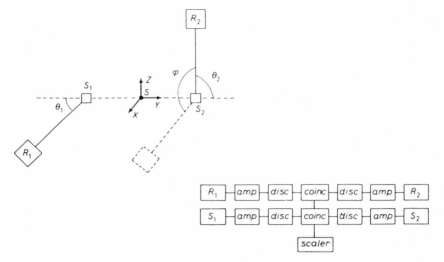

Figure 26. Schematic diagram of the experimental arrangement for the two-photon annihilation experiment to Faraci *et al.*[40] The source S consists of a ^{22}Na positron emitter enclosed in a plexiglas container acting as annihilator; S_1 and S_2 are plastic scintillators acting as Compton scatterers; R_1 and R_2 are NaI scintillators.

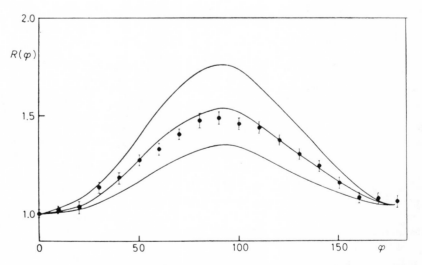

Figure 27. Normalized angular correlation function $R(\phi) = N(60°, 60°, \phi)/N(60°, 60°, 0°)$ for the two-photon annihilation experiment as a function of the relative azimuthal angle ϕ. Upper curve: quantum-mechanical prediction corrected for finite geometry of the experiment. Intermediate curve: largest correlation allowed for Bell's inequality corrected for finite geometry. Lower curve: prediction based upon the Schrödinger–Furry (or Bohm–Aharonov) hypothesis.

al.[40] on the same process. Figure 26 shows their experimental arrangement. The source S is a ^{22}Na positron emitter enclosed in a plexiglas container acting as annihilator. S_1 and S_2 are plastic scintillators acting as Compton scatterers, R_1 and R_2 are NaI(Tl) scintillators. Faraci *et al.* from Catania measured the coincidence count rates $N(\theta_1, \theta_2, \phi)$ between the four scintillators. Figure 27 gives as an example the ratio $R(\phi) = N(60°, 60°, \phi)/N(60°, 60°, 0°)$. The interesting feature of these data is that they significantly disagree with quantum mechanics and tend to agree with Bell's upper limit. The Catania group has also varied the distance between the detectors in order to test any influence possible from distance. Apart from symmetrical flight distances (at 5.5 cm, 10 cm, and 20 cm distances from the source to detector) asymmetrical flight paths of (6 cm, 13 cm) and (5.5 cm, 34 cm) have been reported (see Figure 28) for the anisotropy factor. The surprising "distance effect" (a decrease in the photon polarization correlation) cannot distinguish whether or not the effect depends on the difference of the flight paths or on the relative distance of the scattering events.

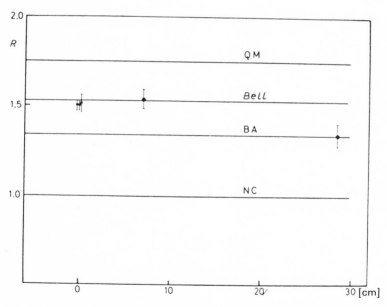

Figure 28. Anisotropy ratio R at $\theta_1 = \theta_2 = 60°$ as a function of the difference in the flight paths of the two annihilation photons in the experiment of Faraci *et al.*[40] Asymmetrical measurements for flight-path differences have been made for the detector positions at (6 cm, 13 cm) and (5.5 cm, 34 cm). The three data points for symmetrical flight paths (abscissa 0) have been taken at 5.5 cm, 10 cm, and 20 cm.

3.3. Wilson, Lowe, and Butt (1976)

The puzzling distance effect in the experiment of Faraci et al.[40] has stimulated further investigations, and Figure ·29 shows the results of this distance effect carried out by Wilson et al.[41] In their measurement, both the symmetrical as well as the asymmetrical anisotropies appear to be distance-independent up to 2.45 m and they are also in agreement with quantum mechanics. The resolving time of the coincidence apparatus of Wilson et al.[41] was 1 ns, which corresponds to a spread of $\Delta l = c\Delta t = 0.3$ m for the photons. Accordingly the two polarization detectors can be considered as "space-like separated" for source–polarimeter separation larger than 0.3 m. More recently, in an interesting extension to this experiment, Paramananda and Butt[42] have shown there is no change in the anisotropy factor for distances between the detectors as large as 24 m. With a resolving

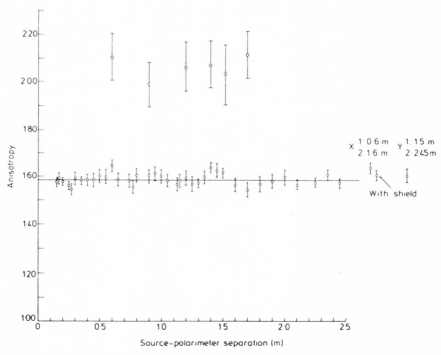

Figure 29. Measured anisotropy in the two-photon annihilation experiment by Wilson et al.[41] as a function of the source–polarimeter separation. The upper results were obtained with the NaI scintillation crystals separated by a distance $d = 11$ cm from the Compton scatterer: the results associated with the straight line were obtained with $d = 4.9$ cm. For both these cases the positron source was placed symmetrically between the Compton polarimeters, while for the results marked X and Y the source was positioned asymmetrically with the separations of the two polarimeters from the source being indicated.

time for the detection electronics which could be made as small as 100 ps, this experiment gave the largest space-like separation so far achieved in such a measurement.

3.4. Bruno, d'Agostino, and Maroni (1977)

Another study of the polarization correlation of annihilation radiation has been reported by Bruno *et al.*[43] applying the ^{22}Na positron source. Again their results are in agreement with quantum mechanics and they did not find a decrease of the polarization correlation in a range of distances between the Compton scatterers as wide as 10 photon "coherence lengths" resulting from the finite resolving time of the apparatus.

3.5. Lamehi-Rachti and Mittig (1976)

In a unique experiment of its kind, low-energy proton–proton scattering has been applied by Lamehi-Rachti and Mittig[44] as a test experiment between quantum mechanics and Bell's inequality. The basic idea behind the experiment is the spin correlation of two spin-$\frac{1}{2}$ particles in a temporary singlet state. During the collision between the two protons the interaction may dominantly be a "singlet" scattering process (two spins of the protons antiparallel). After the collisional interaction the two protons would separate from the singlet state, so that a possible change of the initial-spin antiparallel spin correlation can be tested at a certain distance between the protons. In a reaction process we may express this situation as follows:

$$p_1(\uparrow) + p_2(\downarrow) \xrightarrow[\substack{\text{singlet} \\ \text{state}}]{} p_1(\uparrow) + p_2(\downarrow)$$

In practice such a spin correlation experiment can be carried out by means of a twofold combined double-scattering experiment as indicated in Figure 30. A beam of 13.2 or 13.7 MeV protons hits a target of hydrogen. After scattering at $\theta_{\text{lab}} = 45°$ ($\theta_{\text{cm}} = 90°$) the protons strike a carbon foil and are scattered for a second time. Four detectors for the doubly scattered protons measure the following coincidence count rates: N_{LL}, N_{RR}, N_{LR}, and N_{RL}, where N_{LL} are the coincidences between the left counters L_1 and L_2, etc. The detectors of one left–right scattering analyzer are in the scattering plane while those of the other left–right scattering analyzer are rotated by an angle θ round the axis defined by the protons entering the analyzer. As a test quantity Lamehi-Rachti and Mittig[44] defined the following correlation function which is related to the spin polarization correlation:

$$P_{\text{meas}}(a, b) = \frac{N_{\text{LL}} + N_{\text{RR}} - N_{\text{LR}} - N_{\text{RL}}}{N_{\text{LL}} + N_{\text{RR}} + N_{\text{LR}} + N_{\text{RL}}} = P_1 P_2 T_1 T_2 \cos(a, b) \quad (3.4)$$

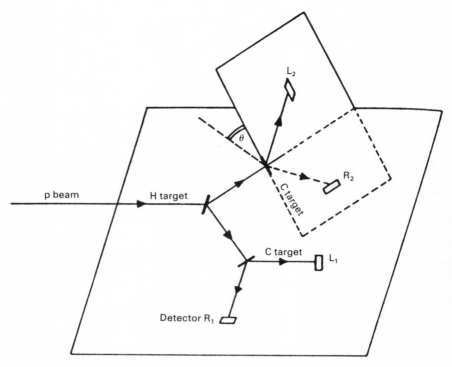

Figure 30. Schematic experimental arrangement for the measurement of the spin correlation in proton–proton double scattering.

where $P(a, b)$ is the probability of one proton having its spin in the direction a and of the other having its spin in the direction b; $P_{1,2}$ and $T_{1,2}$ are the analyzing power and transmission of the analyzers.

Indeed, the left–right scattering asymmetries of each scattering analyzer are directly related to the transverse spin polarization of the protons with regard to the plane of the second scattering. Accordingly, the device of the proton double-scattering scheme in Figure 30 is analogous to the double

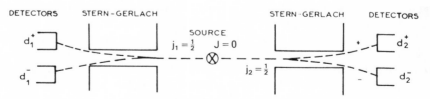

Figure 31. Double Stern–Gerlach experiment for a test of Bell's inequality. The source consists of particles with $J = 0$, decaying into two particles with $j = \frac{1}{2}$.

Stern–Gerlach experiment of Figure 31, which is Bohm's version of the EPR paradox. In this Stern–Gerlach experiment, a source of particles in the intermediate state $J = 0$ disintegrates into two kinds of particles which are moving in opposite direction. After the separation each particle has spin $j = \frac{1}{2}$, so that $J = j_1 + j_2 = 0$ is valid for a correlated pair of disintegrated particles. The corresponding correlation function for detected particles in coincidence after passing the Stern–Gerlach magnets with spin up (+ sign) and spin down (− sign) would be

$$P_{\text{meas}}(a, b) = (N_{++} + N_{--} - N_{-+} - N_{+-})/N$$

where $N = N_{++} + N_{--} + N_{+-} + N_{-+}$.

Quantum mechanics predicts a correlation function of

$$P(a, b)_{\text{QM}} = \langle \sigma_1 \cdot a\, \sigma_2 \cdot b \rangle = -\cos(a, b) = -\cos \theta$$

for the two spins in the direction a or b with an angle θ. This relationship should also be valid for the proton double-scattering experiment apart from a small contribution due to triplet scattering: $P_{\text{exp}}(\theta) = -C_{nn} \cos \theta$ where C_{nn} is the Wolfenstein[45] parameter, which is $C_{nn} = -0.95 \pm 0.015$ according to Catillon *et al.*[46] for the energy in the experiment of Lamehi-Rechti and Mittig.[44] The small difference of C_{nn} from −1 is attributed to the triplet contribution in the proton–proton scattering. To relate P_{exp} to the actual measured correlation $P_{\text{meas}}(a \cdot b)$, one has to take into account the analyzing power of P_1 and P_2 of the two analyzers and a geometric correlation between the two detectors C_g; this gives the final expression

$$P_{\text{meas}}(a \cdot b) = \frac{N_{\text{LL}} + N_{\text{RR}} - N_{\text{RL}} - N_{\text{LR}}}{N_{\text{LL}} + N_{\text{RR}} + N_{\text{RL}} + N_{\text{LR}}}$$

$$= P_1 P_2 P_{\text{exp}}(a \cdot b) + C_g[1 - |P_1 P_2 P_{\text{exp}}(a \cdot b)|^2] \cos(a \cdot b) \quad (3.5)$$

Measuring P_1 and P_2 and C_g gives $P_{\text{exp}}(a \cdot b)$.

Table 1 gives results for the measured correlation function $P_{\text{meas}}(\theta)$ for two target densities. From these data and the measured values for P_1, P_2, and C_g, the quantity $P_{\text{exp}}(\theta)$ has been calculated by means of equation (3.5). Figure 32 shows the results for $P_{\text{exp}}(\theta)$ and a comparison for the quantum-mechanical prediction and the limits following Bell's inequality. It can be seen from this figure that the spin correlation of the proton–proton scattering gave good agreement to quantum mechanics, while the results

Table 1. Final Results for the Measured
Correlation Functions $P_{meas}(\theta)$ as a Function of
the Angle θ for 18.6-mg/cm^2 and 29-mg/cm^2
Targets. Errors Are Statistical One-Standard-
Deviation Errors

θ	$P_{meas}(\theta)$ (18.6 mg/cm^2)	$P_{meas}(\theta)$ (29 mg/cm^2)
0°	−0.40±0.05	−0.38±0.025
30°	−0.38±0.04	−0.27±0.025
45°	−0.29±0.04	−0.26±0.023
60°	−0.24±0.04	−0.17±0.025
90°	−0.01±0.03	−0.03±0.04

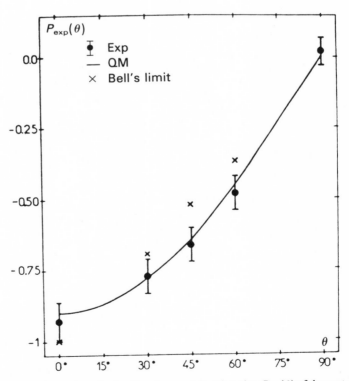

Figure 32. Experimental results for the spin correlation function $P_{exp}(\theta)$ of the proton–proton
double-scattering experiment carried out by Lamehi-Rachti and Mittig,[44] in comparison to
the limits of Bell's inequality (×) and quantum mechanics (QM).

are in contradiction with the limiting values following Bell's inequality. The authors claim that the experimental spin correlation data contradict the limits of Bell with a statistical significance of $\frac{7}{10\,000}$. By applying the prediction of quantum mechanics for $P_{\exp}(\theta)$ the Wolfenstein parameter C_{nn} becomes $C_{nn} = -0.97 \pm 0.05$, in good agreement with the value of $A_{yy} = C_{nn} = -0.95 \pm 0.015$ of Catillon *et al.*[46]

4. Discussion and Proposals for Future Experiments

From the discussions in Sections 2 and 3 it is clear that, with two exceptions, the results of experiment confirm the validity of quantum mechanics and violate Bell's inequality, thus providing *prima facie* evidence against a local realistic view of the world. The experiments also support the conclusion that there is no spontaneous change in the form of the state vector of the kind discussed by Schrödinger and Furry, at least over the distances so far investigated. However, objections to these conclusions have been raised: In the case of the atomic physics experiments (Section 2) where the efficiency of the polarizers is high, these objections have centered on the low efficiency of detectors for photons in the ultraviolet and visible part of the spectrum, which leaves open the possibility that the observed polarization correlation may be the result of some process of enhancement in the detection process itself. On the other hand, for the positronium annihilation and spin experiments, where the efficiency of detection is high, polarization and polarization correlation must be inferred indirectly from the results of a secondary scattering experiment. This fact has allowed the results to be explained easily in terms of *ad hoc* local realistic theories. Also, since in order to analyze the results in this type of experiment the validity of quantum mechanics itself must be assumed, it is questionable, logically, if such experiments can be used to test quantum mechanics.

Given the above objections it would clearly be desirable to carry out an experiment with high-efficiency polarizers and high-efficiency detectors. To achieve such an ideal, in 1981 Lo and Shimony[47] suggested an experiment in which the spin of sodium atoms resulting from the dissociation of sodium dimers (Na_2) would be analyzed by Stern–Gerlach magnets and hot-wire detectors. An experiment of this kind would, of course, be a realization of the type of experiment originally suggested by Bohm[2] in 1951. However, as Shimony[48] himself pointed out, the experiment may not be possible with sodium because of the difficulty in designing the Stern–Gerlach magnets in such a way that the sodium atoms will not precess excessively prior to entering the analyzer and, also, because of the loss of correlations of the electronic spins of the atoms in transit due to coupling with the nuclear spins. In relation to EPR experiments making use of spin, it is probably worth noting at this point that it is not considered possible to use Stern–Gerlach magnets to carry out a spin analysis of free electrons, since the splitting produced by the inhomogeneous magnetic field would be less than the uncertainty in the deflection predicted by Heisenberg's uncertainty relation.

Several proposals have also been made for EPR experiments involving the decay of elementary particles into two components. So far, these experiments are only in preparation or in a state where very preliminary results have been reported. A particularly interesting example, discussed by Selleri[49] and Six,[50] proposes the use of the neutral kaon pair $K^0 \bar{K}^0$ resulting from the annihilation of a proton–antiproton pair according to $p\bar{p} \to K^0 \bar{K}^0 \to K_s K_L$, where $K_s = (K^0 + \bar{K}^0)/\sqrt{2}$ and $K_L = (K^0 - \bar{K}^0)/\sqrt{2}$ represent, respectively, the long-lined and short-lined mass eigenstates of kaons. By applying Einstein locality to the decay of the $J^{PC} = 1^{--}$ state into a pair of neutral kaons, Selleri[49] predicted a rate for double $\bar{K}_0$ observation which is about 12% lower than that predicted by quantum mechanics. Possible experiments in which the neutral pairs $K^0 K^0$, $K^0 \bar{K}^0$, $\bar{K}^0 \bar{K}^0$ are converted into $K^\pm$ by the strong interaction have been discussed by Six.[50] Recently Datta and Home[51] have pointed out that an EPR experiment might be carried out by examining the decay of the spin-1 vector meson $\gamma(4s)$ into a pair of neutral pseudoscalar mesons $B^0 \bar{B}^0$. Yet another suggestion is that of Tixier[52] and Törnquist[53] to observe the reaction $J/\psi \to \Lambda\bar{\Lambda} \to \pi^- p \pi^+ \bar{p}$, in which the polarization properties of the $\bar{\Lambda}\Lambda$ particles are related to the angle between detected $\pi^\pm$ pairs. Preliminary measurements by the DM2 Collaboration[52] have been made of the correlation between the momenta of $\pi^\pm$ pairs and, within the limited statistical accuracy so far obtained, the results are consistent with the quantum-mechanical predictions.

The EPR argument, as we have seen, hinges on the form of the state vector, given in equation (1.1) or (1.2), which is a coherent superposition of two states of joint polarization for the two photons. Thus, any demonstration regarding the existence of such coherent superposition states in quantum mechanics is of considerable general interest. In this context it is worth mentioning the neutron interference experiments of Summhammer, Badurek, Rauch, Kischko, and Zeilinger,[54] the experiment involving the photodissociation of calcium molecules (Ca_2) carried out by Grangier, Aspect, and Vigue,[55] and the proposal by Leggett[56] that such coherent superpositions may also be shown to exist for macroscopic states of superconducting devices. In any event, it is clear that experimental work related to the EPR question and, more generally, the range of validity of quantum mechanics itself will continue to occupy the minds and rouse the passions of physicists everywhere.

References

1. A. Einstein, B. Podolsky, and N. Rosen, *Phys. Rev.* **47**, 777 (1935).
2. D. Bohm, *Quantum Theory*, Prentice-Hall, Engelwood Cliffs, N.J. (1951).
3. D. Bohm and Y. Aharonov, *Phys. Rev.* **108**, 1070 (1957).
4. J. S. Bell, *Physics* **1**, 195 (1964).
5. J. F. Clauser, M. A. Horne, A. Shimony, and R. A. Holt, *Phys. Rev. Lett.* **23**, 880 (1969).
6. E. Schrödinger, *Proc. Camb. Phil. Soc.* **31**, 555 (1935).
7. W. H. Furry, *Phys. Rev.* **49**, 393 (1936).
8. J. F. Clauser and A. Shimony, *Rep. Prog. Phys.* **41**, 1881 (1978).
9. F. M. Pipkin, in: *Advances in Atomic and Molecular Physics* (D. R. Bates and B. Bederson, eds.), Vol. 14, pp. 281–340, Academic Press, New York (1978).
10. J. S. Bell, in: *Proceedings of the International School in Physics "Enrico Fermi", Course 1L* (B. d'Espagnat, ed.), pp. 171–181, Academic Press, New York and London (1971).
11. S. J. Freedman, PhD Thesis, University of California, Berkeley (1972).
12. C. A. Kocher and E. D. Commins, *Phys. Rev. Lett.* **18**, 575 (1967).
13. J. F. Clauser, *Phys. Rev. A* **6**, 49 (1972).
14. S. J. Freedman and J. F. Clauser, *Phys. Rev. Lett.* **28**, 938 (1972).
15. R. A. Holt and F. M. Pipkin, Harvard University preprint (1974); see also Pipkin.[9]
16. J. F. Clauser, *Phys. Rev. Lett.* **36**, 1223 (1976).
17. J. F. Clauser, *Nuovo Cim. B* **33**, 740 (1976).
18. E. S. Fry and R. C. Thompson, *Phys. Rev. Lett.* **37**, 465 (1976).
19. A. Aspect, P. Grangier, and G. Roger, *Phys. Rev. Lett.* **47**, 460 (1981).
20. A. Aspect and P. Grangier, *Lett. Nuovo Cim.* **43**, 345 (1985).
21. A. Aspect, P. Grangier, and G. Roger, *Phys. Rev. Lett.* **49**, 91 (1982).
22. F. Falciglia, L. Fornari, A. Garuccio, G. Iaci, and L. Pappalardo, in: *The Wave-Particle Dualism* (S. Diner, G. Lochak, and F. Selleri, eds.), pp. 397–412, D. Reidel, Dordrecht (1984).
23. A. Aspect, *Phys. Rev. D* **14**, 1944 (1976).
24. A. Aspect, J. Dalibard, and G. Roger, *Phys. Rev. Lett.* **49**, 1804 (1982).
25. T. W. Marshall, E. Santos, and F. Selleri, *Lett. Nuovo Cim.* **38**, 417 (1983).
26. F. Selleri, *Lett. Nuovo Cim.* **39**, 252 (1984).
27. S. Pascazio, *Nuovo Cim. D* **5**, 23 (1985).
28. W. Perrie, A. J. Duncan, H. J. Beyer, and H. Kleinpoppen, *Phys. Rev. Lett.* **54**, 1790 (1985); **54**, 2647(E) (1985).
29. G. Breit and E. Teller, *Astrophys. J.* **91**, 215 (1940).
30. H. Kleinpoppen, in: *Book of Invited Papers, Tenth International Conference on Atomic Physics* (H. Narumi and I. Shimamura, eds.), Tokyo, Japan, August 25–29 (1986), to be published.
31. A. J. Duncan, W. Perrie, H. J. Beyer, and H. Kleinpoppen, in: *Book of Abstracts, Second European Conference on Atomic and Molecular Physics* (A. E. de Vries and M. J. van der Wiel, eds.), p. 116, Free University, Amsterdam, The Netherlands, April 15–19 (1985).
32. A. J. Duncan, in: *Book of Invited Papers, Tenth International Conference on Atomic Physics* (H. Narumi and I. Shimamura, eds.), p. 121, Tokyo, Japan, August 25–29 (1986).
33. J. F. Clauser and M. A. Horne, *Phys. Rev. D* **10**, 526 (1974).
34. T. Haji-Hassan, A. J. Duncan, W. Perrie, H. J. Beyer, and H. Kleinpoppen, in: *Abstracts, Tenth International Conference on Atomic Physics* (H. Narumi and I. Shimamura, eds.), pp. 63–64, Tokyo, Japan, August 25–29 (1986); *Phys. Lett.* **123A**, 110 (1987).
35. A. J. Duncan, in: *Microphysical Reality and Quantum Formalism* (G. Tarozzi and A. van der Merwe, eds.), D. Reidel, Dordrecht (1986), in press.
36. A. Garuccio and F. Selleri, *Phys. Lett.* **103A**, 99 (1984).

37. J. A. Wheeler, *Ann. N.Y. Acad. Sci.* **48**, 219 (1946).
38. C. N. Yang, *Phys. Rev.* **77**, 242 (1949).
39. L. R. Kasday, J. D. Ullman, and C. S. Wu, *Nuovo Cim. B* **25**, 663 (1975); also, L. R. Kasday, in: *Proceedings of the International School of Physics "Enrico Fermi," Course IL* (B. d'Espagnat ed.), pp. 195–210, Academic Press, New York and London (1971).
40. C. Faraci, D. Gutkowski, S. Notarrigo, and A. S. Pennisi, *Lett. Nuovo Cim.* **9**, 607 (1974).
41. A. R. Wilson, J. Lowe, and D. K. Butt, *J. Phys. G* **2**, 613 (1976).
42. V. Paramananda and D. K. Butt, *J. Phys. G* (1987), in press.
43. M. Bruno, M. D'Agostino, and C. Maroni, *Nuovo Cim. B* **40**, 143 (1977).
44. M. Lamehi-Rachti and W. Mittig, *Phys. Rev. D* **14**, 2543 (1976).
45. L. Wolfenstein, *Annu. Rev. Nucl. Sci.* **6**, 43 (1956).
46. P. Catillon, M. Chapellier, and D. Garetta, *Nucl. Phys. B* **2**, 93 (1967).
47. T. K. Lo and A. Shimony, *Phys. Rev. A* **23**, 3003 (1981).
48. A. Shimony, *Phys. Rev. A* **30**, 2130 (1984).
49. F. Selleri, *Lett. Nuovo Cim.* **36**, 521 (1983).
50. J. Six, *Phys. Lett.* **114B**, 200 (1982); also, Contributed Paper, Lear Workshop, Tignes, Savoie, France (1985).
51. A. Datta and D. Home, *Phys. Lett.* **119A**, 3 (1986).
52. M. H. Tixier, in: *Microphysical Reality and Quantum Formalism* (G. Tarozzi and A. van der Merwe, eds.), D. Reidel, Dordrecht (1986), in press.
53. N. A. Törnquist, *Found. Phys.* **11**, 171 (1981); also private communication.
54. J. Summhammer, G. Badurek, H. Rauch, U. Kischko, and A. Zeilinger, *Phys. Rev. A* **27**, 2523 (1983).
55. P. Grangier, A. Aspect, and J. Vigue, *Phys. Rev. Lett.* **54**, 418 (1985).
56. A. J. Leggett, *Prog. Theor. Phys., Suppl.* **69**, 80 (1980).

8

Rapisarda's Experiment: Testing Quantum Mechanics versus Local Hidden-Variable Theories with Dichotomic Analyzers

LORENZO PAPPALARDO AND FILIPPO FALCIGLIA

1. Introduction

Any decision about the validity of different physical theories can be taken only if it is possible to realize an experiment whose results can be directly compared with the different theoretical predictions. Owing to the lack of such an experimental check, the debate between the supporters of the orthodox formulation of the quantum mechanics (QM) and those who believed QM incomplete and therefore trusted in the possibility of completing it causally within a realistic vision of the physical world has been for a long time a comparison between ideologies rather than between different physical theories.

A considerable role against the QM's causal completion has been played by the theorem of von Neumann[1] whose validity, however, rests on some axioms not necessarily required by the properties of physical objects. Indeed Bohm and de Broglie[2-4] showed the existence of physical causal models which reproduced the statistical results of QM, so contradicting the consequences of von Neumann's theorem. However since 1965, thanks to Bell's work,[5] the physicists have the theoretical tools for planning experiments

LORENZO PAPPALARDO • Department of Physics, University of Catania, 95129 Catania, Italy. FILIPPO FALCIGLIA • Institute of Physics, Viale Andrea Doria, 95125 Catania, Italy.

suitable to discriminate between QM and local hidden-variable theories (LHVT).

Bell's original paper gave rise to a considerable number of theoretical works,[6-10] which tackle the problem both in the deterministic and probabilistic hidden-variable theories (DHVT and PHVT) frame, and are today known as "Bell's inequalities" (BIn) or "Bell's theorem." Consequently, some experiments[11-14] were carried out whose interpretation however needs additional hypotheses, experimentally not verifiable, for taking into account the finite efficiencies of the detector. These hypotheses, although "physically reasonable," are not strictly necessary from a theoretical point of view and limit the BIn validity which, at least in an ideal case, should discriminate definitively between QM and LHVT.

These first experiments, based on polarization-correlation measurement on the photons emitted in atomic cascades, although very accurate, suffered both from the additional hypotheses and from experimental problems, e.g., the background noise due to the excitation mechanism of the cascade. That led Rapisarda, in the late 1970s, to propose an experiment requiring only one additional hypothesis, based on the use of dichotomic polarization analyzers, and using the latest experimental techniques.[15]

2. Ideal Experiments with Dichotomic Analyzers

A physical system leading to the Einstein–Podolsky–Rosen paradox,[16] and to which Bell's argument can be applied, is that comprising two parts, having somehow interacted in the past or having a common origin, but "separated" at the time when a measurement is made on either part. A couple of electrons in spin singlet state, as in the Bohm Gedankenexperiment,[2] or a photon couple emitted in an atomic cascade, are examples of such a physical system. The latter is the most suitable for an experimental realization.

We shall now consider the ideal experiment depicted in Figure 1. In the source S photon couples originate in an angular-momentum singlet state created, e.g., by an atomic cascade $J = 0 \rightarrow 1 \rightarrow 0$.

If we consider only photons γ_A, γ_B which travel in opposite directions along the z axis, the state of the system (photon couple before its interaction with the measuring apparatus) is described by QM, in the even-parity case, by the second-kind state vector:

$$|\psi\rangle = \frac{1}{\sqrt{2}}(|R_A\rangle|R_B\rangle + |L_A\rangle|L_B\rangle) \tag{1}$$

where $|R_i\rangle(|L_i\rangle)$ is the right- (left-) circular polarization state along the z axis for the photon $\gamma_i(i = A, B)$.

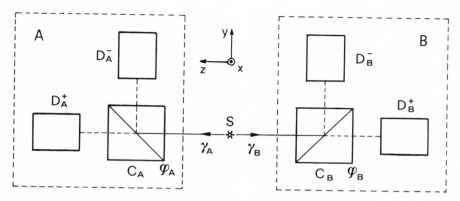

Figure 1. Scheme of the experiment with dichotomic analyzers.

An equivalent description employs linear polarization states along the x and y axes:

$$|\psi\rangle = \frac{1}{\sqrt{2}}(|x_A\rangle|x_B\rangle + |y_A\rangle|y_B\rangle) \tag{2}$$

Let C_i be ideal dichotomic linear polarization analyzers, set at angles φ_i with respect to the x axis. Ideal means that a photon γ_i, linearly polarized at an angle Φ_i with respect to the x axis, will be transmitted along the ordinary path $(+)$ of the analyzer C_i with probability $\cos^2(\varphi_i - \Phi_i)$ or along the extraordinary path $(-)$ with probability $\sin^2(\varphi_i - \Phi_i)$.

We let D_i be photon detectors and ideal too, i.e., they have efficiency equal to one and independent of the impinging photon polarization. Therefore every photon interacting with its measuring apparatus (A or B) will certainly be detected by D_i^+ or D_i^-.

The measured physical observable O_i is the photon "linear polarization": it is a dichotomic random variable with mean value equal to zero if it is assigned the value $+1$ when the photon is detected by D_i^+, and -1 when it is detected by D_i^-. It should be noted that the event "0," namely absence of detection, does not exist in this ideal experiment.

The observable "polarization product" can be defined as the product of the single-photon polarizations and can be described in a similar manner.

Finally, we define the correlation function between the polarizations of photons γ_A and γ_B as the mean value of the polarization products minus the product of the mean values:

$$E(\varphi_A, \varphi_B) = \langle O_A(\varphi_A)O_B(\varphi_B)\rangle - \langle O_A(\varphi_A)\rangle\langle O_B(\varphi_B)\rangle = \langle O_A(\varphi_A)O_B(\varphi_B)\rangle \tag{3}$$

In the QM description the mean value (expectation value in the state $|\psi\rangle$) is

$$E(\varphi_A, \varphi_B) = \cos 2(\varphi_A + \varphi_B) \tag{4}$$

where angles φ_A and φ_B are measured as shown in Figure 2.

In the PHVT it is assumed that:

P1. To each photon couple (φ_A, φ_B) emitted in S is associated the value $\lambda \in \Lambda$, where Λ is the set of values of the hidden variable λ.

P2. $\rho(\lambda)$ is the normalized density function of λ in Λ ($\int_\Lambda \rho(\lambda) \, d\lambda = 1$).

P3. The probability that the value j is obtained as the result of a measurement on the first photon and the value k on the second one is

$$P(j, k \,|\, \varphi_A, \varphi_B, \lambda) = P(j \,|\, \varphi_A, \lambda) \cdot P(k \,|\, \varphi_B, \lambda) \quad (j,k = -1,+1) \tag{5}$$

where the probabilities on the right-hand side refer to measurement on the single photons γ_A and γ_B.

Therefore

$$E(\varphi_A, \varphi_B) = \sum_{j,k} \int_\Lambda P(j \,|\, \varphi_A, \lambda) P(k \,|\, \varphi_B, \lambda) \rho(\lambda) \, d\lambda$$

$$= \int_\Lambda [P(1 \,|\, \varphi_A, \lambda) - P(-1 \,|\, \varphi_A, \lambda)]$$

$$\cdot [P(1 \,|\, \varphi_B, \lambda) - P(-1 \,|\, \varphi_B, \lambda)] \rho(\lambda) \, d\lambda \tag{6}$$

Obviously $|E(\varphi_A, \varphi_B)| \leq 1$, because all the quantities in the square brackets have modulus less than or equal to one.

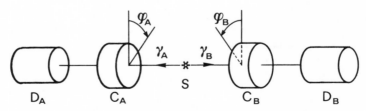

Figure 2. CHSH scheme (single output analyzers experiment).

We note that, given four numbers a, b, a', b' with modulus ≤ 1, it can easily be shown that $|ab - ab'| + |a'b + a'b'| \leq 2$, whence

$$\int (|ab - ab'| + |a'b + a'b'|)\rho(\lambda)\, d\lambda \leq 2$$

Then the function

$$C = |E(\varphi_A, \varphi_B) - E(\varphi_A, \varphi_B')| + |E(\varphi_A', \varphi_B) + E(\varphi_A', \varphi_B')| \qquad (7)$$

satisfies the inequality

$$C_{\text{PHVT}} \leq 2 \qquad (8)$$

Also, using the QM correlation function (4), it is possible to verify that

$$C_{\text{QM}} = 2\sqrt{2} \qquad (9)$$

when $\varphi_A + \varphi_B = \varphi_A' + \varphi_B = \varphi_A' + \varphi_B' = 22.5°$ and $\varphi_A + \varphi_B' = 135°$, so violating inequality (8).

In order to verify PHVT (8) against QM (9), the predicted values must be compared with the results of experimental measurements. In the experimental frame using dichotomic analyzers it is possible to measure simultaneously the four numbers N_{++}, N_{+-}, N_{-+}, N_{--} of coincidences between the outputs of the four detectors.

The experimental frequencies of the event (j, k) are given by

$$f_{jk}(\varphi_A, \varphi_B) = N_{jk} \Big/ \sum_{j,k} N_{jk}$$

and the correlation function will be

$$E(\varphi_A, \varphi_B) = \sum_{j,k} jk f_{jk} = \frac{N_{++} - N_{+-} - N_{-+} + N_{--}}{N_{++} + N_{+-} + N_{-+} + N_{--}} \qquad (10)$$

On performing measurements at various angles, function C in equation (7) can be evaluated and compared with the predictions of PHVT (8) and QM (9).

3. Experiments Based on CHSH and CH Configurations

Clauser, Horne, Shimony, and Holt (CHSH)[6] demonstrated a BIn, valid in the DHVT frame, which takes into account the finite (i.e., < 1) detector efficiency. To do this they introduced the following hypothesis:

> I1. The probability of a joint detection of a photon couple emerging from the analyzers is independent of φ_A and φ_B.

This assumption, though "physically reasonable," is not at all experimentally verifiable but gave an impulse to the development of experimental checks.

The first experiments performed[11-14] were based on the CHSH scheme outlined in Figure 2. The analyzers used have a single output (e.g., polaroids or piles-of-plates), so the only experimental information is: "the photon has been detected by D_A (D_B) on the ordinary channel." The lack of information on the extraordinary channel compels one to perform stunts. Indeed, it is necessary to perform coincidence measurements with and without the analyzers in place and to employ the following hypotheses:

> I2. The density distribution $\rho(\lambda)$ and, in general, the flux of the analyzed photon couples is the same whether or not the analyzers are in place.
> I3. The source is stable during the experiment and emits at a constant rate.

Hypothesis I1 is replaced by the following:

> I4. The detection probability of a photon when the analyzer is in place is always less than or equal to that when the analyzer is removed.

By means of these hypotheses it is possible to derive a BIn that is experimentally testable. We note that both the impossibility of experimentally checking the above assumptions and, moreover, the necessity of performing measurements with and without the analyzers, corresponding to different physical situations, significantly lowers the validity of these experiments.

4. From Ideal to Real Experiments with Dichotomic Analyzers

We shall now analyze the problems to be considered before planning a "real" experiment. They can be divided in two main classes:

C1. The problems concerning the quantum-mechanical purity of the photon-couple state.

C2. The problems raised by the imperfect behavior of the analyzers and of the detectors.

As regards the first class of problems we note that, if the photon-couple state is not a pure singlet state of the type (1), but some statistical mixture of it and other states having a different structure (e.g., odd or undefined parity), the QM correlation value will be lowered so that comparison with the predictions of the LHVT will become more difficult.

An electric-dipole transition with an intermediate state unaffected by nuclear hyperfine or Zeeman splitting is a suitable candidate for the source. The most efficient dichotomic polarization analyzers available are the calcites, and the most efficient detectors are the photomultipliers (PMs), both working in the optical range. Therefore the most suitable choice is the $Ca^{40}I$ atomic cascade $4p^2 \, {}^1S_0 \rightarrow 4s4p \, {}^1P_1 \rightarrow 4s^2 \, {}^1S_0$, in which optical photons are emitted. It is noteworthy that some decorrelation could be caused by rescattering phenomena (see Section 5.1 below).

Another possible choice could be to use an atomic forbidden transition $J = 0 \rightarrow 0$ as in the Stirling experiment.[17] The simultaneous emission of the two photons allows the coincidence time to be reduced, but the overall signal-to-noise ratio does not improve because the photon energy spectrum is very broad, so making more difficult the background-noise filtering.

In any case, the decorrelation introduced by all optical systems used to collect the photons and by the finiteness of the solid angles subtended by them must be taken into account. The evaluation of both effects[9,18] shows that they are negligible when the collecting half-angles are less than $30°$. Obviously, in all the optics (lenses, windows, filters, etc.), birefringent effects like those induced by mechanical stresses must be absent.

The second class of problems, C2, is more intriguing. Owing to the finite efficiency of the detector, only a few of the photon couples emitted from the source initiate a coincident count and so it is necessary to assume that the set of the detected couples, as regards the correlation, is a faithful image of the emitted ones. By definition this is true in the QM description, while problems arise in LHVT. It is then necessary to assume that the eventual hidden variable associated with the photon does not affect its probability to be detected or, from an opposite point of view, to build a model in which the hidden variable appears explicitly in determining this probability.

In this connection we note that the first choice is the "classical" one of CHSH and CH, while the second, which is today of growing importance, leads to the proposal of new experimental configurations.[4,19-21]

A similar problem arises for the analyzers, whose transmittances are not exactly equal to 1 or 0; however, the practical values for the transmittances of a good calcite crystal are not very different from the ideal ones. Hence it is reasonable to neglect their possible effects.

In order to obtain a formulation of BIn which considers the above arguments we shall assume the CHSH hypothesis I1, which is equivalent to the following:

I1.1. The detection probability is independent of the hidden variable associated with the photon.

I1.2. The detection probability of a photon which has passed through an analyzer set at angle φ_i is independent of φ_i.

We wish to stress again that assumptions I1.1 and I1.2 can neither be justified theoretically nor tested experimentally, at least in a direct manner.

If, to these hypotheses, is added that of system symmetry by rotations about the photon propagation axis, experimentally verifiable by the constancy of the total number of counts N_{jk}, then one can obtain the inequality[22,23]

$$Q = |z(\varphi_A, \varphi_B) - z(\varphi_A, \varphi_B')| + |z(\varphi_A', \varphi_B) + z(\varphi_A', \varphi_B')| \leq 2 \quad (11)$$

where the correlation function z is an appropriately defined generalization of the previous one (E) and takes into account the actual values of the analyzer and detector parameters.

Let us assume the following experimental constraints:

E1. The analyzer extinction ratio is small.

E2. The detector sees the photon always with the same polarization.

Then, in the QM description, the above function z becomes

$$z_{QM} = \frac{\beta_A\beta_B + \cos 2(\varphi_A + \varphi_B)}{1 + \beta_A\beta_B \cos 2(\varphi_A + \varphi_B)} \quad (12)$$

where $\beta_i = (T_i^+ - T_i^-)/(T_i^+ + T_i^-)$, T_i^k being the calcite principal transmittances.

We note that equation (12) is true only if the calcites have a small extinction ratio and if the depolarization due to the finite collecting angles is negligible.

Equation (12) reduces to the usual QM correlation function (4) when at least one of the coefficients β_i is equal to zero, i.e., the corresponding calcite has identical transmittance for the ordinary and extraordinary channel.

It can easily be verified[22] that in the interval $0 \leq \beta_A\beta_B \leq 1/\sqrt{2}$ and for the angles which yield the maximum violation of BIn in the ideal case, one has $Q_{QM} > 2\sqrt{2}$, too.

5. Rapisarda's Experiment

We shall now analyze an experiment to measure the correlation function with optical photons and dichotomic analyzers, where the conditions satisfy, as faithfully as possible, the requirements imposed by the theoretical analysis. The experiment was proposed in 1979 by Rapisarda[15] and is presently being carried out in Catania by our group.

The theoretical frame of the experiment is still that depicted in Figure 1, while the experimental setup is outlined in Figures 3 and 4.

5.1. Source of the Photon Couples

The source of the photon couples is the $Ca^{40}I$ atomic cascade shown in Figure 5. Starting from the ground level, the upper level can be populated

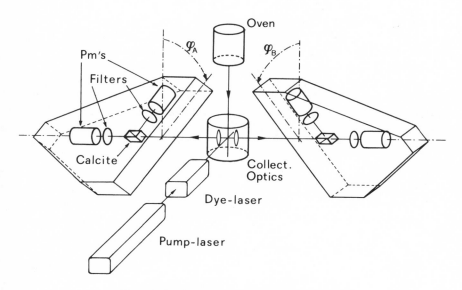

Figure 3. Rapisarda's experiment: general setup.

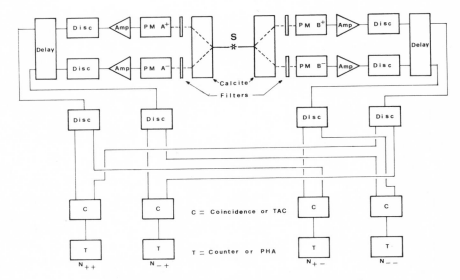

Figure 4. Rapisarda's experiment: scheme of the electronics.

only by a two-photon absorption process, where the single-photon transition is forbidden. The extremely low yield of this process is proportional to the square of the incident-radiation intensity; in practice it can be obtained only by using a tunable laser.

In effect an oven generates thermally a beam of neutral Ca atoms onto which the radiation of a dye-laser is directed. The interaction between the two beams takes place at the interior of a "scattering chamber" endowed with optical windows at Brewster's angle for the laser beam; the two beams are mutually orthogonal so as to minimize the Doppler effect.

The calcium is contained in a tantalum cylindrical crucible, about 100 cm^3 in volume, with an output hole 3 mm in diameter. The operating temperature (650–750°C) is maintained by a heating element electronically fed back so as to attain a temperature stability of ±0.5°C. Suitable

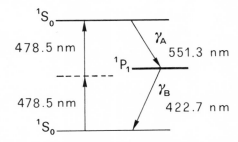

Figure 5. Partial level scheme for atomic calcium: excitation and decay paths.

diaphragms reduce the transversal dimensions of the beam so that, in the interaction region, about 60 cm from the Ca output hole, it is nearly a cylinder (divergence approximately 3.3 mrad) of diameter 1 mm.

The atomic beam density can be regulated by varying the crucible temperature between the above-mentioned limits; the resulting experimental values in the interaction region are $(1.38-5.46) \cdot 10^8$ atoms/cm^3. The measurements were performed by depositing the Ca atoms on a silicon substrate and measuring the surface density by the ion backscattering method. Working at these density values ensures negligible rescattering phenomena[24] (i.e., the absorption and consequent re-emission of some γ_B photons from the atoms, in the ground state, present in the interaction region).

In the interaction region a lens focuses the 478.5 nm radiation from a CR 599-21 dye-laser, using Coumar in 480 as dye. The dye-laser frequency is electronically stabilized and, when the laser is pumped with 2.5 W on the all-violet lines from a krypton-ion laser (CR 3000K Coherent), outputs about 50 mW in single frequency. An external feedback loop stabilizes the dye-laser against the long-term frequency drift by using, as reference, the decay rate of the excited atoms, monitored by a photomultiplier.

5.2. Analyzing and Detecting Apparatus

The photons emitted in the atomic cascade orthogonally with respect to the two beams are collected by two optical systems placed on the opposite sides of the source. Each optical system, comprising antireflection-coated lenses, subtends a solid angle of about 1.12 sr around the propagation direction: in this situation the depolarization effects are negligible.[18]

The polarization analysis is conducted by means of two Foster-type cut calcites of useful aperture approximately 12 mm and length about 30 mm. One output ray has the same direction as the input one, while the other is orthogonal to it. The calcite optical parameters were measured by means of single-photon counting techniques, under conditions similar to those in actual use.[25]

Typical measured values of the principal ($T_{i\|}^k$) and secondary ($T_{i\perp}^k$) transmittances are

$$T_\|^+ = 0.9095 \pm 0.0023, \qquad T_\|^- = 0.7625 \pm 0.0024$$

$$T_\perp^+ = 0.0044 \pm 0.0002, \qquad T_\perp^- = 0.0041 \pm 0.0003$$

These experimental values warrant the above assumptions regarding the extinction ratio in the determination of QM correlation function (12).

Behind the calcite outputs are placed interference filters which, besides selecting the photons of the "correct" wavelength for each couple of detectors, reduce the background noise due to the laser radiation diffused inside the scattering chamber. We note that the filter transmission can be taken into account by a corrective factor of the PM efficiency, if assumption I1.1 is made.

Between each filter and its relative PM is placed a lens which focuses the incident radiation on a small area of the photocathode so reducing the time-spread of the output signals. The aperture time of the concidence circuits can then be maintained at a minimum value, thereby improving the signal-to-noise ratio.

The photomultipliers, suitable for single-photon detection, are placed in cooled housings rigidly connected to the relative ones containing the calcite and the filters; therefore each photomultiplier sees "its" photon always at the same point of the photocathode and always with the same polarization. We note that this is a requirement imposed by hypothesis I1.2.

The photomultipliers were chosen on the basis of their spectral sensitivity, background noise, and time resolution. For the γ_A photons they are EMI 9863B with S20 spectral response, whose efficiency is about 10 % at 551.3 nm, while for the γ_B they are RCA 8850 with a bialkali-type photocathode and efficiency about 27 % at 422.7 nm.

The two systems (analyzer + detectors) can obviously rotate about the photon-propagation axis and, moreover, can be placed at various distances from the source so as to vary the space–time geometry of the experiment. In this manner some of the hypotheses regarding the "luminal"[26] or "superluminal"[27,28] mechanisms of the correlation propagation can be verified (see Section 6.3).

5.3. Data Collection and Analysis

The output signals from the four PMs, suitably treated (see Figure 4), are sent to four double-input coincidence circuits whose outputs, totalized by four counters, give the four numbers N_{jk} appearing in the experimental correlation function (10). The choice of coincidence-circuit resolving time should be made as a function of the intermediate-level mean life τ of the atomic cascade (about 5 ns).

Indeed the probability of a coincident count (apart from the geometrical factors, PM efficiencies, etc.) is $1 - \exp(-t/\tau)$, where t is the coincidence resolving time. A too large t renders this probability near 1, but deteriorates the signal-to-noise ratio; a too narrow t makes this probability unacceptably small. A good choice (taking into account the PM jitter, the amplifier noise, etc.) is $t = 20$ ns.

A more comprehensive analysis can be performed by using, instead of the coincidence circuits, four time to amplitude converters (TACs) whose output signals are directed to four pulse height analyzers (PHAs). A single TAC together with its PHA works as a large number (that of PHA channels) of coincidence circuits whose resolving time, depending on the TAC and PHA characteristics and settings, is normally much smaller than that of a standard coincidence circuit (about 0.1 ns against approximately 2 ns).

Therefore it is possible to study more accurately the time behavior of the system by choosing, among the events stored in the PHA memory, those with which to calculate the correlation function. We can say that a single measurement run is equivalent to a series of experiments performed in the same spatial configuration of the apparatus, but in different time situations.

6. Interpretation of Results and Further Experimental Tests

It is well known that almost all the experiments performed so far in the CHSH configuration have yielded results seemingly favorable to QM predictions. The same holds for the experiments performed more recently by Aspect and co-workers both in the old CHSH configuration[29] and in the Rapisarda one.[30] We note that the interpretation of these experiments is not as straightforward as it may seem; this has already been stressed by Selleri.[4,21]

Indeed, a result seemingly favorable to QM could imply two things: the QM description is true, or one of the auxiliary assumptions is false. In any case one must search for the validity of the assumptions and for the possible mechanisms of the nonlocality propagation. A result seemingly favorable to LHVT, moreover, does not imply the correctness of the assumption, in that it could be due to poor experimental apparatus. Therefore it is necessary to rule out any possible source of decorrelation or, better, to plan a variant of the experiment which can give prediction quite distinguishable from QM ones. Again this implies the necessity of abandoning I1.

We show how in the basic experimental configuration just described some of the hypotheses adopted in order to explain the "mechanism" of the correlation propagation can be tested with only minor experimental changes, besides the effects of magnetic fields on the photon-couple state and the validity of hypothesis I1.

6.1. On the Kinematics of the Correlation Propagation

Various mechanisms of the correlation propagation have been proposed. Two of them can easily be tested: both are connected to the propagation of a certain type of "information" generated when the first photon (γ_A) is detected, going back toward the source[26] or the other detection apparatus[27,28] and conditioning the emission of the second photon of the couple or its detection.

The first mechanism should occur at a (sub)luminal velocity and can be normally ruled out, provided that the two detection events are separated by a space-like interval. On the other hand, the second mechanism requires a superluminal velocity and can be analyzed by measuring the correlation function at various space–time configurations of the apparatus. In practice, if L_A (L_B) is the distance of detector D_A (D_B) from the source, a set of measurements is performed by keeping L_A fixed and by varying L_B. One should find a minimum value L_B below which the correlation descends below that expected from QM. Then it can be shown[27] that the velocity of the superluminal signal is $v = c(L_A + L_B)/|L_A - L_B|$.

6.2. Magnetic Field Effects on the Polarization Correlation

If in the basic experimental configuration an external magnetic field **B** is applied,[31] the intermediate level $^1P_1$ is split into three levels $m = -1$, $0, +1$ whose energy difference is $\Delta E = \mu_B B$. If the analysis is conducted along the **B** direction (coincident with the direction of the z axis), the only contributions are due to the two decays shown in Figure 6.

When the level separation is sufficiently larger than their natural width, then it is possible, at least in principle, to distinguish which of the two decay paths has been taken. This implies that the photon-couple state is no longer described by equation (1), but by a statistical mixture (with almost the same probability) of the following two states:

$$|\psi^+\rangle = |R_A\rangle|L_B\rangle \qquad \text{and} \qquad |\psi^-\rangle = |L_A\rangle|R_B\rangle \qquad (13)$$

For such a system $f_{jk} = \frac{1}{4}$, which in the main satisfies the BIn (11).

When the magnetic field is switched on the correlation function should decrease abruptly, and that would indicate transition from a nonlocal situation (strong correlation) to a local one (no correlation). Besides, such an experiment should verify, in an indirect way, the capability of the experimental apparatus to discriminate between situations satisfying or not satisfying the locality.

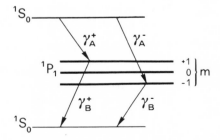

Figure 6. Magnetic field effect on the Ca atomic cascade.

If the natural width of the intermediate level is $\Delta E_{nat} \simeq \hbar/\tau \simeq 1.3 \cdot 10^{-7}$ eV, an easily obtainable magnetic field $B \simeq 200$ G would give $\Delta E \simeq 20\Delta E_{nat}$, allowing one to perform such an experiment. We note that the separation due to the Earth magnetic field (about 0.4 G) is considerably smaller (approximately $4.6 \cdot 10^{-9}$ eV) than the natural width of the level. Therefore, it is not normally necessary to shield the interaction region.

6.3. Test of Additional Assumptions

We saw that to obtain experimentally verifiable inequalities it is necessary, in addition to the locality hypothesis P3, to assume hypothesis I1 about the real detector's behavior. In all the experiments to which we have so far referred, this hypothesis could not be verified.

The only experimentally verifiable statement is that the detection probability of a linearly polarized photon is statistically independent of the polarization. This means that, given a huge number of photons, all with the same polarization and with hidden variables statistically distributed over a certain ensemble, the mean value of the detection probability does not depend on the polarization value. Nothing can be deduced from that about the detection probability of an individual photon to which a particular λ value is associated.

The only way around the problem is to dispense with the hypothesis and assume that the detection probability is a particular function both of the polarization and of λ whose average is equal to the detector's quantum efficiency.

Various models of this class, recently proposed,[4,19-21] lead to predictions experimentally distinguishable from the QM ones. These models can be verified in the above basic experimental configuration by simply adding auxiliary polarizers or half-wave plates between the calcites and the photomultipliers.

References

1. J. von Neumann, *Die Mathematische Grundlagen der Quantenmechanik*, Springer-Verlag, Berlin (1932).
2. D. Bohm, *Phys. Rev.* **85**, 166, 180 (1952).
3. L. de Broglie, *J. Phys. Radiat.* **20**, 963 (1959).
4. F. Selleri, in: *Fundamental Processes in Atomic Collision Physics* (Kleinpoppen, Briggs, and Lutz eds.), pp. 421–451, Plenum Press, New York (1985).
5. J. S. Bell, *Physics* **1**, 195 (1965).
6. J. F. Clauser, M. A. Horne, A. Shimony, and R. A. Holt, *Phys. Rev. Lett.* **23**, 880 (1969).
7. M. A. Horne, *Experimental Consequences of Local Hidden Variables Theories*, Ph.D. Thesis, Boston University (1970).

8. J. F. Clauser and M. A. Horne, *Phys. Rev. D* **10**, 526 (1974).

9. E. S. Fry, *Phys. Rev. A* **8**, 1219 (1973).

10. A. Garuccio and F. Selleri, *Found. Phys.* **10**, 209 (1980).

11. S. J. Freedman and J. F. Clauser, *Phys. Rev. Lett.* **28**, 938 (1972).

12. R. A. Holt and F. M. Pipkin, *Quantum Mechanics vs. Hidden Variables: Polarization Correlation Measurement on an Atomic Mercury Cascade*, preprint unpublished, Harvard University (1974).

13. J. F. Clauser, *Phys. Rev. Lett.* **36**, 1223 (1976).

14. E. S. Fry and R. C. Thompson, *Phys. Rev. Lett.* **37**, 465 (1976).

15. V. A. Rapisarda, *Lett. Nuovo Cim.* **33**, 437 (1982).

16. A. Einstein, B. Podolsky, and N. Rosen, *Phys. Rev.* **47**, 777 (1935).

17. A. J. Duncan, *Tests of Bell's Inequality and No-Enhancement Hypothesis Using an Atomic Hydrogen Source*, Proc. of the "Microphysical Reality an Quantum Formalism—Urbino (1985)," Plenum Press, New York (in press).

18. M. Baldo, F. Falciglia, M. E. Oliveri, and V. A. Rapisarda, *Nuovo Cim. B* **63**, 679 (1981).

19. T. W. Marshall, E. Santos, and F. Selleri, *Phys. Lett. A* **98**, 5 (1983).

20. A. Garuccio and F. Selleri, *Phys. Lett. A* **103**, 99 (1984).

21. F. Selleri, *Variable Photon Detection as an Explanation of EPR Experiments*, Proc. of the "New York Conference—N.Y. (1986)," New York Academy of Sciences Ed. (in press).

22. A. Garuccio and V. A. Rapisarda, *Nuovo Cim. A* **65**, 269 (1981).

23. F. Falciglia, L. Fornari, A. Garuccio, G. Iaci, and L. Pappalardo, in: *The Wave-Particle Dualism* (Diner *et al.*, eds.), pp. 397–412, D. Reidel, Dordrecht (1984).

24. S. Pascazio, in this book.

25. F. Falciglia, A. Garuccio, G. Iaci, and L. Pappalardo, *Lett. Nuovo Cim.* **37**, 66 (1983).

26. L. Pappalardo and V. A. Rapisarda, *Lett. Nuovo Cim.* **29**, 221 (1980).

27. F. Falciglia, G. Iaci, and V. A. Rapisarda, *Lett. Nuovo Cim.* **26**, 327 (1979).

28. A. Garuccio, V. A. Rapisarda, and J. P. Vigier, *Lett. Nuovo Cim.* **32**, 451 (1981).

29. A. Aspect, P. Grangier, and G. Roger, *Phys. Rev. Lett.* **47**, 460 (1981).

30. A. Aspect, P. Grangier, and G. Roger, *Phys. Rev. Lett.* **49**, 91 (1982).

31. F. Falciglia, A. Garuccio, G. Iaci, and L. Pappalardo, *Lett. Nuovo Cim.* **38**, 52 (1983).

9

Nonlocality and the Einstein-Podolsky-Rosen Experiment as Understood through the Quantum-Potential Approach

D. Bohm and B. J. Hiley

1. Introduction

The Einstein, Podolsky, and Rosen (EPR) experiment was originally suggested as a criticism of the conceptual completeness of the quantum theory.[1] However, with the advent of Bell's theorem[2] it became possible to look on it as a test between locality and nonlocality in the basic properties of matter.

For several centuries there has been a feeling that nonlocal theories are not acceptable in Physics. It is well known, for example, that Newton felt uneasy about action-at-a-distance and that Einstein regarded this action as "spooky." However, until the development of field theory there was no way to avoid such an assumption. But, as is well known, field theories explain interaction entirely through local forces.

When these theories were made relativistic, the requirement of locality became even stronger because relativity demanded that no signal should be propagated faster than light. When field theories were quantized, it appeared at first sight that the question of locality was not fundamentally changed. For the field operators were now still local in the sense that the

D. Bohm and B. J. Hiley • Physics Department, Birkbeck College, University of London, London WC1E 7HX, England, United Kingdom.

operators at different points on a space-like surface either commuted or anticommuted. However, the question of locality versus nonlocality cannot be fully treated merely by considering the commutation or anticommutation of operators. For it was the essential meaning of the EPR argument that, even though the operators for different places commuted, quantum mechanics could still imply a nonlocal relationship between the measurements.

Within the context of Bohr's[3] discussion of the EPR experiment, however, it is not necessary to go into this issue. What Bohr did was to show that even if the EPR experiment is done at macroscopic orders of distance, the actual process of the movement of the particles is unanalyzable so that no detailed conception of the means by which interaction takes place, has any meaning. Therefore it would be pointless to raise the question as to whether the forces were local or nonlocal.

It is significant to note, however, that most physicists do not follow the Bohr interpretation consistently, but for the most part tend to use the von Neumann interpretation in terms of the quantum states, as represented by the wave function. As we shall see, a certain kind of nonlocality can already be discerned through the von Neumann approach. However, since this has not generally been clearly realized, it has not thus far had much influence on the development of the subject. Rather, the main source of such development was an interest in hidden variables. As a result of this interest Bell[2] was led to suggest a criterion that had to be satisfied by a local hidden-variable theory. Since that time a number of experimental tests have been performed, and the majority of these support the conclusion that if there are to be hidden variables they must be nonlocal.[4,5] Of course, from Bohr's point of view all of this discussion is irrelevant, but if one wishes to pursue the question of whether "quantum reality" is objective and not fundamentally dependent on observation, then the meaning of the experiments aimed at testing Bell's inequality must be considered seriously.

One possibility is to open our minds to the suggestion that there may after all be nonlocal interactions in nature. However, there seems to be a very strong aversion to this idea. Some people regard it as aesthetically unattractive, others feel that the possiblity of doing science itself depends on restricting scientific thought to purely local theories in space-time.[6] There is no way to answer these objections, which are personal judgments. However, a more significant objection would be that nonlocal theories imply that the transmission of impulses faster than light violate special relativity. But, as we shall see later, this objection can be answered with the aid of our own proposals in terms of the quantum-potential approach.

This whole discussion generally has suffered from the fact that people have not agreed on a clear space–time model, which could illustrate what is meant by nonlocality and show what the so-called nonlocal hidden variables would have to be doing in order to explain the implications of the quantum theory. Evidently the meaning of nonlocality cannot even be put except on the basis of some kind of space-time structure. We feel that the main value of the quantum potential in this context is that it is able to do just this.

In this paper we shall show in detail how the quantum-potential approach deals with the EPR experiment in a way that is free of confusion and paradoxes. No insoluble problems arise and the only objections that remain are those based on personal judgments as to what is a suitable theory. In particular, we discuss the question of transmission of impulses faster than light. We show that the quantum potential is so highly unstable and fragile, that any attempt to impose a form on it will change it radically and in such an unpredictable and uncontrollable way that no preassigned meaning can be given to this form. And so the quantum potential cannot carry a signal. Therefore, while the quantum potential may make possible connections that are faster than light, these will not, as we have indeed already indicated earlier, violate relativity.

We are suggesting this approach not as a definitive theory but rather as something to explore. Although it certainly does not disagree with experiment there is, at present, no way to tell whether it is right or wrong. However, we feel that such an exploration will bring about insight into meaning of the quantum theory which is not available in other approaches and which will, perhaps, suggest new directions of research.

2. The Experiment of Einstein, Podolsky, and Rosen

To begin the discussion, let us recall the salient features of the EPR experiment.[1] Consider a quantum state of two particles, with coordinates $\mathbf{x}_1$ and $\mathbf{x}_2$, and which is given by

$$\Psi(\mathbf{x}_1, \mathbf{x}_2) = S(\mathbf{x}_2 - \mathbf{x}_2 - \mathbf{a}) \tag{1}$$

where S is a very sharply peaked function with width much less than a. If the position of the first particle, $\mathbf{x}_1$, is measured, then we know that the position of the second particle is $\mathbf{x}_2 - \mathbf{a}$, with a negligible uncertainty. Therefore by measuring $\mathbf{x}_1$, we are simultaneously able to know $\mathbf{x}_2$. However, the wave function may be Fourier analyzed as

$$\Phi(\mathbf{k}_1, \mathbf{k}_2) = \sum_k S_k \, e^{ik(x_2 - x_1)} \, e^{-ik \cdot a} \tag{2}$$

where S_k is the Fourier coefficient of $S(x_2 - x_1 - a)$ and where $\mathbf{k} = \mathbf{k}_2 + \mathbf{k}_1$. If the momentum of particle 1 is measured, then particle 2 will have exactly the opposite momentum. Thus a measurement of k_1 enables us to know k_2.

The above results will hold regardless of how large **a** is. This led EPR to conclude that quantum mechanics as it now stands cannot be conceptually complete. To show how they came to this conclusion, we recall that they defined what they called "elements of physical reality." They did not attempt to provide a necessary criterion for what such an element is. However, they proposed that if a variable corresponding to an element can be measured precisely without disturbing the system, then this is a sufficient criterion that this variable represents an element of reality. As is well known, when we measure the properties of particle 1, they are generally disturbed, so that we cannot obtain simultaneous values of x_1 and p_1 precisely. However, the measurement of x_1 enables us to know x_2 as precisely as we please without disturbing particle 2 itself. Likewise, the measurement of p_1 enables us to know p_2, again without disturbing particle 2. Of course we cannot know x_2 and p_2 precisely at the same time, but nevertheless, the implication of the above argument is that they must exist in the second particle with precisely defined values. (For, since there is no disturbance, they cannot be made uncertain.) But in the usual interpretation of the quantum theory, all properties are assumed to be contained in the wave functions and the operators. However, this provided no way to define x_2 and p_2 simultaneously. EPR therefore argued that some further concepts are needed to describe the elements of reality corresponding to the simultaneous definition of x_2 and p_2.

It must be emphasized once again that this conclusion depends upon the fact that there is no interaction between the particles. Of course, we are presupposing that the classical potential $V(x_1, x_2)$ is zero so that, at least in terms of the language of current quantum theory, it would follow that there is no such interaction. However, it would seem natural when presented with the argument of EPR to ask whether or not there may be some unknown interaction that would connect the particles. If there were such an interaction, particle 2 might be disturbed when particle 1 is measured and so one of the basic assumptions of EPR concerning this experiment would be called into question. EPR do not seem to have considered this as a serious possibility, probably because it would imply a nonlocal interaction even at macroscopic distances. Such a nonlocal interaction could perhaps be avoided by postulating an interaction propagated at a speed not greater than that of light. But this would have implied that the EPR correlations would not be obtained if measurements on both particles are carried out simultaneously. EPR seemed to have sufficient confidence in the current formalism so that they did not even raise the question of whether there is such a propagated interaction (which clearly would have led to predictions contradicting those of the current quantum theory). Their criticism concerned only the lack of an adequate conceptual basis and was not directed at the statistical predictions of the theory itself.

Schrödinger[7] was one of the earliest to understand the full implications of the argument of EPR. He realized that many-body wave functions, such as that given in equation (1), represent a new type of "entanglement" of systems which is significant even when the potential energy of interactions between them is zero. Such an entanglement is, in fact, the rule in quantum mechanics and is behind all the quantum-mechanical treatments of molecules, chemical properties, properties of solids, metals, etc. Few physicists would feel uncomfortable with this sort of nonlocality if it took place only at very short distances. What was hard to accept, however, was the implication of nonlocality at large distances.

Bohr saw that the experiment of EPR was an important challenge to his own interpretation of the theory and he was quick to provide what he regarded as an adequate response.[3] Essentially this consisted in pointing out that his notions of complementarity still applied, even when particles are separated by large distances and when the results of experiments are separated by large intervals of time. He concluded therefore that the attempt to discuss in detail how the "disturbances might be carried from one particle to another" has no meaning. This is because the form of the experimental conditions and the content (meaning) of the experimental results are, in Bohr's view, a whole that is not further analyzable. Therefore there is no way, properly, to think about the properties of particle 2 apart from the experimental contexts within which they are measured. But the context needed in a position measurement is not compatible with that needed for a momentum measurement. This signifies that even though we can infer either the position or the momentum of particle 2 without disturbing the latter, there is no experimental situation with regard to particle 2 in which both of these inferences could have meaning together. Therefore the questions raised by EPR have no place to which they could be relevantly applied.

It also follows from the above that there can be no meaning to discussing in detail any underlying process by which the experimental results could be brought about (since there is no experimental situation in which this process could reveal itself). According to Bohr all that is available is the overall phenomenon which is, as we have said, an unanalyzable whole and all that we can do is to use what he terms the "quantum algorithm" to calculate the probabilities of the various results.[8]

Originally, there was no technically feasible way to test whether the predictions of the quantum theory with regard to the EPR correlations are correct. However, one of us[9] proposed an equivalent experiment involving a molecule of zero spin consisting of two particles of spin one-half. The wave function for this system is

$$\Psi(1, 2) = (1/\sqrt{2})[\phi_+(1)\phi_-(2) - \phi_-(1)\phi_+(2)] \qquad (3)$$

This implies that if the spin of particle 1 is measured in any given direction, that of particle 2 measured in the same direction will come out opposite no matter how far apart they are. A very similar situation arises with two photons whose combined spin is zero. A test was proposed that would distinguish between the predictions of the quantum theory and another theory[10] that was essentially an extension of the hypothesis proposed by Furry.[11] Basically this consisted in assuming that when the two photons separate, the individual states of polarization in *some* direction are well defined, but both photons have states of opposite polarization. It was further assumed that the polarization directions of individual pairs would be random.

It is fairly evident that with this model there will be less correlation than is implied by quantum theory. Thus, suppose we measure the polarization in a certain direction, clearly almost all the photons will in general be polarized in other directions. It will therefore be impossible to guarantee that when photon 1 is polarized in a certain direction, the other will be opposite as demanded by the quantum theory. It was found that the experiments were in agreement with the predictions of the quantum theory and not with those of this model.[12-14]

We emphasize, however, that the Furry model was really based on the idea that after the photons separated, each developed certain properties which were carried along with the movement of the photon. In this way correlations were explained basically in the classical manner. However, quantum mechanics was still being used for the individual particles themselves, but the two-body wave function of the type shown in equation (3) was no longer used once the photons separated.

The experimental results ruled out this rather natural way of trying to treat the problem. But because the experiments used the Klein–Nishina formula to analyze the scattering of photons off the plastic scatterers, this conclusion has been criticized. For it is not simply a test of the long-range correlations, but rather of a combination of the long-range correlations and the Klein–Nishina formula, both of which are based on the quantum theory.

The above model clearly provides a rather restricted test of this nonlocal feature of the quantum theory. Bell later proposed a far more general test. The basic idea may be considered as follows. A molecule disintegrates into two particles 1 and 2, associated with which are the parameters λ_1 and λ_2 which determine, at least statistically, the results of all measurements on these respective particles. We may say, therefore, that λ_1 and λ_2 are able to carry information about the molecule from which the particles emerged, in a purely local way. This information is represented by some probability function $f(\lambda_1, \lambda_2)$ which can explain the correlated properties of the two particles. We then assume that the probability of a given result of a measurement on the first particle, for example, the spin in the direction $\hat{\mathbf{n}}_1$, is

$g_1(\lambda_1, \hat{\mathbf{n}}_1)$. The corresponding probability of the second particle is $g_2(\lambda_2, \hat{\mathbf{n}}_2)$. The total probability of the combined result is

$$\int g_1(\lambda_1\hat{\mathbf{n}}_1)g_2(\lambda_2\hat{\mathbf{n}}_2)f(\lambda_1, \lambda_2)\, d\lambda_1\, d\lambda_2$$

The essential feature of the above is that the probability of any measurement in the direction $\hat{\mathbf{n}}_1$ is independent of the measurement in the direction $\hat{\mathbf{n}}_2$ in which the second particle is measured. This seems to cover the most general type of hidden-variable theory one can consider, in which the correlations are carried or propagated locally along with the movements of the particles.

Using essentially this assumption, Bell arrived at his well-known inequality which has to be satisfied by such a local theory, and he demonstrated that the quantum mechanics does not satisfy this inequality. However, without the assumption that the probability of the two measurements contained the product $g_1(\lambda_1\hat{\mathbf{n}}_1)g_2(\lambda_2\hat{\mathbf{n}}_2)$, the Bell inequality would not have been obtained. This product function represents the relative independence of the two distributions implying no direct contact between the two particles. Therefore the failure to satisfy the Bell inequality would mean that measurements made on one of the particles would directly affect those made on the other, even though the distance between them is large. The quantum mechanics would therefore still be compatible with a model having some kind of parameters that were nonlocally related.

The Furry model is actually a special case of the class of (local) models considered by Bell. However, Bell's approach is more general in that it may include "hidden variable" theories in which even the individual particles do not obey quantum mechanics.

A number of experiments have been done to test not only the extension of the Furry model to a pair of photons, but also the more general models implied by Bell's inequality.[4] On the whole the results strongly favor the quantum theory. Moreover, the experiments of Aspect *et al.*[5] have been done under conditions in which the properties of both photons have been measured at the same time. In this way any possible explanation in terms of the propagation of an unknown force at a speed not greater than that of light is ruled out.

The above experiments have been criticized by several authors,[15] who point out that they all depend on assumptions concerning the functioning of the apparatus that they feel can be questioned. However, it appears that it will be difficult in the present state of the art to close these possible loopholes. Nevertheless, we feel that the available evidence makes a reasonable *prima facie* case for the conclusion that if there is an explanation of the results, it must be nonlocal (recalling, of course, that in Bohr's approach such an explanation would have no meaning).

3. The von Neumann Theory of Measurement

We are going to give an explanation of the EPR experiment in terms of the quantum potential which, in general, is nonlocal in its action. However, before doing this we want to emphasize further that nonlocality as such does not depend on the assumption of a quantum potential or, indeed, of any kind of hidden-variable theory. For nonlocality is already implied by the usual interpretation, and this implication is especially clear, as we have pointed out earlier, in the form proposed by von Neumann.[16] To show this we shall begin by summarizing some of the essential features of von Neumann's theory of measurement as extended along the lines described by Bohm.[9]

An essential step is to divide the measurement process into two parts. In the first part, different quantum states are distinguished, but this distinction has not yet been made irreversible. In the second step, amplification to a macroscopic scale of the outcome of the measurement is made and in this way the result is fixed irreversibly.

Let the initial wave function of the observed system be $\psi(\mathbf{x})$ and let the significant apparatus variable, such as a pointer, be represented by $\mathbf{y}$. Initially the state of the apparatus will correspond to a packet $\phi_0(\mathbf{y})$. It is sufficient for our purposes to deal with impulsive measurements such that, during the interaction of the observed system and the measuring instrument, the self-Hamiltonians of the both systems may be neglected. During this period we assume an interaction Hamiltonian

$$H_I = i\lambda O \, \partial/\partial y \tag{4}$$

where O is the operator to be measured. Its eigenvalues and eigenfunctions are, respectively, O_n and $\psi_n(\mathbf{x})$. We express the initial wave function of the measured system as

$$\psi_0 = \sum_n C_n \psi_n(\mathbf{x}) \tag{5}$$

The initial wave function for the whole system is

$$\Psi_0 = \sum_n C_n \psi_n(\mathbf{x}) \phi_0(\mathbf{y}) \tag{6}$$

The use of the Hamiltonian (4) then gives rise to

$$\Psi = \sum_n C_n \psi_n(\mathbf{x}) \phi_0(\mathbf{y} - \lambda O_n T) \tag{7}$$

where T is the duration of the interaction. If λT is much larger than the width of the wave packet ϕ_0, then, clearly, there will be no overlap of the packets corresponding to different possible results O_n, of the measurement.

However, as we have already indicated, this process is still, in principle, reversible. What we do now is to assume an interaction with a detecting apparatus, D, which magnifies the differences in O_n to macroscopic orders of magnitude. Such an apparatus will have a vast number of internal coordinates Z_s corresponding to all of its constituent particles. It is evident that this process will be irreversible in the thermodynamic sense.[9]

After the detection process is over, the overall wave function will be

$$\Psi = \sum_n C_n \psi_n(\mathbf{x}) \phi_0(\mathbf{y} - \lambda O_n T) \chi_n(\mathbf{Z}_1, \ldots, \mathbf{Z}_s, ..) \tag{8}$$

where χ_n represents the state of the apparatus that has registered the eigenvalue n. Since the χ_n represents states of the detecting device that are clearly distinct for different states of the observed system, it follows that the χ_n are orthogonal. The set of $\phi_0(\mathbf{y} - \lambda O_n T)$ is evidently also orthogonal because they do not overlap. Using this orthogonality we obtain for the average value of an arbitrary operator, M, belonging to the observed particle

$$\bar{M} = \int \sum_n C_n^* C_n \psi_n^*(\mathbf{x}) M \psi_n(\mathbf{x}) \, d\mathbf{x} \tag{9}$$

Thus far the wave function (8) still spreads over all values of n. However, the average value M given by expression (9) can also be obtained if we assume that the system falls into one of the states, $n = m$, with probability

$$P_m = C_m^* C_m$$

This is equivalent to the assumption that wave function collapses from equation (8) to

$$\Psi_m = \psi_m(\mathbf{x}) \phi_0(\mathbf{y} - \lambda O_m T) \chi_m(\mathbf{Z}_1, \ldots, \mathbf{Z}_s, \ldots) \tag{10}$$

Statistically, this collapse is equivalent to going from a pure state to a mixed state. Of course, such a collapse violates what has been assumed to be a basic quantum law, namely, Schrödinger's equation. Nevertheless, von Neumann has assumed this step to be a basic feature of the measurement process which is, indeed, now called the projection postulate. (As we have indeed already pointed out, this approach seems to be the one commonly adopted, at least tacitly, by most physicists.)

To see what the projection postulate implies, consider two alternative kinds of measurement, that of $\mathbf{x}$ and that of $\mathbf{p}$. When we measure $\mathbf{x}$, then the original wave function $\psi_0(\mathbf{x})$ undergoes the projection

$$\psi_0(x) \rightarrow \psi_n(\mathbf{x})\delta(\mathbf{x} - \mathbf{x}_n) \tag{11}$$

where $\mathbf{x}_n$ is the actual result of the measurement. If, however, we measure $\mathbf{p}$, then the projection is

$$\psi_0(x) \rightarrow \phi_n e^{i\mathbf{p}_n\cdot\mathbf{x}} \tag{12}$$

where ϕ_n is the Fourier coefficient of ψ_0. The effect of measurement is thus to transform the original wave function into something very different—in this case, either into a δ-function, or into a plane wave. Therefore it seems inappropriate even to call this process a measurement, because the result does not in the least resemble the state in which the system started. Rather, as has also been explained by Bell,[17] it would be better to call this process a *transformation* in which the apparatus and the system of interest are left in correlated states.

Let us now go on to discuss the EPR experiment in terms of the above analysis. The system to be "measured" now contains the two particles $\mathbf{x}_1$ and $\mathbf{x}_2$ with initial wave function

$$\Psi_0 = S(\mathbf{x}_2 - \mathbf{x}_1 - \mathbf{a})$$

However, measurements will be carried out on only one of the particles, say, $\mathbf{x}_1$. Initially the total wave function is

$$\Phi_0 = S(\mathbf{x}_2 - \mathbf{x}_1 - \mathbf{a})\phi_0(\mathbf{y})\chi_0(\mathbf{Z}_1, \ldots, \mathbf{Z}_s, \ldots) \tag{13}$$

After the position-measuring Hamiltonian has operated (on the variables, $\mathbf{x}_1$ only), the wave function will be

$$\Lambda_A = \int S(\mathbf{x}_2 - \mathbf{x}_1 - \mathbf{a})\delta(\mathbf{x}_1 - \boldsymbol{\xi})\phi_0(\mathbf{y} - \lambda\boldsymbol{\xi}T)\chi_0(\mathbf{Z}_1, \ldots, \mathbf{Z}_s, \ldots)\,d\boldsymbol{\xi} \tag{14}$$

and after the momentum-measuring Hamiltonian has operated the wave function will be

$$\Lambda_B = \sum_{\mathbf{k}} S_{\mathbf{k}}\, e^{i\mathbf{k}(\mathbf{x}_2-\mathbf{x}_1)}\, e^{-i\mathbf{k}\cdot\mathbf{a}}\phi_0(\mathbf{y} - \lambda\mathbf{k}T)\chi_0(\mathbf{Z}_1, \ldots, \mathbf{Z}_s, \ldots) \tag{15}$$

The next step is to bring in the detecting device. When this operates, the wave function $\chi_0(\mathbf{Z}_1, \ldots, \mathbf{Z}_s, \ldots)$ in equation (14) will change to $\chi_{\xi_i}(\mathbf{Z}_1, \ldots, \mathbf{Z}_s, \ldots)$ and in equation (15) it will change to $\chi_\mathbf{k}(\mathbf{Z}_1, \ldots, \mathbf{Z}_s, \ldots)$. We then use the projection postulate which gives the final states

$$\Phi_A \to S(\mathbf{x}_2 - \zeta_1 - \mathbf{a})\phi_0(\mathbf{y} - \lambda\zeta_1 T)\chi_{\zeta_1}(\mathbf{Z}_1, \ldots, \mathbf{Z}_s, \ldots) \qquad (16)$$

and

$$\Phi_B \to e^{i\mathbf{k}\cdot\mathbf{x}_2} e^{-i\mathbf{k}\cdot\mathbf{x}_1}\phi_0(\mathbf{y} - \lambda\mathbf{k}T)\chi_\mathbf{k}(\mathbf{Z}_1, \ldots, \mathbf{Z}_s, \ldots) \qquad (17)$$

The key new feature here is that the measurement of particle 1 alone leads to the simultaneous projection of both particle 1 and particle 2, either into δ-functions of position, or into plane waves. One might perhaps accept that the apparatus could transform particle 1 to bring about its projection into an appropriate state. But we have no understanding *at all* of how particle 2 is projected in a similar way without any interaction between them or with the measuring apparatus. It is thus clear that something highly nonlocal is implied, especially in view of the fact that the state into which particle 2 is projected depends on the apparatus with which particle 1 has interacted. We therefore emphasize again that von Neumann's interpretation implies a certain kind of nonlocality. Since, as we have already indicated, this is at least tacitly the interpretation most commonly adopted in physics, it would follow that the implications of nonlocality of the quantum theory are not to be avoided simply by refraining from using "hidden variables." And, as we have seen, this kind of nonlocality is one of the most essential features of the quantum theory and therefore cannot be dismissed as a mere side issue.

It must be emphasized, however, that von Neumann's approach and Bohr's approach are different in certain key ways. In Bohr's approach there is no attempt to analyze the process of measurement as von Neumann has done. Rather, as we have already pointed out earlier, Bohr expressly enjoins us to regard the entire experimental situation as an unanalyzable whole which he considers to be a *phenomenon*, i.e., an appearance. Bohr does not clearly commit himself as to whether or not there is a "quantum reality" underlying this phenomenon. But it seems to be implied by him that even if there is such a reality, we can say nothing about it. Therefore it follows in Bohr's view that analyses of the type given by von Neumann, along with the necessary assumption of the "collapse," are not really relevant to the issues under discussion.

In such an analysis von Neumann's notion of the quantum state, represented by the wave function $\psi(\mathbf{x}, t)$, plays a key role. In effect von

Neumann assumes that the quantum state is the most complete description of reality that is possible. Indeed it would be fair to say that for von Neumann the quantum state *is* the basic element of reality. Thus, for example, in a typical experiment, one begins with the initial quantum state $\psi_0(\mathbf{x}, t)$ and after the interaction Hamiltonian has operated this "collapses" into a new state $\psi_f(\mathbf{x}, t)$.

Bohr, however, never even refers to the quantum state, and, in fact, for him this concept can have no meaning. For, as we have already pointed out, all that we can properly talk about in this point of view is the one unanalyzable whole phenomenon which is described in terms of the form of the experimental conditions and the content of the experimental results. In this description there is no room for any element of "quantum reality" such as would be implied by the term quantum state. That is to say, it has no meaning to think, for example, of an electron in a quantum state that is acted on by an apparatus that gives rise to another quantum state. Rather we must begin with one total initial experimental phenomenon and use the quantum algorithm to compute the probability that this will give rise to a certain experimental result. As we have already emphasized throughout this chapter, not only is there no way to discuss a process by which this happens, but also we cannot relevantly discuss even the initial and the final states of the quantum system. The terms appearing in the solution of Schrödinger's equation, such as the initial and final wave functions, are merely parts of the quantum algorithm and have no more physical significance than, for example, do the terms in the power-series expansion of a function in classical physics. (Feynman[18] has come out with a very similar view.)

Since nothing can thus be said either about the state of "quantum reality" or about its detailed process of movement and interaction, it follows in Bohr's view that there is no way relevantly to use either the concepts of locality or of nonlocality (which after all could refer meaningfully only to some kind of "quantum reality"). One simply has to be silent on this issue except, perhaps, to say that we have gone beyond the domain in which this distinction has any meaning.

Most physicists who reject nonlocality would not be satisfied, thus, to have to reject locality as well and to be required to conclude that quantum reality is so mysterious that nothing at all can be said about it.* In some ways this view should perhaps be even more repugnant to such physicists, since the inability to speak either of locality or of nonlocality is surely even "spookier" than would be the need to speak of mere nonlocality. To be sure it seems to be logically consistent to talk as Bohr does, but anyone who is interested in the question of the locality versus the nonlocality of

---

* In this connection d'Espagnat[19] for example discusses a "veiled reality" about which nothing can be known.

the details of the actual process (and not merely of the formal properties of the commutation relations of operators at different points) cannot use Bohr's approach as a foundation or even as a point of departure.

Most physicists, however, accept Bohr's interpretation, at least verbally. Nevertheless, they generally also accept von Neumann's notion of the quantum state which is, as we have seen, not compatible with Bohr's approach at all. As we have already suggested earlier, it does not seem to be commonly known that the two views are so different. It would perhaps be better for physicists to acknowledge this difference and state which, if any, of these views they prefer. It seems to us that if such a choice were made in most cases, von Neumann's interpretation would be favored. And, as we have seen, in this view one can already see a certain physical meaning of the kind of nonlocality that is present in the many-body Schrödinger equation. Moreover, as we shall show in subsequent sections, this feature of nonlocality can be brought out more clearly in the causal interpretation in terms of the nonlocal quantum potential.

4. Brief Resumé of Measurement Theory in the Causal Interpretation

We shall now summarize the essential features of measurement in the causal interpretation.[20] First we recall that in the causal interpretation, we assume the particle has a well-defined position, $\mathbf{x}(t)$.

In addition the particle has associated with it a new kind of "Schrödinger wave" satisfying the wave equation

$$i\hbar\frac{\partial\psi}{\partial t} = -\frac{\hbar}{2m}\nabla^2\psi + V\psi \tag{18}$$

The momentum of the particle is assumed to be $\mathbf{p} = \nabla s$ and its equation of motion will then be

$$\frac{dp}{dt} = -\nabla(V + Q) \tag{19}$$

where V is the classical potential and Q is the quantum potential given by

$$Q = -\frac{\hbar^2}{2m}\frac{\nabla^2 R}{R} \tag{20}$$

and where ψ is expressed in polar form, $\psi = R\,e^{iS/h}$.

It has been shown that the main new features of the quantum theory can be explained in terms of the quantum potential.[21,22] However, one must add to the assumptions that we have already given, a statistical postulate, namely, the probability density of particles in states having the same Schrödinger wave ψ is

$$P = |\psi|^2 \tag{21}$$

This postulate has been shown to be consistent with the equations of motion, so that if it holds at any one time, it will hold at all times. We may therefore simply take it as an assumption. However, this assumption can be justified in various ways, for example, by postulating a new random Brownian-type motion to the particle[23] leading eventually to a stable distribution of the type given in equation (21).

In the many-body system, the wave function $\psi(\mathbf{x}_1, \ldots, \mathbf{x}_N, t)$ depends on the configuration space of all the particles. This cannot be understood as a wave in a three-dimensional space but, as we have shown,[22] it can be interpreted as a kind of active information in that it determines the many-body quantum potential,

$$Q(\mathbf{x}_1, \ldots, \mathbf{x}_N, t) = -\frac{\hbar^2}{2m} \sum_n \frac{\nabla_n^2 R}{R} \tag{22}$$

where $\psi = R\, e^{iS/\hbar}$.

We show that, in general, this quantum potential determines a nonlocal interaction which may have an indefinitely long range. Moreover, it is not a preassigned function of the particle coordinates but depends on what is commonly called the quantum state of the whole system. However, if the wave function factorizes into a product of functions of the coordinates of each particle, then the quantum potential reduces to a sum of independent terms

$$Q(\mathbf{x}_1, \ldots, \mathbf{x}_N, t) = \sum_j Q_j(\mathbf{x}_j, t) \tag{23}$$

In this case there is no nonlocal interaction and the behavior of the particles that are far from each other is independent.

We show that, generally speaking, in the large-scale limit and at high temperature, such factorization takes place, at least to the extent that large-scale phenomena arising at large distances generally do exhibit the kind of independence that is observed. Nevertheless, at the atomic level and at low temperatures, e.g., in superconductivity, the quantum potential is seen to bring about a characteristic new kind of wholeness of the entire system. This is similar in certain key ways to Bohr's notion of wholeness, but differs in that the process is analyzable in thought, while for Bohr it is essential, as we have already pointed out, that the process is completely unanalyzable even in thought.

To deal with the measurement process, we use a treatment of the wave function which is essentially the same as that given in Section 3 for the von Neumann interpretation. The new feature is that we assume the observed system to be a particle with a well-defined position, $x(t)$, and that the apparatus variables $y(t)$ and $Z_i(t)$ are also well defined. Because of the stochastic process responsible for the probability distribution, $P(x, y, Z_i)$, there is no way to predict or control the initial positions of these particles apart from saying that they are somewhere in the region of configuration space in which the wave function of the system is appreciable.

Initially, the quantum potential breaks into a sum of separate terms and therefore $x(t)$ and $y(t)$ move independently. When the interaction Hamiltonian, H_I, as described in equation (4), is operating, the wave function becomes a sum of products. The quantum potential is then a very complex and rapidly fluctuating function of all the relevant coordinates. However, when the interaction is completed the apparatus packets corresponding to different eigenvalues of the operator, O_n [as given in equation (7)] no longer overlap. Because the probability density is proportional to $|\Psi|^2$ no particle can enter the region between the packets (where $\Psi = 0$). Every particle enters one of the packets and thereafter stays there. The unoccupied packets make no contribution to the quantum potential acting on the particle.

We can obtain a helpful image of what happens by saying that each of the apparatus packets determines a channel into which the particle may enter. During the period in which the interaction Hamiltonian is operating, the quantum potential will contain a set of bifurcation points such that particles on one side enter a particular packet, while those on the other do not. The motion is highly unstable and, indeed, chaotic in the sense of modern chaos theory.[24] Even if we could predict or control the initial positions of the relevant particles to a fair degree of accuracy, we could not determine beforehand which channel the particle will enter. And if one assumes the statistical distribution (21) of initial positions of the particles, one can show[20] that the probability of entering the channel, n, is $|C_n|^2$. This is, of course, the same as the probability of obtaining the corresponding result in the usual interpretation.

It is clear that the interaction Hamiltonian has induced a transformation of the whole system such that the net result is to have the x particle in what is commonly called the quantum state $\psi_n(x)$ while the apparatus particle y is in a correlated channel. However, as we have already pointed out in

Section 3, this transformation is still reversible. This implies that, by means of a suitable interaction Hamiltonian, the apparatus packets could be made to overlap, and thus interfere again. In the causal interpretation, this means that the packets not occupied by the particle can still come to affect the quantum potential at a later time. Therefore the result of this operation is not yet irrevocable. But, of course, as we have already indicated in Section 3, after the system interacts with the detecting device no more interference of this kind is possible. The quantum potential in any one channel therefore cannot later be affected by the empty wave packets and the result of the process is irreversible.

In this way we explain how the measurement process produces a definite result without invoking any assumption of collapse of the type needed in the von Neumann interpretation. The definite result follows from the theory itself, without further assumptions.

Some physicists have found the notion of empty wave packets disturbing, but we have given a further explanation of their significance in our paper.[22]

Our treatment also gives a simple explanation of Heisenberg's uncertainty relationships. For example, to measure x we may use an interaction Hamiltonian

$$H_{I_1} = i\lambda X \frac{\partial}{\partial y} \tag{24}$$

and to measure **p**, we may use

$$H_{I_2} = i\lambda P \frac{\partial}{\partial y} \tag{25}$$

These two operators do not commute. From this it follows that they cannot be carried out at the same time, in the sense that the conditions needed to set up one of them are incompatible with those needed to set up the other. What this implies in essence is that, as we have indeed already pointed out, what has commonly been called a measurement is actually not a measurement but a transformation. We shall therefore call this a measurement operation. We can thus understand in a simple way why x and p cannot be "measured together."

In this connection we could say that the value of a quantity that comes out of a measurement operation does not, in general, correspond to some preexistent "element of reality" within the system itself. To be sure in this interpretation the system is assumed to possess elements of reality which include, for example, a position x and a momentum $\mathbf{p} = \nabla s$ as well as the wave function $\psi(\mathbf{x}, t)$. However, in the measurement operation these elements undergo transformations. Thus, as we have seen, the wave function is transformed into an eigenfunction of the interaction Hamiltonian. Moreover, in most cases, both the position and the momentum change in a way, as we shall show, that is unpredictable and uncontrollable (though in special cases, such as measurement of position, it is, in principle, possible to carry out a measurement operation which reveals the value that the corresponding element of reality had before the operation). Owing to the incompatibility of the conditions needed to bring about an eigenfunction of momentum and an eigenfunction of position, only one of these potentialities can be actualized on any given occasion. (A similar conclusion has been drawn in the usual interpretation by Bohm[9] and by Heisenberg[25].) This is, in essence, the way in which the causal interpretation treats the same phenomena which Bohr treats in terms of complementarity.

To discuss how the causal interpretation deals with the Heisenberg relationships $\Delta x \Delta p \sim \hbar$, we consider the fact that the initial coordinates of the apparatus particles and those of the observed system have a statistical scatter. As a result, and because the quantum potential is chaotically unstable and full of bifurcation points during the period of operation of the interaction Hamiltonian, it follows that the final channel that the particle enters is unpredictable and uncontrollable. But since the probability of each result is the same as it is in the usual interpretation, it also follows that the scatter in the results will satisfy the Heisenberg relationships.

It can be seen from the above that in the causal interpretation the uncertainty relations are explained in a way that is basically similar in spirit to the original explanation proposed by Heisenberg, i.e., that they come about through unpredictable and uncontrollable disturbances taking place during the "measurement." The key difference is that, in Heisenberg's approach, there was no clear concept of what the trajectory would have been without such a "measurement" and therefore no clear meaning could be assigned to the disturbance. In the causal interpretation, however, all these notions are clearly defined and even the unpredictability and uncontrollability of the disturbance is a consequence of the theory and is not merely an *a priori* assumption.

5. The Causal Interpretation of the EPR Experiment

We now proceed to discuss the causal interpretation of the EPR experiment. Here the treatment of the wave function is also the same as in the von Neumann approach described in Section 3. The main difference from our discussion of Section 4 is that the observed system contains two particles $\mathbf{x}_1$ and $\mathbf{x}_2$. However, only one of these particles, $\mathbf{x}_1$, is directly observed. Nevertheless, as we shall see, the unobserved particle goes into a definite state correlated to that of the observed particle.

In Section 3 we have worked out how the wave function changes in the "measurement" process. This is shown in equations (13), (14), and (15). All that we have to add is that these wave functions are now determining the quantum potential and that during the period of operation of the interaction Hamiltonian on particle 1, this becomes a very rapidly fluctuating function of all the variables x_1, x_2, and y that is full of bifurcation points implying unstable and chaotically complex motions.

However, after the interaction Hamiltonian ceases to operate, the system enters one of its possible channels and, as explained in Section 4, the wave function representing the other channels can thereafter be neglected (as if the wave function had "collapsed" into the actual channel). During the interaction period, particles 1 and 2 are closely connected owing to the nonlocal character of the quantum potential, but afterwards the effective wave function (representing the channel actually occupied by the whole system) is just a product of factors. From then on all interactions are local and each system behaves independently. Nevertheless, the properties of each particle are left correlated with those of the other in the way demanded by the quantum theory.

Thus we have explained the correlations which are, in fact, produced by the action of the quantum potential, the latter giving rise to a nonlocal interaction between the particles.[26,27] In this way we deny one of the basic assumptions of EPR, that is, no interactions between the particles. Once we admit that these particles interact, we no longer conclude with EPR that the properties of particle 2 can be measured without disturbing that particle. Therefore we also no longer say that the properties of particle 2 are already "existent elements of reality" before interaction with the apparatus has occurred. Rather we have shown that, as we have indeed already pointed out in Section 3, they are potentialities which are actualized not only through the effects of the interaction Hamiltonian acting on particle 1, but also through the overall quantum potential which acts on both particles. That is to say, the properties actualized in quantum measurement operations are potentialities of the combined system (including the apparatus). However, even the particle that does not interact directly with the apparatus still has its potentialities realized in a measurement operation. Through such an analysis the EPR experiment can be made intelligible in a relatively simple way.

Since the experimental tests of the EPR correlation have all involved spin or polarization in some way, we shall discuss briefly how these considerations can be extended to include the latter. Beginning with spin we point out that there are several ways of extending the causal interpretation to include this. One of these was proposed by Bohm *et al.*[28] but this was only possible in the case of the one-particle system. Another was suggested by Bell[29] which could, in principle, be developed so as to treat the N-body system. However, to deal with this would involve a long and complex series of questions. Instead we shall discuss the case of a pair of atoms each with spin one (i.e., with internal orbital angular momentum $L = 1$).

If the center of mass of the atom is denoted by **R** and the coordinate of the electron relative to the nucleus by x, then a state of unit angular momentum has the wave function

$$\psi_m(\mathbf{R}, \mathbf{x}) = F(\mathbf{R})f_m(\mathbf{x}) \tag{26}$$

where $f_m(\mathbf{x})$ represents a state of unit angular momentum of the internal electron and $F(\mathbf{R})$ represents the state of the center of mass.

A molecule of total spin zero will then have the wave function

$$\Psi_0 = F(\mathbf{R}_1)G(\mathbf{R}_2)[f_0(\mathbf{x}_1)f_0(\mathbf{x}_2) + f_1(\mathbf{x}_1)f_{-1}(\mathbf{x}_2) + f_{-1}(\mathbf{x}_1)f_1(\mathbf{x}_2)] \tag{27}$$

We have chosen F and G separately to allow the centers of mass to be in different places.

Clearly the above wave function implies that if the angular momentum of particle 1 is measured, that of particle 2 will come out opposite. Thus we have an EPR correlation similar to that discussed for spin $\frac{1}{2}$, but different in that each particle has three possible states rather than two. But it is clear that the same treatment that we have used for the spin system will go through here and so no further discussion of this case is required.

There remains the polarization states of photons. The proper treatment of this requires the quantum-potential treatment of the electromagnetic field. We have already given such a treatment elsewhere and indicated how it works out for the EPR paradox.[22]

6. The EPR Experiment and Relativity

Our treatment has so far been nonrelativistic and so no problems arise with the notion that the quantum potential acts instantaneously. Moreover, we have already shown that in the classical limit (i.e., large scale and high temperatures) the quantum potential becomes essentially factorizable, so that there will be no significantly nonlocal interactions. Therefore there will be no contradiction between our approach and relativity physics applied in the classical domain.

In the quantum domain, however, we might, at first sight, conclude that the instantaneous connection of distant systems brought about by the nonlocal action of the quantum potential could lead to contradictions with the requirement of relativity. But here let us recall that these requirements will be satisfied if no *signal* can be transmitted faster than light. However, as we shall see, the quantum potential cannot be made to carry a signal, and so on this score there need be no violation of the requirements of relativity theory.

For example, the simplest way to send a signal between two particles by means of EPR correlations would be to put one of the particles in a definite state, e.g., of spin, and then to see if this state is changed when a measurement is carried out on the other particle. However, the interaction that puts the first particle into a definite state will bring about a product wave function for the two particles. Thus the quantum potential would now be local and so no signal could be propagated between the particles.

The above illustrates a feature of the quantum potential that is very significant in this context, i.e., that anything one would do to it to make a signal possible changes it so radically that no signal can actually be transmitted. For example, suppose one were to try to modulate the quantum potential as one does with a radio wave so that it could carry a signal. To treat this process one would have to take into account the modulating system to produce a combined wave function which includes the latter as well as the particles. The interaction Hamiltonian between the modulating system and one of the particles would produce complex changes in the overall wave function similar to those in a measurement operation. This would result in a quantum potential with many bifurcation points leading to chaotic motions of both particles. In this context the order of the signal would be totally scrambled up in an unpredictable and uncontrollable way and so no meaningful signal could be transmitted.

More generally, the nonlocal type of quantum potential is so fragile and its effects are so unpredictable and uncontrollable that it can never be used to carry a signal. From this we conclude that the causal interpretation of the EPR correlation will not lead to a violation of the theory of relativity.

One can see this in another way which is based on the fact that the causal interpretation and the usual interpretation give the same statistical results for all measurement operations no matter which operator is "measured." But the usual interpretation has been shown to be capable of a covariant extension in which field operators at different positions either commute or anticommute. From this it follows that the causal interpretation, which gives the same statistical results, must be covariant in this purely statistical context.

What about the context of a single measurement operation? As long as the initial conditions are unpredictable and uncontrollable in the way we have assumed, nothing more can be concluded about a single measurement operation. And so no further question of this kind can arise in this context. However, one of us[21] has elsewhere discussed the possibility of going beyond the laws of current quantum theory in such a way that more might be known about the individual process than is possible in the present statistical theory. In this case the instantaneous connections brought about by the quantum potential could lead to a definite violation of the theory of relativity (which would, however, still be statistically valid). If this happens, then we would have to admit that there is a preferred frame in nature and the covariance of the laws of physics would apply only in some limited context which includes that of the statistical results discussed in the current theory. (The classical limit is a particular and limiting case of the statistical context in the sense that the statistical fluctuations can be neglected for this domain.)

The above would imply that the theory of relativity is limited, just as Newtonian mechanics is. This sort of notion is not entirely foreign to the current development of physics. Thus special relativity is limited by general relativity and the latter is, in turn, limited by some of the cosmological implications of the theories dealing with the origin of the universe (e.g., there must have been a time so close to the origin that, through quantum fluctuations, the gravitational tensor $g_{\mu\nu}$ would fluctuate so much that there would be no way even to express what could be meant by the notion of the covariance).

We are proposing that if we go in the opposite direction to reach a domain beyond that of the quantum theory, this too will carry us beyond the range of applicability of the theory of relativity. In some ways our proposals might be similar to those of Lorentz, who explained the Lorentz covariance of the laws of physics in terms of an "ether" which constituted a preferred frame. Nevertheless, Lorentz invariance was shown to follow from the changes of the material structures as they moved through this ether. We are also suggesting there is some ultimately non-Lorentzian substructure which, however, implies Lorentz invariance of all processes at the statistical level treated by the current quantum theory.

7. Conclusion

We have shown that the quantum potential provides an adequate explanation of the EPR experiment. The nonlocality that it implies was seen to be a perfectly rational idea leading to no insoluble physical or mathematical problems. Its implications may eventually require us to revise our basic concepts of physics, to go beyond both those of relativity theory and those of quantum theory as these are now understood.

References

1. A. Einstein, B. Podolsky, and N. Rosen, *Phys. Rev.* **47**, 777 (1935).
2. J. S. Bell, *Physics* **1**, 195 (1964).

3. N. Bohr, *Phys. Rev.* **48**, 696 (1935).
4. J. F. Clauser and A. Shimony, *Rep. Prog. Phys.* **41**, 1881 (1978).
5. A. Aspect, J. Dalibard, and G. Roger, *Phys. Rev. Lett.* **49**, 1804 (1982).
6. T. W. Marshall, Private communication.
7. E. Schrödinger, *Proc. Camb. Phil. Soc.* **31**, 555 (1935); **32**, 446 (1936).
8. N. Bohr, *Atomic Physics and Human Knowledge*, p. 71, Science Editions, New York (1961).
9. D. Bohm, *Quantum Theory*, Prentice-Hall, London (1960).
10. D. Bohm and Y. Aharonov, *Phys. Rev.* **108**, 1070 (1957).
11. W. H. Furry, *Phys. Rev.* **49**, 393 (1936); **49**, 476 (1936).
12. C. S. Wu and I. Shaknov, *Phys. Rev.* **77**, 136 (1950).
13. H. Langhoff, *Z. Phys.* **160**, 186 (1960).
14. A. R. Wilson, J. Lowe, and D. K. Butt, *J. Phys. G* **2**, 613 (1976).
15. T. W. Marshall, E. Santos, and F. Selleri, in: *Open Questions in Quantum Physics* (G. Tarozzi and A. van der Merwe, eds.), pp. 87–101, (1985).
16. J. von Neumann, *Mathematical Foundations of Quantum Mechanics*, Princeton University Press, Princeton (1955).
17. J. S. Bell, in: *Quantum Implications: Essays in Honour of David Bohm* (B. J. Hiley and F. D. Peat, eds.), Routledge and Kegan-Paul, London (1986).
18. R. P. Feynman, *The Strange Theory of Light and Matter*, Princeton University Press, Princeton (1985).
19. B. d'Espagnat, *In Search of Reality*, Springer-Verlag, New York (1983).
20. D. Bohm and B. J. Hiley, *Found. Phys.* **14**, 255 (1984).
21. D. Bohm, *Phys. Rev.* **85**, 166, 180 (1952).
22. D. Bohm and B. J. Hiley, *Phys. Rep.* **144**, 323 (1987).
23. D. Bohm and J.-P. Vigier, *Phys. Rev.* **96**, 208 (1954).
24. Hao Ban-Lin, *Chaos*, World Publications, Singapore (1984).
25. W. Heisenberg, *Physics and Philosophy*, Harpers, New York (1959).
26. D. Bohm and B. J. Hiley, *Found. Phys.* **5**, 93 (1975).
27. B. J. Hiley, in: *Determinism in Physics* (E. I. Bitsakis and N. Tambakis, eds.), pp. 201–216, Gutenberg Publ. Co., Athens, Greece (1985).
28. D. Bohm, R. Shiller, and J. Tiomno, Nuovo Cim., Suppl. **1**, 48 (1955); D. Bohm and R. Shiller, Nuovo Cim., Suppl. **1**, 67 (1955).
29. J. S. Bell, *Epistemol. Lett.* **20**, 1 (1978).

10

Interpretation of the Einstein-Podolsky-Rosen Effect in Terms of a Generalized Causality

JERZY RAYSKI

1. Introduction

After many years of discussion of the Einstein–Podolsky–Rosen (EPR) effect (and of similar effects which appear in the consideration of a pair of correlated systems which were once coupled with one another and subsequently separated), a majority of physicists have come to the conclusion that the quantum phenomena, revealed by measurements performed on such systems, exhibit a manifestly nonlocal character, even on a macroscopic scale. A measurement performed on one of two widely-separated systems affects the other system—as it seems—instantaneously, or in other words, information about the result of the measurement appears to propagate with superluminal velocity, in apparent violation of the laws of relativity and causality. Quantum physicists explain this situation by pointing out that quantum phenomena are characterized by a certain "discreteness" and integrity due to the existence of the quantum-of-action h. Some of them even believe that quantum phenomena cannot be completely accommodated within the realm of space-time and, therefore, surpass its framework.

The question arises as to whether these nonlocal features are, in effect, a violation of relativity. They certainly violate our usual concept of causality, but may still be reconcilable with a generalized concept of causality which accepts also a backward-in-time propagation of information. Whether this

JERZY RAYSKI • Institute of Physics, Jagellonian University, Cracow, Poland.

implies that we must put causality on an equal footing with anticausality will be the main subject of our analysis and the principal problem of investigation in this chapter.

2. Causality within Classical Physics

The notion of causality in classical physics differs from the Aristotelian concept of causality as well as from our everyday intuition. The popular and intuitive understanding of causality gives decisive preference to one direction along the time axis: cause must precede effect, or in other words, the effect should be a consequence that follows from a precedent—an earlier cause. However, the temporal sequence of cause and effect is absent from classical physics, the latter being expressible in terms of differential equations (usually of the second order, but occasionally of the first order in the time derivatives) wherein the "initial" conditions, at an arbitrary time instant $t = 0$, determine not only the future of the physical system but also, equally well, its past ($t < 0$).

If the physical state at the initial instant of time is understood to constitute a "cause" then its "effects" are determined not only for the future, but also for all previous instants of time. Thus, in classical physics, causality is on an equal footing with anticausality, and ordinary action with retro-action.

The conflict with everyday intuition may be removed by assuming a consequently "deterministic" viewpoint: according to such determinism our free will is to be regarded as an illusion, explicable by the simple circumstance that our actions often coincide with our wishes and desires. It appears to us that we act freely, while in reality we are unable to make any free decision but are simply passive participants tightly captivated in a chain of events, by causal bonds on the one side and anticausal bonds on the other. Being unable to change our fate or to influence the course of events, we must therefore also be passive spectators in any act of measurement, its type and result being predetermined by our own (i.e., the observers') fate.

3. The EPR Effect and Anticausality

If the same fatalism would also hold true in quantum mechanics, then the EPR paradox could be solved immediately. We could imagine that the effect of a measurement which we executed upon one of a pair of systems (and that we were forced to perform by command of our destiny) would first propagate backward in time, until the instant when both systems constituted a single whole; would be transferred to the other system; and then the relevant information would propagate forward in time. Both fragments of information propagated in this roundabout way would be carried along by the particles themselves (of both systems) whose velocities do not surpass the velocity of light *in vacuo*. Thus, nonlocality of interaction, or the superluminar propagation of information, would only be illusory because, in reality, information would be transferred partly anticausally and partly causally, in accordance with relativity, but allowing for both retarded as well as advanced interactions. This would remove the necessity of introducing action-at-a-distance and an intrinsic nonlocality of quantum phenomena.

The paradox of being able to "kill our grandfather" by means of a retroaction could be avoided by assuming that our destiny generally forbids actions that form closed loops, or knots, along the two-way causality chain. However, it may be doubted whether quantum mechanics (QM) is reconcilable with such a fatalistic understanding of our role as being merely that of passive observers, unable to decide what kind of action or what sort of experiment we are going to perform.

Nevertheless, some workers[1-4] pursued investigations along these lines and accepted the existence of retroaction in order to remove paradoxes of the EPR type. In particular, it was pointed out by Costa de Beauregard[1] that the equations of relativistic quantum-field theories are invariant under the charge–parity–time (CPT) transformations and are certainly applicable in both directions along the time axis. Paradoxes that might appear in consequence of admitting causality with anticausality, symmetrically, must be—according to him—restricted to microscopic domains of space-time and cannot affect macroscopic and classically describable systems for which only classical determinism holds true.

4. Is Quantum Mechanics Reversible?

The explanation of symmetry under time-reversal, described in the previous section, gives rise to serious doubts. Several authors have pointed out that reversibility of the equations of motion, or of the Schrödinger equation, is not sufficient to put past and future on an equal footing because, in quantum physics, one must also take account of the problems of measurements that involve manifestly irreversible features. The most radical position in this connection was assumed by Weizsäcker[5] who was of the opinion that the mere concept of probability, in quantum theories, applies only to future, as yet unaccomplished events. As long as an event is still virtual, it remains undecided and therefore subject to a probabilistic treatment in terms of a state vector or Schrödinger wave function. On the other hand, past events have already been decided and therefore cannot be subject to a probabilistic treatment in terms of the concept of state functions. The author does not agree with these views and would like to point out that physical science, does not distinguish between an accomplished past and a prospective future and nor does it admit the existence of a single time-instant between them which denotes "now." All points along the time axis are equivalent and a point $t = 0$ is merely a convention. It could be shifted arbitrarily by a transformation, i.e., a translation of the time coordinate. Moreover, physical science has included neither the concept of time flow nor that of an uninterrupted streaming of our consciousness distinguishing a present instant, a "now" (ever-changing), between the approaching future and escaping past. Those are categories of psychology, but not of physics. Consequently, the concept of probability has nothing to do with the question of whether an event has already happened or is still virtual. The point is only whether information is, or is not, available. Whenever full information is missing there exists the possibility of introducing the concept of probability, regardless of whether the lack of information concerns the future or past.

In physical science there exists a sharp distinction between theories of a statistical type, that privilege one direction along the time axis (e.g., the theory of diffusion), and dynamical theories that are symmetric under time reversal (or, more generally, under the CPT transformation). This is quite independent of the questions of which time instant denotes an actual "now," or of whether there is a constant flow of time. The question arises as to whether QM is a dynamical theory or a theory of a statistical type. If the first alternative holds true, QM should be applicable both to predictions and to "retrodictions" (of an unknown past), symmetrically. If the second alternative holds true, the situation is less clear: it might still be applicable in both directions (although the results of predictions and retrodictions would not be symmetrical) or—in view of the asymmetry—even the mere applicability backward in time might be forbidden by some *a prioristic* reason. In this case Weizsäcker would be right.

The simple fact that, in the Heisenberg picture, the equations of motion assume exactly the same form as in classical mechanics, assures one that QM is a dynamical theory. The same applies to quantum field theory: since quantized-field equations admit retarded as well as advanced solutions, probabilistic retrodictions are admissible. Nevertheless, there also exist contrary arguments. It could be pointed out that the equations of motion (in the Heisenberg picture) or Schrödinger's equation (in the Schrödinger picture) do not yet represent the whole quantum theory. The latter also involves the problem of measurements and their repeatability, while each quantum measurement entails an irreversible element. This circumstance, as well as the fact that quantum theories yield only probabilistic assertions, may be regarded as arguments in favor of their statistical, rather than dynamical, character. In this case retarded potentials would be preferred to advanced ones and anticausality would have to be dispensed with.

In order to solve this dilemma, a more detailed discussion of the problems of measurements in the world of quanta is needed.

5. An Analysis of the Measurement Process

In most papers devoted to interpretational problems of QM, one finds statements to the effect that measurements always introduce an uncontrollable perturbation which leads to uncertainty in the measurement's result and thus only probabilistic, and not deterministic, assertions are possible. It is also stressed that the perturbation caused by the act of measurement is irreversible. This last assertion seems to imply that probabilistic statements of QM refer only to predictions, and not to retrodictions. Such statements are not precise and constitute only half-truths.

Three phases may be distinguished in the measuring process:

1. An initial phase—a preparation for the act of measurement by the installation of a suitable apparatus and the securing of suitable measuring conditions.
2. The main phase—interaction between the apparatus and the object of measurement.
3. The final phase—registration of the result.

In the case of a position measurement one has to fix, in the initial phase, a certain domain Δx in order to be able to answer "yes" or "no" to the question of whether the particle is situated in that domain. In the case of a measurement of an angular-momentum component one first has to distinguish a certain axis in space (e.g., by introducing a constant magnetic field) so as to specify which of the components is to be measured. The following interpretation is to be recommended: The initial phase is an active phase while the proper measurement (second phase) is passive with respect to the quantity that is to be measured. By this we mean that already in the introductory phase a decision has been made as to the value of the respective observable (e.g., whether the particle is or is not in the previously fixed domain). The value of the observable to be measured has already been fixed in the first phase. This value has been decided, but nevertheless remains unknown. In the proper, second phase of the act of measurement one secures information about the value which was predetermined in the first phase. It is really only a confirmation of a preexisting, but hitherto unknown, fact (or property). In this sense the second phase is a passive securing of information about a preexisting, objective reality. The reader is here referred to the Appendix for an explanation of what constitutes a "perfect" measurement.

In connection with the main topic of this book, it should be stressed that not all measurements cause a perturbation of the object of measurement, namely those that constitute "indirect" measurements, e.g., those encountered in the EPR effects. But even in the case of ordinary, i.e., "direct," measurements, such statements are misleading. They do not distinguish between the act of measurement, itself, and what may be deduced from it with regard to the results of other possible measurements. The result of a measurement need not be uncertain: if one is measuring an observable possessing a discrete set of eigenvalues, a good measurement will yield a specific eigenvalue. If, on the other hand, an observable Q possesses a continuous spectrum, we may always perform a measurement with an arbitrarily small uncertainty ΔQ. In this sense, the result of a measurement may be arbitrarily exact. An inexactitude, or rather a *dispersion of the results of possible future measurements*, refers to the results of possible future measurement of *another* quantity described by an operator that does not commute with the former one (otherwise we could apply an axiom of repeatability of measurements.)

Now, it would be incorrect to conclude that, due to an element of irreversibility inherent in any measurement, QM is only applicable to predictions, and not to retrodictions.

First of all, we must distinguish this irreversible element, which is unavoidably connected with the act of measurement itself (both its first and second phases), from that connected with the necessity of registering the result. Although in practice these stages of the process of measurement are often intermingled, this does not mean that in better-planned measurements they cannot be separated. The contrary opinion is reminiscent of the belief of those critics of Newton who maintained that free motion must always, sooner or later, come to rest. The assumption that it is possible to perform measurements that completely separate the act of measurement (and its consequences for the object of measurement) from that irreversible process which is inherent in the act of registration of its result, is a fruitful one just as was Newton's idealization and extrapolation consisting of the complete neglect of friction.

This does not mean that during the act of measurement itself (i.e., its second phase), there is no other irreversible process involved. On the contrary, every measurement in quantum physics, even an ideal measurement, possesses two aspects which one may compare to two faces of the god Janus (or, if one likes, to the two Chinese characters Yang and Yin).

On the one hand the measurement increases one's degree of knowledge about the quantity actually being measured, and on the other hand it produces an unwanted side effect in the form of uncontrollable and unpredictable perturbations of all other observables whose operators do not commute with that one describing the quantity actually measured. These side effects diminish one's level of information about these observables (if such information was, indeed, formerly available) thereby increasing the total entropy, and thus the process becomes irreversible.

We may therefore distinguish two sides, or aspects, of the act of measurement: an "instructive" and a "destructive" side. The instructive (or constructive) side increases ones information, while the destructive side produces an uncontrollable change in the values of other, complementary, observables. It decreases information and increases entropy. In order to properly interpret quantum phenomena one must be aware of this "double-faced" character of measurements: their constructive and destructive aspects comparable to Janus' peace-like and war-like faces.

Since the two aspects of the measurement process concern different and complementary properties, there is no contradiction between the possible applicability of QM backward in time, and the irreversibility of the perturbation effects unavoidably connected with the act of measurement itself.

6. Applicability Backward in Time

A clear understanding of the "double-faced" character of the act of measurement solves the problem of whether QM can be applied symmetrically to past and future. It should be stressed that this problem has nothing to do with the discovery that some physical objects (i.e., kaons) are describable by a Hamiltonian which is asymmetric under time reversal. The problem is not whether there exists symmetry (even under a CPT transformation) but, more generally, whether it is possible to draw conclusions about the past and make probabilistic retrodictions. Inasmuch as evolution in time is described by the Schrödinger equation, a differential equation of the first order in time, knowledge of the value of the wave function at t_0 enables one to find a solution not only for $t > t_0$ but also for $t < t_0$. Thus, the Schrödinger equation is also applicable backward in time. It does not contradict our previous interpretation of the act of measurement, since the Schrödinger equation is only valid when the physical system is decoupled from the measuring device and since the wave function ψ is a function of information (not of "de-information"), i.e., it refers to the aspect which does not increase the entropy and, consequently, is equally meaningful in both directions along the time axis.

The other aspect of the act of measurement, or metaphorically speaking, the other face of the god Janus, only reveals itself when a new measurement takes place, thereby canceling the existing information, i.e., when the old wave function is invalidated and validity conferred on a new function (which is an eigenfunction of the actually measured observables).

However, it should be noted that if we admit the applicability of QM backward in time, then we will find that for one and the same time interval $\{t_1, t_2\}$ there in general apply two different wave functions, one computed from the initial values at t_1, and the other from the final values at t_2. But this does not imply any contradiction since neither of the two "state functions" is an intrinsic property, as such, of the physical system (in other words, neither of them is its "state"), rather they are functions (or states) of information that serve different purposes: one serves to predict and the other to retrodict. The state should not be called a "state of the physical system" but a "state of information" (about the system). This makes a great difference, according to the modern philosophy of languages.

7. The EPR Effect and Realism

While analyzing the Einstein–Podolsky–Rosen effect, several authors have tried to convince the reader that the correlations exhibited in this effect are in full accord with the formalism of QM and, consequently, must be realistic because QM is a realistic theory, since it shows a satisfactory agreement with experiment. Such an explanation does not seem very satisfactory. It is not sufficient to prove that QM does not violate the requirements of consistency or that it is not incomplete. The nonlocality exhibited by the EPR effect is of a macroscopic character, far beyond the nonlocality of position Δx appearing in the uncertainty relations of Heisenberg, and requires some explanation, not only a formal demonstration that it follows as a straightforward consequence of the formalism of QM.

Inasmuch as Einstein *et al.* have cast doubt on the realistic character of the quantum-mechanical description of physical phenomena, we should—first of all—define what we mean by "realistic" in this context. In our opinion, a realistic theory must satisfy two requirements: (i) repeatability of measurements, and (2) causal propagation of perturbations as well as causal transfer of information. The EPR effect appears to violate the latter requirement, well beyond the limits imposed by the uncertainty relations, i.e., it violates not only microcausality but macrocausality as well.

However, as we mentioned earlier, the action-at-a-distance exhibited by the EPR effect can be regarded as simply an illusion because, in effect, the information obtained from a direct, spin or momentum, measurement performed upon the first of the two particles, is to be viewed as propagating first backward in time until the instant when the particles formed a common system, and then—after transfer to the second particle—as propagating, together with it, forward in time. Neither of the two segments of information exceeds the velocity of light because they are conveyed, in a roundabout way, by the two particles themselves.

Our intuition revolts against the introduction of anticausality since it would mean that the effect precedes its cause. But a closer examination of the measurement process in QM may convince us of its admissibility and intelligibility. As discussed earlier, there are two different aspects to quantum measurement processes, comparable to the two faces of the god Janus. One acquires information about a quantity A that is going to be measured, and the other produces a loss of information about those observables $B, C, \ldots$ which are represented by operators that do not commute with A. Such a loss is an unavoidable side effect of any quantum measurement. It is only this unwanted side effect that produces an increase in entropy and thereby privileges one direction along the time axis, i.e., propagates only forward in time, while the main (and positive) role of the act of measurement (the peace-like face of Janus, in our metaphor) does not prefer future to past or *vice versa*. Instead it is strictly symmetric, in agreement with the fact that dynamical equations of physics admit retarded solutions on an equal footing with advanced solutions. The other (war-like) face of the act of measurement is quite different: by introducing a decrease of information, i.e., an increase of entropy, it results in an irreversible process.

We conclude by stating that anticausality is on an equal footing with causality in classical physics, but is not on quite so equal a footing in quantum physics. A generalized causality ("ambicausality") admits causal as well as anticausal propagation of information, but not of de-information, the latter being an inherent property of the quantum measurement process.

In this way the paradox of a future action actively disturbing the past is avoided. As regards the possibility of an influence acting backward in time, it should be stressed that a measurement may, at most, contribute to making the past better understood, or more definite, but can never change or destroy it.

We believe that the above explanation of the EPR paradox would satisfy both Einstein and Bohr.

8. The Grandfather Paradox

By admitting advanced actions along with retarded actions, we are faced with the "grandfather paradox," mentioned earlier, which consists of the possibility of "killing, by means of an advanced action, our own grandfather in his earliest childhood." This paradox is avoided by classical physics (where advanced and retarded interactions appear on a perfectly equal footing) by denying the existence of free will. Classical observers and classical experimentalists are not situated beyond the world of classical physics, but constitute an intrinsic part of it—an "ingredient"—so that their actions are also predetermined. If we are thus treated classically, then we are unable to change either our future or our past. We cannot produce knots or closed loops in the bicausal chain of events, either.

The situation as regards quantum causality appears quite different. There is not a complete symmetry between causality and anticausality. While—metaphorically speaking—classical physics admits only one face of the god Janus: the peace-like face (which looks rather passive), quantum physics admits both faces simultaneously so that we are not only spectators of, but also actors in, the act of measurement. In quantum physics the "peace-like face" reveals itself to be less passive than it appears in classical physics: we choose which (complete set of) observables we are going to measure and our measurement creates some order—it is constructive, while the complementary "war-like" face produces disorder and deinformation as unavoidable side effects of the measurement act.

The two complementary aspects of physical reality reveal themselves not only in acts of observation and measurement, but also in other actions that we might be involved in. The appearance of complementary aspects of physical reality introduces an element of indeterminism and a certain tolerance, which together constitute a margin for the existence of a free will. If looked upon from the point of view of an external observer, our actions appear subject to an unpredictable quantum uncertainty; but from our personal viewpoint the same actions appear to be controlled by our free will. A free will—if it exists—acts in one direction only: toward the future, otherwise we would merely be passive spectators as in classical physics, rather than actors as is admitted by quantum physics. Indeed, both aspects of reality are essential for us to be regarded as active participants rather than mere spectators, while only one of the aspects (that one tending to increasing entropy) admits propagation toward the future only. This is a sufficient reason for us to be unable to "kill our grandfather."

9. Conclusions

A thorough analysis of the quantum measurement process reveals the existence of two complementary aspects of quantum phenomena: one is symmetric, the other asymmetric under time reversal (or a CPT transformation). The propagation of information with a velocity that does not exceed that of light *in vacuo* is symmetric under time reversal, while its counterpart induced by unavoidable and uncontrollable perturbations, and carrying deinformation and increasing entropy, is asymmetric and propagates only toward the future.

The symmetric-under-time-reversal propagation of information guarantees the realistic character of the EPR effect, in agreement with the requirements of relativity and (a generalized) causality. A roundabout transfer of information (partly anticausal) demonstrates that the nonlocality inherent in the EPR phenomenon is only apparent. On the other hand the other aspect, connected with an increase of entropy, explains why there is no danger of disturbing the past by means of a retroaction (or killing our grandfather).

The completely symmetric-under-time-reversal classical dynamics exhibits only one side (or aspect) of physical reality and is strictly deterministic, while quantum mechanics with its two complementary aspects, the one symmetrically causal and anticausal and the other only causal, admits some tolerance as regards determinism and allows for something subjectively interpretable as free will. The problem of free will is strongly connected to the EPR paradox and appears unavoidable if we ask the question: what will happen if we choose freely this or that experiment to be performed on one of two objects that once interacted with one another and then separated.

Thus, a solution to the EPR controversy necessitates a thorough discussion of the fundamental ideas of contemporary physics: quantum measurements, uncertainty relations, relativity, complementarity, as well as the introduction of a generalized causality concept, including both anticausality and retroaction, in order to remove the apparent nonlocality of quantum phenomena. The EPR debate involves the deepest philosophical arguments, pro and contra realism and neopositivism, and even touches on the questions of fatalism and free will which are so deeply rooted in both Eastern and Western cultures.

Appendix: Perfect Measurements

In the last two or three years interest in the foundations of quantum mechanics and its philosophical implications has increased considerably. The main reasons for this renewed interest are twofold. First, we have recently witnessed a considerable advance in the technology of quantum measurements, enabling us to actually perform some crucial experiments which in earlier times could only be discussed as the so-called "Gedankenexperiment." The second reason was the 50th anniversary of the EPR debate which was marked by the organization of several symposia, devoted to the foundations and philosophy of QM, in particular those in Finland (Joensuu, June 1985) and Italy (Urbino, September/October 1985).

Our aim is to discuss the EPR effect, but in order to do so we must discuss, more thoroughly than hitherto, the problem of observation of events on a microscale and what is usually called a "theory of measurement."

Several physicists have complained that the originators of QM had not paid enough attention to the process of measurement on the microscale and tried to fill this gap by formulating a "theory of measurement" for quantum physics. There seems, however, to have existed—from the very beginning—considerable confusion and misunderstanding surrounding this point. We must be aware of the fact that in the modern, highly complex world specialization is unavoidable, otherwise we would not make any real progress in our endeavors. In particular, science and technology form two distinct domains and physicists should be, and in fact are, classified as theoreticians, experimentalists, and technical physicists. Obviously, the concept of measurement is viewed by each of them from a different angle, and different aspects of the problem of measurement are of particular interest to each of these three categories of physicists.

One should not forget that QM is a branch of *theoretical* physics. Consequently, neither the design of measurement apparatus, nor the practical methods of conducting measurements, belong to the main domain of interest of a theoretician but rather to that of the experimental and technical physicists. If that is so, the question arises as to what a "theory of measurement" implies for a theoretical physicist specializing in QM. Is a theory of measurement a meaningful concept at all? It has often been argued that a "theory of measurement" is essential to the interpretation of QM, or even that it constitutes its most fundamental part. However, what does "interpretation of QM" mean? Quantum mechanics itself consists of two parts: the formalism and its interpretation. The latter establishes links between the formalism itself and the results of measurements performed on the physical objects of the microworld, as well as establishing relationships between the results of different measurements. Therefore it is incorrect to speak of the "interpretation of quantum mechanics" because both the formalism and its interpretation constitute a single whole: a physical theory called "quantum mechanics." If, despite this, theoreticians decide to develop what they call a "theory of measurement" they begin by producing some mathematical formulas. But, by so doing, they are merely building up to an even greater extent, the formalism of QM. This does not seem to be very legitimate because it amounts to explaining and interpreting the formalism of QM by means of the formalism of QM itself. To be more specific, some theoretical physicists practice a "theory of measurement" by trying to apply a mathematical (quantum) description to the act of measurement itself. In this way they only complicate the problem by including the apparatus, or at least its essential parts, into the physical system that is going to be observed, forgetting that the enlarged system, itself, needs another apparatus for its observation. Including the apparatus into the system is thus an unending process, the system meanwhile becoming more and more complicated. Such a procedure seems to contradict the spirit of a theorem in mathematics, known as Goedel's theorem, which states that it is impossible to construct a complete and closed logical system without the help of some elements from outside the system. There is no reason why this theorem should not also apply to QM. Thus we conclude that QM divides physical reality into two distinct parts: the physical phenomena and the experimental apparatus, the latter constituting an external element with respect to the physical system, itself subject to observations and measurements.

Contrary to experimental physics and technology, theoretical QM physics is not so much interested in the process of measurement itself, or in detailed methods and recipes of measurements, but rather in the results of possible measurements and in their consistency with the existing formalism of QM. To this end, a concept of the "perfect" (or ideal) measurement is needed. This concept should not be confused with what is usually understood by an imperfect but practical measurement, i.e., one which is limited by the available technology.

The characteristics of perfect measurements must be introduced axiomatically. They are not direct consequences of the formalism but constitute its completion and, of course, must not be inconsistent with the mathematical formalism.

Let us define what is meant by perfect measurements by means of the following two postulates:

1. Inasmuch as the observables (described by self-adjoint operators) possess either discrete eigenvalues or continuous spectra, the perfect measurement yields either an exact eigenvalue of the measured quantity represented by this operator, or inexact values belonging to the continuous spectrum with an inexactitude that reaches a minimum. For instance, a perfect measurement of a pair of canonically-conjugate quantities q and p is characterized by an uncertainty given by

$$\Delta q \cdot \Delta p = \tfrac{1}{2}\hbar \tag{1}$$

in agreement with Heisenberg's uncertainty relation. If this uncertainty were larger than $\tfrac{1}{2}\hbar$ we would qualify this measurement as imperfect.

2. Perfect measurements are measurements of complete sets of observables (described by commuting operators), enabling one to determine the ψ-function (or equivalently, a normalizable state vector in Hilbert space) valid for a statistical description of the future *as well as* the past of the physical system, as long as the latter remains disentangled from interaction with a similar or different apparatus.

Most practical measurements turn out to be imperfect. They either do not reach the limit of exactitude imposed by the Heisenberg relation (1) or they introduce a violent disturbance that completely cuts the system off either from its past (i.e., destroys any memory about its past state) or from its future. In that case the measurement serves merely as a preparation of the state for the future (and is unable to say anything about its past) or confirms (statistically) predictions (supplied by another state of the system) valid prior and up to the instant of the measurements, but destroys the system or makes it useless for predictions of its future fate (e.g., measuring a particle's position by stopping it in a photographic emulsion).

Many physicists deny the validity of the concept of a perfect measurement as defined above. They distinguish sharply between the act of preparing the state of the system for the future and the actual act of measurement, and believe that in order to repeat a measurement one must necessarily use another specimen from an ensemble of identical systems or prepare the initial state of the same system anew. Basing themselves on a superficial analysis of some imperfect measurements they speculate in the spirit of Aristotle (who said that every motion tends to a state of rest) instead of following the example of Newton, who boldly extrapolated everyday experience and introduced an idealization of it for the case of motion free of any sort of friction (and said that every body not subject to external forces moves with constant velocity). Similarly to Newton's idealization of zero friction, the idealization of perfect measurements constitutes a very useful concept. Perfect measurements are to be regarded as the limiting case, to which realistic measurements may approach arbitrarily closely.

The consequences of the existence of perfect measurements (even if they constitute only a limiting case) are crucial to the interpretation of QM. One consequence is the repeatability of measurements in the "strong" sense, i.e., a repeatability of measurements on one and the same object, and not that one need necessarily make each measurement on another specimen from an ensemble. Such strong repeatability follows from the fact that a perfect measurement also constitutes [according to postulate (2)] a preparation of the system for a subsequent measurement. Strong repeatability is evident in the tracks observed in Wilson or bubble chamber wherein each droplet or bubble constitutes a new measurement of position of the same particle which is performed without any (other) previous preparation.

The applicability of QM backward in time, that is permitted (*ex definitione*) by the concept of perfect measurements—i.e., those that allow not only predictions but also retrodictions or, in other words, the conveying of information in both directions along the time axis—will be shown to be of importance for a "realistic" explanation of the EPR effect and for a satisfactory conclusion to the discussion between Einstein and Bohr.

References

1. O. Costa de Beauregard, *Nuovo Cim. B* **51**, 267 (1979).
2. C. W. Rietdijk, *Found. Phys.* **8**, 815 (1978).
3. H. P. Stapp, *Found. Phys.* **9**, 1 (1979).
4. J. A. Wheeler, presented at a meeting of the American Philosophical Society and the Royal Society, London (June 5, 1980).
5. C. v. Weizsäcker, in: *Proceedings Joensuu Symposium: 50 Years of the Einstein–Podolsky–Rosen Gedankenexperiment* (16–20 June 1985).

11

Quantum Action-at-a-Distance: The Mystery of Einstein-Podolsky-Rosen Correlations

A. Kyprianidis and J. P. Vigier

1. Introduction

Among Einstein's many remarkable papers, one of the most astonishing was published in 1935[1] (in collaboration with Rosen and Podolsky) with the objective of establishing the incomplete character of the Copenhagen Interpretation of Quantum Mechanics (CIQM). Far from being obsolete or refuted it is still now, in 1986, at the very center of the current crucial epistemological confrontation on the physical and philosophical implications of quantum mechanics. Its discussion, originally limited to the question of the complete (or incomplete) character of the quantum-mechanical description (Bohr believed in completeness; Einstein, Podolsky, and Rosen did not), has indeed blossomed, in successive steps, into questions concerning the space–time coordination of quantum events and their objective existence beyond measurements; the causal nature of physical laws; the wave–particle dualism; the existence of hidden variables; the nature of quantum correlations; and so on. All these questions were already present in the Bohr–Einstein debate in which strongly antagonistic positions were defended: Einstein tended to a causal–deterministic conception while Bohr's

A. Kyprianidis and J. P. Vigier • Laboratory of Theoretical Physics, Henri Poincaré Institute, 75231 Paris Cedex 05, France.

approach led in the opposite direction. This debate has now arrived at a new stage following the realization of the Bohm spin version of the original Einstein–Podolsky–Rosen (EPR) experiment by Aspect[2] that established the validity of quantum predictions over those of local models on the basis of Bell's inequalities (see Figure 1). But the most interesting implication of Aspect's experiments is, perhaps, the emergence of quantum nonlocal correlations and their relation to causality in microprocesses. This point is, of course, of great importance for the interpretation of the quantum formalism.

Confronted with this situation, various attitudes are possible: One could still follow Bohr's line of reasoning, deny the objective existence of microphenomena unless measurement intervenes, and explain the EPR effect by means of nonseparability. On the other hand, one could try to reconcile these results with local-realistic models which could reproduce, sufficiently accurately, the experimental results.[3] A third possibility would be to introduce nonlocal hidden-variable models which, while being equivalent to the quantum-mechanical formalism, would satisfy causality and space–time coordination despite the presence of nonlocal interactions.

It is precisely this approach that we will follow in the presentation of this chapter. In the next section we will discuss the possible implications of the quantum formalism and we will show that what is often presented in the name of the Copenhagen school as a vigorous "sticking to the facts," is nothing more than a set of arbitrary, unjustified, irrational philosophical assertions. Sections 3 and 4 treat a specific action-at-a-distance quantum-potential model (in Section 3 the scalar, and in Section 4 the spin version of it) that we shall construct in order to account for quantum nonlocal correlations. It is demonstrated that this specific form of quantum nonlocality satisfies the criteria of Einstein causality, and does not produce quantum paradoxes. In Section 5 a simple example is presented so as to illustrate

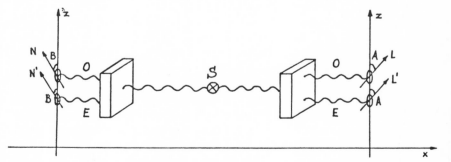

Figure 1. Schematic representation of the Aspect experiment on the Bohm spin EPR-correlation. Random switches orient the photons in the ordinary (O) or extraordinary (E) ray in order for them to be detected through pairs of linear polarizers L, L' and N, N'.

the shortcomings of the Copenhagen interpretation of EPR correlations and quantum-measurement theory. This example also serves as a test of the validity of our nonlocal quantum-potential model in this context. Finally, in Section 6 we will briefly discuss the physical origin of quantum nonlocality in terms of the existence of a random subquantal medium, i.e., Dirac's "covariant aether" model.

2. The Quantum Formalism and Acausal Deviations

It would be very useful, from the point of view of our further presentation, to be clear on the meaning we attach to causality in Physics (see Figure 2). In our opinion there is no better way to clarify this concept than by presenting Einstein's and de Broglie's views on the subject. These can be summarized by the following few points:

1. Microprocesses exist objectively, i.e., independently of any measurement or act of consciousness. Matter, space, and time are objective realities and not merely mathematical or physical conventions. The laws of nature must remain invariant under the causality group $G = T_4 \otimes \mathscr{L}{\uparrow} \otimes D + P$, where T_4 denotes space–time translations in the objective four-dimensional space–time, $\mathscr{L}{\uparrow}$ is the orthochronous Lorentz group (which preserves the time ordering of time-like separated events and therefore the order of causal chains), D is a scale dilatation, and P represents space parity.
2. The causality group implies the absolute conservation of energy-momentum P^{μ} and angular momentum $M_{\mu\nu}$ in all macroscopic and microscopic phenomena.

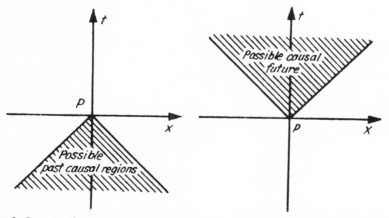

Figure 2. Causal origins and causal consequences of an event are limited to its backward and forward light cone, respectively.

3. All physical processes consist of positive energy: $E = P_0 > 0$ propagating in positive time with a velocity $\leq c$. Antimatter can be represented as negative energy moving backward in time, this being a mathematical (and not a physical) convention.

4. For all isolated systems one can solve the Cauchy problem, i.e., one can determine the future evolution of a system from a set of initial conditions, defined on a space-like surface at some initial time.

5. Probabilistic theories reproduce real chaotic motions which are, in principle, analyzable in terms of "hidden parameters." Any probability distribution resulting from acts of observation on physical systems reflects an objective limit of real frequencies which result from a deeper random causal behavior of the system, inextricable at the observation level.

Why do we lay so much stress on the causality concept and its unambiguous definition? The reason is because it is very frequently claimed that this concept is in contradiction with the quantum formalism; moreover, it is claimed that the quantum formalism implies an acausal behavior of microsystems. Advocates of the Bohr standpoint[4] have recently developed an approach to EPR correlations, based on a relativistic S-matrix scheme, which claims to exhibit an "isomorphism between the formalism and its interpretative discourse."[5] Let us briefly state the basic ingredients of this approach:

1. If $|\phi_0\rangle$ is the initial prepared state and $|\psi_0\rangle$ the measured state, then the system's state $|S\rangle$ is neither in retarded evolution: $|\phi_i\rangle = |U_{i0}\phi_0\rangle$, nor in advanced evolution: $|\psi_i\rangle = |U_{i0}^{-1}\psi_0\rangle$, but only in a transient state located beyond space-time.[5] This combined advanced and retarded action (which is PT invariant) manifests an arrowless causality on the microlevel.[6] The macroscopic irreversibility is "fact-like" but not "law-like," i.e., not a strict consequence of the formalism.

2. Physical causality should merely be identified with conditional expectations: $(A/C) = |\langle A|C\rangle|^2$, where $\langle A/C\rangle$ is a transition conditional amplitude which is, in fact, a timeless definition, invariant with respect to time zigzagging[7] (see Figure 3).

3. The solution to the Cauchy problem can be written as

$$\langle x'|a\rangle = \langle x'|x\rangle\langle x|a\rangle$$

where $\langle x'|x\rangle$ is the Pauli–Jordan propagator which is invariant under $P = -T = -C = 1$ separately, contrary to the Feynman propagator, invariant under $P = CT = 1$.[5] The latter entails an exponential

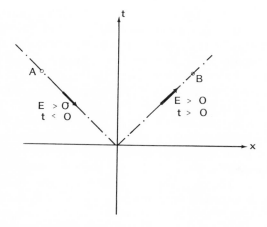

Figure 3. In the retroaction-in-time model, nonlocal correlations result from a time zigzag where $E > 0$ is always propagated.

decay of higher energy levels in a predictive calculation, a fact not present in the structure of the former.

4. EPR correlations should be explained by taking account of the fact that adjustable parameters exist at the measurement positions, and not at the place of common preparation of the system.[8] The link that establishes the correlation is the time zigzag, consisting of two time-like vectors with a relay at the source in the past,[9] a mechanism which is insensitive to the space-like or time-like character of the separation of the two measuring processes.[10]

5. Finally, the following are some ontological consequences of this point of view[9,11]:

> ... both "reality" and objectivity concepts must yield in favor of "inter-subjectivity," thus entailing a world-view very akin to the Hindu "maya" concept ... a sort of a common daydream, the illusory character of which is pinpointed by the so-called "paranormal phenomena."

> ... Fact-like irreversibility does not mean suppression but repression of advanced waves and decreasing probabilities; it does not mean suppression but repression of the lower against the upper arrow in the reversible negentropy $\rightleftarrows$ information, or $N \rightleftarrows I$ transition. Taken as basic, the I–N transition has a name: psychokinesis. The formalism not only allows, but suggests its existence.

All these assertions are not, of course, implications of the formalism but only arbitrary deviations from it. A simple criticism is sufficient to recognize this fact: No need exists for a "collapse plus retrocollapse" mechanism for quantum measurement since Bohm[12] and Cini[13] have shown that "fact-like" irreversibility results from a realistic model for the measurement process, where the latter is conceived as a spectral-decomposition procedure of the wave packets, the particle entering one of the resulting subpackets (see Figure 4). The CPT invariance of the formalism simply denotes the existence of antiparticles in the relativistic frame. Feynman zigzags are a mathematical picture of combined particle/antiparticle-creation/annihilation processes, and the use of Feynman propagators (instead of acausal Pauli–Jordan ones) implies an irreversible increase of entropy in future evolution, as well as the prohibition of particle evolution towards the past. The shooting of positive energies backward in time, advocated by this model, violates the conservation laws at the source, since energy appears from the future without any apparent cause. Finally, let us stress that this "antitelegraph toward the past" mechanism is not a consequence of the formalism, nor a consistent extrapolation of it, but simply an arbitrary and unallowed adjunction to it.

If one wishes to examine what the quantum formalism really implies then one has to limit the arbitrary assertions to a minimum and, using a physical model, try to approach the quantum correlations by checking, at every step, the consistency of an attempted explanation (and its congruence) with formal apsects of quantum theory itself. We intend to follow precisely this line of approach by using an action-at-a-distance scheme as an explanatory pattern, and we hope to demonstrate that it consistently reproduces

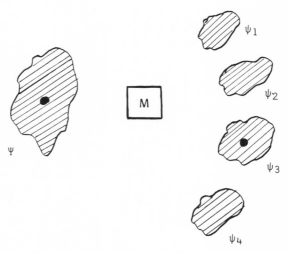

Figure 4. A wave packet Ψ interacting with a measuring device M splits into nonsuperimposed packets $\psi_1, \ldots, \psi_4$ and a subsequent measurement will always find the particle in ψ_3.

the formal quantum aspects as well as satisfies the usual causality conditions in the frame of a space–time approach to quantum mechanics.

3. A Scalar Action-at-a-Distance Mechanism

The essential difficulty with nonlocal correlations (or action-at-a-distance mechanisms) arises in the context of relativistic theories. Classical theories and dynamics (which do not contain a limiting-velocity concept and obey Galilean invariance) are devoid of any similar problem. If one takes, as an example, Newtonian mechanics, this is in fact an action-at-a-distance theory and no formal difficulty arises in connection with it. However, since in relativity theory all interactions propagate with a specific velocity smaller than or equal to the velocity of light (i.e., are retarded interactions), this concept of action-at-a-distance seems, at first sight, to be in conflict with basic relativistic assumptions. In fact, it seems that the retarded propagation and the direct interaction properties are mutually exclusive on the simple ground of causality: The former guarantees a causal evolution, independent of any frame characteristic, while the latter makes the causal chain of cause–effect connection a frame-dependent property and, moreover, authorizes the existence of causal loops and retroaction in time.

Let us clarify this point in the case of a simple example. We consider two relatively moving observers O_1 and O_2 with respective rest frames S and S'. At the event ε_1, observer O_1 sends a superluminal signal in its relative future which is absorbed by O_2 at ε_2. Then observer O_2 waits until the event ε_3 happens and sends, again in its relative future, a superluminal signal which is absorbed at ε_4. It has been shown[14] that one can always arrange things in such a way that ε_4 precedes ε_1, so that we find ourselves confronted by the following paradox: Starting from ε_1, we can use superluminal signals in order to modify the past of O_1 by performing a criss-crossing of space-like paths and effectively transporting positive energy backward in time from ε_1 to ε_4 (see Figure 5). If a direct interaction inevitably implies such a criss-crossing mechanism, then it is clear that it is in severe conflict with causality. Moreover, the action-at-a-distance mechanism is then acausal and, whenever we reveal the presence of this mechanism, the underlying behavior of the related system must violate causality.

But first of all let us determine the precise physical origins of possible space–time criss-crossing mechanisms. By examining the problem closely, we can extract from the above qualitative description two major causes for the existence of a causal paradox. The first is linked to the fact that the relativistic potentials which mediate the mutual interdependence of the constituent elements of a system of particles are, in general, nonlocal, i.e.,

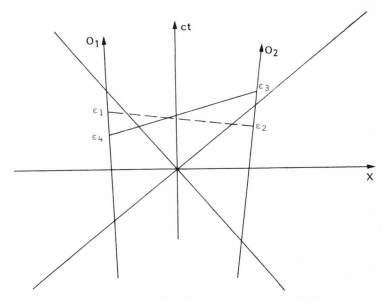

Figure 5. An example of space–time causal anomalies.

they depend on the characteristic properties (positions, momenta, and so on) of all the particles in a direct (nonretarded) way. The second cause is related to the basic difference between relativistic and nonrelativistic mechanics: The latter possesses a universal evolution parameter, the physical time t, while in the former this role is played by the proper times τ_i of each particle, which figure as nonuniversal particle-dependent parameters.

One would therefore be tempted to ascribe an acausal behavior to action-at-a-distance theories, if it were not for the fact that this problem has been satisfactorily resolved in the frame of the Hamiltonian formulation of relativistic constraint dynamics.[15] Let us briefly state the main lines of this demonstration in the simplest case of two scalar relativistic particles.

Here we can define a 16-dimensional phase space for the canonical coordinates q_i^μ, p_i^μ ($i = 1, 2$, $\mu = 0, 1, 2, 3$) which satisfy the standard Poisson bracket relations $\{q_i^\mu, p_{j\nu}\} = \delta_{ij}\delta_\nu^\mu$. In this phase space we define the two covariant Hamiltonians which consist of free terms and additive interaction terms which are, in general, nonlocal potentials:

$$H_i = \tfrac{1}{2}p_{i\mu}p_i^\mu + V_i(q_1^\mu, p_1^\mu; q_2^\mu, p_2^\mu) = \tfrac{1}{2}m_i^2 \tag{1}$$

As can be deduced from the above equation, the Hamiltonians are constants of the motion but, unlike in the nonrelativistic case where they are equal to the particle energy, here they are identified with the squared masses. Furthermore we see, for example, that the potentials V_i are momentum dependent with respect to both p_i and p_j, a dependence which is, in general, nonlocal. These are two major differences from nonrelativistic dynamics but there is still a significant third one: This formalism is a multitime formalism with parameters τ_i which are suitable generalizations of the proper times.

In general, if we allow an unconstrained evolution for the system, the Hamiltonians are functions of both parameters. In that case, however, no world lines can be defined for any of the particles and the system's behavior is unpredictable. It is, therefore, necessary to impose a predictivity constraint on the system, which mainly states that the Hamiltonian of system i is a function only of the parameter τ_i and not of $\tau_j (j \neq i)$, or equivalently:

$$\frac{\partial H_i}{\partial \tau_j} = \{H_i, H_j\} = 0 \tag{2}$$

This predicting condition ensures the existence of the world line $l_i(\tau_i)$ and the time-like character of the paths, thus avoiding the emergence of the causal paradoxes discussed above. To get a clearer idea of how this can be ensured, we now proceed with a brief quantitative treatment of the two-particle system.

It is sufficient for our purposes to treat the single-potential case, i.e., $V_1 = V_2 = V$ because it simplifies the calculations considerably and, furthermore, because it is the case relevant to our later discussion. First, by introducing the center-of-mass (CM) total momentum $P^\mu = p_1^\mu + p_2^\mu$, we can rewrite the difference between equations (1) as follows:

$$P^\mu (p_{1\mu} - p_{2\mu}) = \tfrac{1}{2}(m_1^2 - m_2^2) \tag{3}$$

A transformation to the CM rest frame, where $P = (P^0, \mathbf{P} = 0)$ because $P^0 = p_1^0 + p_2^0$ and $p_1^k = -p_2^k$, yields the following two equations for the particle energies:

$$p_1^0 = \frac{P^0}{2} + \frac{m_1^2 - m_2^2}{2P_0} \qquad p_2^0 = \frac{P^0}{2} - \frac{m_1^2 - m_2^2}{2P_0} \tag{4}$$

which implies that, in this frame, no exchange of energy occurs between the particles since both p_1^0 and p_2^0 are constants of the motion.

We can then evaluate the implications of the predictivity condition of equation (2), which is, in fact, a condition on the nonlocal potential V. Equation (2) now takes the form

$$\{P^\mu (p_{1\mu} - p_{2\mu}), V\} = 0 \tag{5}$$

This condition has been shown to restrict the general (p, q) dependence of the nonlocal potential V to the following functional dependence on a projection of the relative variable $z^\mu = q_1^\mu - q_2^\mu$:

$$V(\tilde{z}) = V(z^\alpha \Pi_\alpha^\beta) = V\left(z^\beta - \frac{(z^\alpha P_\alpha) P^\beta}{P_\lambda P^\lambda} \right) \tag{6}$$

This is a suitable form from which the nonlocal property of the relativistic potential can immediately be derived. Consider, e.g., a transformation to the rest frame, i.e., $P_\mu = (P_0, 0)$. Then the projected relative coordinate is $\tilde{z}^\mu = (0, z^k)$ so that the potential is only a function of the spatial distance z^k, i.e., $V_{CM} = V(z^k)$. Since the fourth component of the relative coordinate vanishes, we see that the potential does not depend on the relative time but only on the space interval between the two particles. It represents an instantaneous form of action-at-a-distance which, because p_1^0 and p_2^0 are constants of the motion [cf. equation (4)], does not imply any exchange of energy in the CM rest frame. There is, however, a minor detail that we have not, as yet, commented upon: The relative coordinate z^μ is a difference between the canonical position coordinates which, owing to the existence of the so-called "no interaction theorem,"[17] cannot be identified with the physical position coordinates in the presence of an interaction. However, it has been shown[15] that the identification $q^\mu = x^\mu$ can be performed, even in the case of interacting particles, on the "equal-time surface," i.e., on a three-dimensional surface in Minkowski space where the physical times t_i for each particle coincide. This is precisely the case in the CM rest frame where the relative time coordinate of V vanishes, so that we find ourselves on the "equal-time surface." Consequently, in this frame, the potential must be a function of the form $V(x_1^k - x_2^k)$, i.e., a function of the relative physical distance between the particles.

Finally, using the obvious relation $P^\mu \tilde{z}_\mu = 0$, we immediately deduce that the action-at-a-distance mechanism between the two particles exists only on a space-like surface perpendicular to the CM total momentum P^μ, which is a time-like vector. The action-at-a-distance mechanism is thus always confined to a space-like direction perpendicular to P^μ owing to the constraint introduced by the predictivity condition. Furthermore, at any instant, an exchange of energy/momentum between the two particles in an arbitrary frame occurs in the way prescribed by equation (3), i.e., by preserving a constant angle between the CM momentum P^μ and the relative momentum $p_1^\mu - p_2^\mu$ (which becomes perpendicular in the case of $m_1 = m_2$). Any such exchange occurs in a ladder scheme and space–time criss-crossings are excluded (see Figure 6).

Up to now we have been treating a system of two point particles, tied by nonlocal potentials, in the scheme of relativistic constraint dynamics.

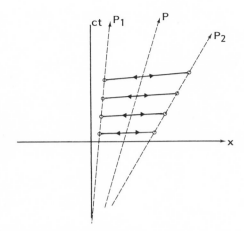

Figure 6. The relative four-momentum of two particles correlated by a nonlocal potential forms a constant angle with the CM four-momentum. No causal anomalies are therefore possible.

The question which naturally arises now is to what extent this fictitious system has any relation to the system of two free Klein–Gordon particles which we have set out to examine. As we shall now show, this question can be answered by demonstrating that the two systems can be mapped one upon another, so that a clear correspondence can be established between them. To this end, we consider a system of two Klein–Gordon equations

$$\left(\Box_i + \frac{m_i^2 c^2}{\hbar^2}\right)\psi(x_1^\mu, x_2^\mu) = 0 \tag{7}$$

which describes two free spin-zero relativistic particles obeying the laws of quantum mechanics. We write the wave function as $\psi = \exp(P + i/\hbar S)$ and consider the case where the CM propagates freely, i.e., is an eigenstate of the total four-momentum operator $\hat{P}^\mu = i/\hbar(\partial_1^\mu + \partial_2^\mu)$. In this case we can separate the motion by writing

$$\psi = \exp\left[\frac{i}{\hbar} K^\mu \frac{x_{1\mu}m_1 + x_{2\mu}m_2}{m_1 + m_2}\right]\phi(x_1^\mu - x_2^\mu, K^\mu) \tag{8}$$

where K^μ is a constant time-like vector representing the CM motion and ϕ is the wave function of the relative motion:

$$\phi(x_1^\mu - x_2^\mu, K^\mu) = \exp\left[P(x_1^\mu - x_2^\mu, K^\mu) + \frac{i}{\hbar} S'(x_1^\mu - x_2^\mu, K^\mu)\right] \tag{9}$$

We can now show that the following relations hold:

$$(\partial_1^\mu + \partial_2^\mu)P = 0$$

$$(\partial_1^\mu + \partial_2^\mu)S' = 0 \quad \text{or} \quad (\partial_1^\mu + \partial_2^\mu)S = K^\mu \tag{10}$$

which imply the independence of the relative motion from the CM motion.

To proceed further, we can decompose equation (7) into real and imaginary parts, thereby obtaining a Hamilton–Jacobi type equation:

$$-\frac{1}{\hbar^2}\partial_{i\mu}S\partial_i^\mu S + \frac{m_i^2 c^2}{\hbar^2} + (\Box_i P + \partial_{i\mu}P\partial_i^\mu P) = 0 \tag{11a}$$

and a continuity equation:

$$\partial_\mu(e^{2P}\partial^\mu S) = 0 \tag{11b}$$

From equation (11b) we can justify an assumption $\partial_i^\mu S = p_i^\mu$. Furthermore, by introducing the quantum potential in its relativistic form $U_i = -\frac{1}{2}\hbar^2(\Box_i P + \partial_{i\mu}P\partial_i^\mu P)$ and using equation (10), we can show that the quantum potentials in the two Hamilton–Jacobi equations are equal, i.e., $U_1 = U_2 = U$. Then equations (11b) can be rewritten in the following ($c = 1$) form:

$$\tfrac{1}{2}p_{i\mu}p_i^\mu + U = \tfrac{1}{2}m_i^2 \tag{12}$$

where $U = U(x_1^\mu - x_2^\mu, K^\mu)$. This result enables us to identify equation (12) [which is the real part of equation (7)] with equation (1) of relativistic constraint dynamics, under two conditions: (1) K^μ is put equal to P^μ; and (2) the quantum-potential dependence on the relative coordinate and the CM time-like vector is reduced to a projection of the relative coordinate of the form presented in equations (6), namely

$$U(x_1^\mu - x_2^\mu, K^\mu) = U\left((x_1^\mu - x_2^\mu) - \frac{[(x_1^\nu - x_2^\nu)K_\nu]K^\mu}{K_\lambda K^\lambda}\right) \tag{13}$$

Our two systems can thus be mapped one upon another and what has been deduced for the relativistic Hamiltonian system holds equally well for the system of two Klein–Gordon particles.

Let us therefore summarize the information gained from this comparison. The system consisting of two free noninteracting Klein–Gordon particles is, in reality, a system of two relativistic scalar particles coupled by nonlocal potentials. The role of coupling potential is played by the quantum potential which is a nonlocal action-at-a-distance mechanism. This direct interaction, restricted to space-like three-surfaces in Minkowski space, becomes instantaneous on the equal-time surface of the CM rest frame of the system. It does not violate causality, in the sense of Einstein, because it is subject to specific constraints (the predictivity condition) implying a specific dependence on the space–time coordinates and the CM four-momentum. What appears in the quantum-mechanical formalism as a free-particle system is therefore, in reality, a system of nonlocally, directly interacting particles which nevertheless preserves Einstein causality and does not imply causal paradoxes.

4. Spin-Dependent Action-at-a-Distance

In the previous section we gave as a simple example of a scalar action-at-a-distance mechanism a system of two relativistic quantum particles. How can we now demonstrate the corresponding effect in the case of spinning particles, in particular for two spin-1 particles, as are involved in the experimentally tested version of the original EPR gedanken experiment? It is evident that we must extend the preceding treatment to the spin-1 case, and we can use as a basis for this treatment de Broglie's theory of light,[18] which is essentially a nonzero-mass photon model ($m_\gamma \neq 0$). This is justified by the fact that the nonzero-mass theory (Proca theory) has a zero-mass limit which is physically indistinguishable from the usual Maxwell wave: the main difference between the two models lies in the appearance of an extra longitudinal mode in the Proca field theory which, for $m_\gamma \to 0$, decouples from the usual transverse modes and describes a Lorentz-scalar Coulomb field.[19] The Proca nonzero-mass theory is thus equivalent to the usual photon theory provided m_γ is very small ($m_\gamma \ll 10^{-48}$ g).

The generalization of the calculus of the preceding section is now straightforward if one substitutes a second-rank tensor $A_{\mu\nu}$ in place of the scalar field ψ in the usual Klein–Gordon equations[20]:

$$\left(\Box_i + \frac{m_\gamma^2 c^2}{\hbar^2}\right) A_{\mu\nu} = 0 \tag{14}$$

where the Proca field tensor can be written as

$$A_{\mu\nu}(x_1^\lambda, x_2^\lambda) = \omega_{\mu\nu} \exp\left(P + \frac{i}{\hbar} S\right) \tag{15}$$

with $\omega_{\mu\nu}\omega^{\mu\nu} = 1$ and where the amplitude and phase effects have been separated out on the right-hand side. As we know this most general form of a compound state of two spin-1 particles can be split into three separate contributions: A symmetric part $A_{(\mu\nu)}$ with total spin $J = 2$, a skew part $A_{[\mu\nu]}$ with total spin $J = 1$, and a trace $A_{\mu\mu}$ corresponding to our present singlet state $J = 0$. Therefore we can write, for this state, the simplified Proca equations:

$$\left(\square_i + \frac{m_\gamma^2 c^2}{\hbar^2}\right) A_{1\mu}(x_1) A_2^\mu(x_2) = 0 \tag{16}$$

where $A_{1\mu}(x_1) A_2^\mu(x_2) = \exp[P'(x_1, x_2) + (i/\hbar)S(x_1, x_2)]$, the factor $\omega_{\mu\mu}$ being included in the amplitude part of the wave function. This system, as can be easily seen, is formally equivalent to the scalar Klein–Gordon case and therefore the whole demonstration of Section 3 can, without difficulty, be carried over to our present formalism. We can recover the nonlocal character of the quantum potential by decomposing equation (16) into real and imaginary parts; we can show that the direct interaction is limited to space-like directions and becomes instantaneous in the CM rest frame on the "equal-time" three-surface; and, finally, we can again demonstrate the specific dependence of the quantum potential on the projected relative variable $\Pi^\alpha_\beta x^\beta = z^\alpha - (z_\nu P^\nu) P^\alpha / P^2$ where P^μ is the time-like four-momentum of the system's CM motion. We therefore see that the Proca singlet state can be treated on exactly the same footing as the scalar Klein–Gordon system, but that its different formalism obscures the essentially new feature of the system of two spin-1 particles, namely, the spin–spin correlation and its nonlocal characteristics. One can easily see that this is the case if one preserves the distinction between the contributions due to the scalar and spin-dependent amplitudes, introduced in equation (15), by writing

$$A^\mu_\mu(x_1^\lambda, x_2^\lambda) = \omega^\mu_\mu \exp\left[P + \frac{i}{\hbar} S\right] \tag{17}$$

where $\omega^\mu_\mu = a_{1\mu}(x_1) a_2^\mu(x_2)$ and $a_{i\mu} a_i^\mu = 1$ ($i = 1, 2$). Then one can calculate explicitly the form of the quantum potential for this wave function:

$$U_i = -\frac{\hbar^2}{2}\left[\square_i P + \partial_{i\mu} P \partial_i^\mu P + \frac{\square_i \omega^\mu_\mu}{\omega^\nu_\nu} + 2\frac{\partial_{i\mu}\omega^\nu_\nu}{\omega^\lambda_\lambda} \partial_i^\mu P\right] \tag{18}$$

From this expression we can deduce that the usual quantum potential of the relativistic scalar case is supplemented by the two spin-dependent contributions which together manifest the spin–spin correlation between the spin angular momenta of the two Proca particles. Since we have deduced the direct-interaction form of the total expressions for the quantum potential, this property, i.e., nonlocality, evidently holds for the spin-dependent contribution to U_i as well.

We can try to make this point a little clearer, without extending our discussion to complicated formal patterns, by presenting the following argument: The Proca equation (16) can be evidently derived from a Lagrangian formalism by performing the usual variations with respect to the field variables and writing the Euler–Lagrange equations of motion. The appropriate Lagrangian for this approach has been shown to be[21]

$$\mathcal{L} = \frac{m_\gamma^2 c^2}{\hbar^2} \Phi^* \Phi + \partial_{1\mu} \Phi^* \partial_1^\mu \Phi + \partial_{2\mu} \Phi^* \partial_2^\mu \Phi \tag{19}$$

with $\Phi(x_1, x_2) = A_{1\mu}(x_1) A_2^\mu(x_2)$. The advantage of this approach rests on the fact that, on the basis of this Lagrangian, we can perform a classical relativistic hydrodynamical analysis[22] and build the usual field magnitudes, such as the energy–momentum tensors for each particle $t_{1\mu\nu}$, $t_{2\mu\nu}$. Of interest to us is the fact that we can hereby define a spin-density tensor (with $u_{1\lambda} u_1^\lambda = 1$):

$$\tfrac{1}{2} S_1^{\alpha\beta} = (A_1^\alpha A_2^\beta - A_1^\beta A_2^\alpha) u_1^\lambda \partial_{1\lambda} \Phi^* + \text{c.c.} \tag{20}$$

and from it a spin vector:

$$S_{1\mu} = \frac{i}{2} \varepsilon_{\nu\alpha\beta\mu} u_1^\nu S_1^{\alpha\beta} \tag{21}$$

From this expression we see that the spin vector of photon 1 does not depend simply on the field magnitude $A_1^\alpha(x_1)$, which would imply a perfectly local theory, but also on the field magnitude $A_2^\beta(x_2)$, present in the spin-density tensor $S_1^{\alpha\beta}$ in a nonlocal way, as we have previously established. An explicit calculation of the spin-vector variation along a line of flow manifests the specific form of this nonlocal spin correlation between S_1 and S_2 but, even at the level of this general formalism, one recovers the essential features of this action-at-a-distance property. In order not to complicate the discussion with tedious calculations, we shall from now on restrict our presentation to a semiqualitative approach by showing the direct and indirect consequences of the spin–spin nonlocal correlations in a photon singlet state, as is present in the EPR spin experiments.

To get an idea of what is, in reality, the consequence of spin action-at-a-distance, one has to consider the effect of a one-sided measurement act, i.e., measurement of the spin of particle 1 and calculation of the spin of particle 2 which can, of course, be verified by a second measurement act. In order to be clear about the local classical prediction, let us consider an example from classical mechanics (see Figure 7). Two particles leave the source A with antilinear and equal momenta: $\mathbf{p}_1 + \mathbf{p}_2 = 0$, and remain so all the way from A to O and O', respectively. At O, particle 1 rebounds on a wall and reverses its momentum: $\mathbf{p}_1 \to -\mathbf{p}_1$, by transferring a quantity $2\mathbf{p}_1$ to the wall. Meanwhile, the second particle propagates unperturbed: $\mathbf{p}_2 \to \mathbf{p}_2$. Total-momentum conservation holds since $\mathbf{p}_1 + \mathbf{p}_2 = 0$ at every point between A and O, and A and O', respectively, before particle 1 hits the wall. But even after passing O and O', respectively, momentum conservation is still guaranteed, at every instant, if the momentum transfer to the wall is taken into account, since $\mathbf{p}_1' + \mathbf{p}_w + \mathbf{p}_2 = 0$. Therefore if we think of the wall as a momentum-measuring device then we can say that, in our example, local conservation laws (i.e., at the source A and at the collision point O) imply total-momentum conservation if the change in momentum of the measuring device is taken into account. And, furthermore, total-momentum conservation implies, in its turn, that the conserved quantity of one part of the system is unchanged, unless a local application of the conservation law intervenes, no matter what changes the other part of the system undergoes: Particle 2 propagates with momentum $\mathbf{p}_2$, unless it collides with another particle or system, no matter what happens to particle 1. Total conservation is thus guaranteed by a succession of local conservation laws.

What, then, is the situation in quantum mechanics? There are some major differences from the classical case. Let us consider a spin singlet state, as shown in Figure 8, where the spin-up state is measured in a measuring device M (a Stern–Gerlach apparatus), while the other state is

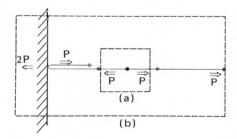

Figure 7. Momentum conservation holds for the two-particle system in case (a). It also holds in case (b) if the momentum transfer to the wall is taken into account.

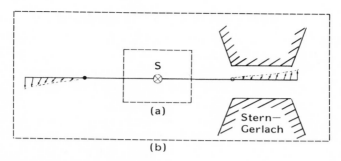

Figure 8. Angular-momentum conservation holds before measurement (case a). After measurement this is no longer the case due to the local angular-momentum change of the apparatus (case b). Conservation can be reestablished if one takes the quantum torque into account.

not subjected to any measurement at all. The remarkable result, given in a recent publication,[23] is the following: If one calculates the total angular-momentum change after the measuring interaction, i.e., the change of spin of particle 1 and of particle 2 and, additionally, the angular-momentum change of the apparatus M, then one finds that the total angular-momentum change is not zero, contrary to the classical case discussed above. What does this imply? Since the process of twisting the spin of particle 1 to the "up" position occurs under local angular-momentum conservation, this results in a net change of zero. Therefore the spin orientation of particle 2 has changed. This is precisely the prediction of the quantum calculation. By twisting the spin of particle 1 to the "up" position at M, the spin of particle 2 is, without any measuring intervention, twisted to the "down" position by means of the spin action-at-a-distance. There is no classical analog of this situation because it would imply, in our first example, that particle 2 reverses its momentum when particle 1 rebounds on a wall. The difference between the quantum and classical cases lies, according to our approach, in the existence of an action-at-a-distance mechanism in the quantum case, only. Of course, there exists an explanation of this phenomenon in the usual Copenhagen context, by claiming that the spin expectation value becomes a definite spin value only upon quantum measurement. Before a quantum measurement there is no significance attached to spin variables, twisting of spin, and action-at-a-distance. "Reality" of observables is only created by a measurement. This is a consistent view, which is not a necessary implication of the formalism although often presented as such, but one which can run into serious difficulties in certain unusual situations. In the next section we shall discuss one such interesting configuration which reveals the shortcomings of ordinary measurement theory and which was recently presented by Sutherland[24] in connection with a specific EPR-type measurement in a singlet-state system.

5. The Sutherland Paradox and Its Causal Resolution

Interestingly, a simple and elegant test of the measurement theory, in the context of EPR spin correlations, was proposed by one of the former proponents of the "retroaction model": Sutherland[24] proposed a "paradox" involving a measurement on EPR-correlated pairs whose result depends on the time-like or space-like character of their separation. The paradox holds equally well in the quantum calculus, unless additional restrictions are introduced. Let us summarize the Sutherland paradox as follows: In Figure 9 we show the space–time picture of a decay, at O, of a spin-zero state into two spin-$\frac{1}{2}$ particles. Spin measurements are performed at M_1 and M_2, or M_1' and M_2, any observed differences depending on the space-like or time-like character of the intervals. According to the quantum formalism, the results at M_1 and M_2, as well as those at M_1' and M_2, are always correlated since there are no predictions that depend on the specific space-like or time-like character of the separating interval. Then for a time-like separation of M_1' and M_2 the following reasoning can be advanced: (1) The result at M_2 depends on the choice at M_1'. Therefore, for some pairs of particles, it is true that choosing a direction ω_1 for the M_1' measurement would yield $+\frac{1}{2}\hbar$ at M_2, and choosing ω_2 at M_1' would yield $-\frac{1}{2}\hbar$ at M_2. (2) Since the interval between M_1' and M_2 is time-like we can, in principle,

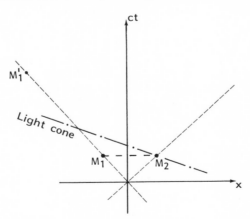

Figure 9. Space-like ($\overline{M_1 M_2}$) and time-like ($\overline{M_1' M_2}$) separated measurements performed on a two-particle system.

signal from M_2 to M_1' and, at will, instruct the measuring device at M_1' to choose a particular direction. (3) Then when $+\frac{1}{2}\hbar$ is observed at M_2 we can instruct M_1' to choose ω_2 and for $-\frac{1}{2}\hbar$ at M_2 we can instruct M_1' to choose ω_1. Using point (1) above, we conclude that neither $+\frac{1}{2}\hbar$ nor $-\frac{1}{2}\hbar$ is a consistent outcome for the particle at M_2', this paradox invariably emerging for all pairs of directions.

This paradox raises a serious difficulty in the standard Copenhagen interpretation as well as in the "retroaction in time" approach, as we shall presently discuss. However, let us first approach this problem in the frame of the quantum-potential model and the resulting action-at-a-distance mechanism that we presented in the previous sections. Our resolution of the paradox rests on the following two points:

1. In the "retroaction approach" and the standard Copenhagen school, it is usually claimed that composite states are nonseparable and unanalyzable and reality is created *iff* a measurement process is actually performed. Contrary to that approach, our model rests on an incommensurable physical assumption: The system plus context (e.g., experimental arrangement, measuring devices) are a unified whole owing to permanently active nonlocal action-at-a-distance correlations between existing real physical (i.e., observable) variables of its constituent elements, these correlations being independent of any attempted or performed measurement process. Nevertheless, the system is analyzable, at least in principle.

   To gain a clearer understanding let us examine the Sutherland paradox in our scheme. The paradox implies: (i) the existence of spin values for each particle which are independent of any measurement. We gave a brief sketch of a calculation, in the relativistic frame, in the preceding section. In a nonrelativistic approach, by means of a causal interpretation of the Pauli equation, similar calculations have been carried out in the case of different experimental arrangements and can be found elsewhere.[25] (ii) The spin vectors of the two particles in a singlet state are correlated by the spin–spin action-at-a-distance, and the spin motion by a quantum torque, which forces them to remain in opposite directions.[26] If we twist the spin of one particle, by means of an interaction, so as to lie in a particular direction, then the nonlocal quantum torque twists the second particle's spin vector to a corresponding direction so that the singlet state is preserved, despite the external interaction.

2. The quantum-measurement results are assumed to correspond to real interactions between the apparatus and the measured particles.

No wave-packet collapse exists in this model. Since the two particles are permanently nonlocally spin correlated (i.e., in the singlet state) for space-like separations, if space-like separated measurements are performed on the two particles then, owing to the existence of a "quantum torque," the measurement results at M_1 and M_2 must possess this correlated feature depending on the choice of polarization directions.

The situation is quite different if the two measurements are separated by a time-like interval, M_2 preceding M_1' in the time ordering. Then at M_2, due to the interaction with the device, the spin S_2 rotates and acquires a certain value for its z component, say $+\frac{1}{2}\hbar$, and consequently the "quantum torque" twists S_1, which is space-like separated from S_2, into a corresponding value: $S_{z1} = -\frac{1}{2}\hbar$ which is not measured. The measurement at M_2 is an objective procedure, for example, a Stern–Gerlach splitting of the original wave packet into two packets where only the spin-up packet is occupied by particle 2. This measurement result rests on the action of the "quantum torque" which depends on the system as a whole (i.e., the two particles and the apparatus).[25] The system remains in the correlated singlet state, i.e., particles 1 and 2 interact nonlocally, after particle 2 has emerged from the measuring device. The subsequent measurement at M_1' on particle 1 can be conceived along the same scheme and the results again depend on the context of the experiment and the "quantum torque" between the two particles. In any case, the spin values S_1 and S_2 remain singlet correlated. However, this is not the point envisaged by the "paradox," because the latter is concerned with the correlation of the measurement results at M_1' and M_2.

In connection with the spin values remaining singlet correlated we can state the following: If the relaxation time of, e.g., the $2(\uparrow)-1(\downarrow)$ state is big with respect to the time separation of the events in the observer frame, then the separation of the spin-up and spin-down packets after passage through M_2 persists and a subsequent M_1' measurement will yield a result correlated with the M_2 measurement. No direct correlation exists between the measuring devices, but the correlation of results is mediated via the fixing of the S_2 value after the passage through M_2. On the other hand, if the relaxation time is very small with respect to the time separation of the two measurements, or if the separated spin-up and spin-down wave packets of particle 2 recombine, then the subsequent M_1' measurement will not be correlated to the M_2 measurement owing to the loss of the splitting characteristics after the M_2 measurement although, at M_1', S_1 and S_2 are still singlet correlated.

Let us once more stress that since action-at-a-distance is limited to space-like directions, any nonlocal influence exists between space-like separated particles and not between time-like separated measurements. It is really incomprehensible to think of measurement results as being influenced by the preexisting settings of a measuring device supposed to operate in the future light cone of an actually occurring measurement on one of the particles.

The Sutherland paradox is really a paradox of the quantum measurement theory and its shortcomings; it is not real but only conceptual. It is tied to the basic philosophical assertion that reality is created by a measurement which also lies at the origin of the completely unfounded, and in our opinion unjustified, extrapolations of the time-retroactive type. On the other hand, our model meets the requirement imposed by Sutherland, namely, a different behavior for time-like and space-like separations between the measurement processes, because it is not the measurement processes which are nonlocally connected but the particles themselves. Moreover, our model is plausible because it does not rest on an *a priori* assumption but deduces the results out of a nonlocal causal action-at-a-distance mechanism. Finally, it does not contradict quantum mechanics, as wrongly implied by Sutherland,[24] but only certain assertions of the highly controversial topic of "quantum measurement theory"—a common label for a set of mostly controversial, and in any case incoherent, calculatory recipes with one common denominator: the "reality creation by measurement" assumptions.

6. The Physical Origin of Nonlocality: Stochastic Motions in the Dirac Aether

The notion of the existence of nonlocal correlations between physical observables and, moreover, the idea of attributing a physical reality to quantum ψ waves, advocated by the pilot-wave interpretation of de Broglie, inevitably introduces the concept of an underlying medium as a possible carrier of these phenomena. The problem is to reconcile this need for a subquantal medium, with the relativity requirement of space–time isotropy in the light cone. The solution was found by Dirac.[27] He observed that the rejection of the aether on the basis of relativity theory applies only to the classical concept of an aether. A quantum aether is necessarily subject to uncertainty relations and its velocity, at a certain space–time point, will not be a well-defined quantity but, rather, will obey a probability-distribution law according to $|\psi|^2$ of the aether. Dirac then assumed the existence of a wave function which makes all values of the aether velocity equiprobable and showed that the perfect vacuum state, which results from it, is in accordance with the principle of relativity. One can, thus, construct this covariant aether vacuum by requiring that the four-momenta of the particles of Dirac's vacuum are uniformly distributed over mass hyperboloids: $P_\mu P^\mu = m^2 c^2$ and, in particular, by assuming that in the ground state all negative-energy states are filled while the positive-energy hyperboloid contains empty states (see Figure 10). Any Lorentz transformation thus leaves both this hyperboloid, and the assumed uniform-state distribution on it, unchanged. Futhermore, it has been shown[28] that the uniform-state distribution implies a nonuniform energy distribution, in the sense that it favors the momenta close to the light cone: In Dirac's aether distribution the weight of the almost light-like four-momenta is thus predominant.

The main problem in this context is to specify the reaction of Dirac's vacuum when a positive-energy particle is introduced into it. One has, of course, to introduce a specific mechanism for this interaction; but one already knows the results of this interaction: it is reflected in the behavior predicted by quantum laws for microphenomena which reproduces, on a higher level, the action of the subquantal medium on quantum entities. As for an explicit model for this interaction, several proposals already exist in the literature: A mechanism of momentum exchange by simultaneously conserving energy was proposed by Cufaro and Vigier,[28] and a complex aether model, in the form of a superfluid state of particle/antiparticle pairs, was introduced by Sudarshan, Sinha et al., and Vigier et al.[29-31] Futhermore, the structure of the subquantal medium has been shown to provide us with

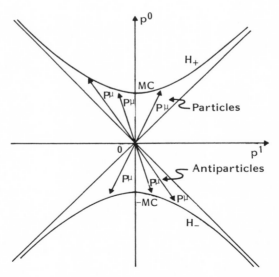

Figure 10. The momentum distribution for particles and antiparticles in Dirac's covariant aether for a given mass value (or mass shell).

an H-type theorem which ensures the stability of quantum mean distributions despite temporary deviations from it due to perturbative effects.[32,33]

Apart from having opened this new direction of research (which is still in its early stages but holds great promise), the hypothesis of an underlying random covariant vacuum has one very important consequence. It is apparent that a particle, inserted into this medium, is subjected to a random process which is a consequence of random collision processes, analogous to classical Brownian motion. This is not merely conjecture but a point of fact, because Nelson[34] has shown that the Schrödinger equation can be derived from a stochastic process, provided that specific assumptions are made about this process. These specific assumptions deviate from the self-evident scheme of Brownian theory, and imply a strange behavior of the random component of the motion which does not fit into an ordinary random process scheme. In ordinary Brownian theory, drift and random forces are added together in the equation of motion, i.e.,

$$(D_d D_d + D_r D_r)x = 0 \qquad (22)$$

while in the present case they must be subtracted, so that we get

$$(D_d D_d - D_r D_r)x = 0 \qquad (23)$$

where D is a derivative and the subscripts d and r denote "drift" and "random," respectively. One further puzzle remains in the following sense: If one assumes a local arbitrary collision model for a particle ensemble, then the statistical behavior of the ensemble is reproduced by Maxwell–Boltzmann statistics.[35] These puzzles, although formally resolved, cannot be answered in a physically satisfactory way, in the frame of Schrödinger quantum mechanics.

The real breakthrough in the understanding of the physical context of the quantum stochastic process was achieved in parallel research in the relativistic domain.[36] In particular, a nontrivial extension of the Markov process property,[37] according to which the future and past are disconnected if the present is known, was only achieved at the price of introducing apparent space-like motions. Furthermore, Vigier[38] showed that the specific form of the drift- and random-force addition law in equations (22) and (23) is due to a requirement of sign inversion under proper-time reversal, i.e., a characteristic particle/antiparticle symmetry, and that the difference between the additive and subtractive forms of equations (22) and (23) lies in a restriction to light-cone processes in the former form, while the latter form contains space-like contributions as well (see Figure 11). What is the consequence of all this? The evident implication is that quantum stochastic

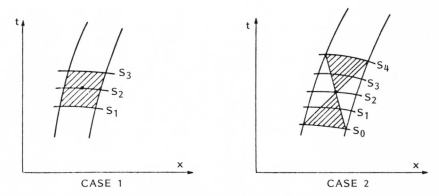

Figure 11. In case (a) the stochastic processes contributing to the force law come from space-like and time-like separated regions while in case (b) the contributions are restricted to the volume inside the light cone.

processes contain apparent space-like contributions which are due to particle/antiparticle transition processes (see Figure 12). A Feynman zigzag due to a creation/annihilation process yields this apparent space-like motion while, at the same time, no element of the process leaves the light cone. Finally, the quantum-statistics puzzle has been clarified because it has been shown[39] that Bose–Einstein or Fermi–Dirac statistics can be deduced from Maxwell–Boltzmann distributions if the probability weight of the phase-space states are not constant but random, a fact established by the quantum potential. It therefore seems that the quantum-mechanical formalism can be deduced from a generalized stochastic mechanics endowed with nonlocality, and no serious obstacle has been reported (up to now) in the efforts to extend this approach to different applications in the relativistic domain.[40]

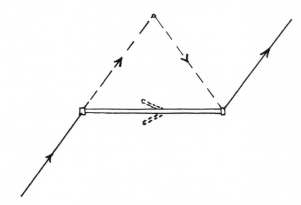

Figure 12. Decomposition of an apparent space-like motion in a particle/antiparticle transition process (dashed line).

It is now quite clear that the vacuum, in the form of Dirac's random covariant aether, is a mixture of particles and antiparticles and that this additional degree of freedom lies at the origin of nonlocal correlations and action-at-a-distance phenomena. Furthermore, it allows us to understand the quantum evolution as a generalized stochastic process where particle/antiparticle symmetries play an essential role in its constitution. Of course, several problems still remain unsolved, for example, a specific model for the superfluid medium or the introduction of some alternative mechanism. But we believe that the fundamental steps have already been taken toward a real space-time approach to quantum mechanics, i.e., in the field that we tend to label, schematically, as the "causal stochastic interpretation" of quantum mechanics.

7. Conclusions

The EPR paradox, originally proposed to demonstrate the incompleteness of quantum mechanics, has thus followed a quite remarkable trajectory. After being thought of as yet another solved problem, by the Copenhagen orthodoxy, immediately following Bohr's reply in 1935, it has reemerged several decades later to find itself in the middle of unceasing scientific debate. Apart from the refined experiments that have been performed in connection with it and the problem of nonlocality or action-at-a-distance that it revealed, it has already achieved the major goal of every scientific argument. Namely, it initiated a discussion just where the matter seemed to be settled, and forced physicists to analyze a situation which went beyond the "nonseparability" curtain. While quantum incompleteness is generally, explicitly or implicitly, admitted, the proponents of strict quantum agnosticism now try to avoid the main consequence of the EPR argument by insisting on the confirmation of quantum predictions by the Aspect experiments. They thus attempt to avoid the new reality created by the research of the last few decades (and the EPR-type experiments), namely, that alternative theories and concepts have emerged which do not refute quantum mechanics but are able to reproduce it and can still go beyond it in the limit. This Copenhagen "impossibility" dogma was first successfully opposed (in the form of an erroneous theorem of Von Neumann) in connection with the de Broglie pilot-wave theory but it failed to prevent the revival of the subject in the form of the quantum-potential model of Bohm, its relativistic extension and nonlocal implications, the new vacuum theory of Dirac's aether, the stochastic quantization method, as well as other alternative approaches. This recent revival which is so scientifically encouraging and theoretically promising, owes much to the EPR trajectory and a basic element contained in it which we could summarize as follows: *Scientifically*

complete physical theories are the (temporary) product of (philosophically)
completely self-satisfied physicists.

References

1. A. Einstein, B. Podolsky, and N. Rosen, *Phys. Rev.* **47**, 777 (1935).
2. A. Aspect, G. Grangier, and G. Roger, *Phys. Rev. Lett.* **47**, 460 (1981), **49**, 91 (1982); A. Aspect, J. Dalibard, and G. Roger, *Phys. Rev. Lett.* **49**, 1804 (1982).
3. T. W. Marshall, E. Santos, and F. Selleri, *Phys. Lett.* A **98**, 5 (1983); T. W. Marshall, *Phys. Lett.* A **99**, 163 (1983); A **100**, 225 (1984); M. Ferrero and E. Santos, *Phys. Lett.* A **108**, 373 (1985); A. Garuccio and F. Selleri, *Phys. Lett.* A **103**, 99 (1984); F. Selleri, *Phys. Lett.* A **108**, 197 (1985).
4. O. Costa de Beauregard, *Compt. Rend. Acad. Sci. Paris* **236**, 1632 (1953); *Nuovo Cim.* B **42**, 41 (1977), B **51**, 267 (1979); H. Stapp, *Nuovo Cim.* B **29**, 270 (1975); W. C. Davidon, *Nuovo Cim.* B **36**, 34 (1976); J. Rayski, *Found. Phys.* **9**, 217 (1979); J. G. Gramer, *Phys. Rev.* D **22**, 362 (1980); C. W. Rietdijk, *Found. Phys.* **8**, 615 (1978); **11**, 783 (1981); R. I. Sutherland, *Int. J. Theor. Phys.* **22**, 377 (1983).
5. O. Costa de Beauregard, Proc. Int. Symp. on Foundations of Quantum Mechanics, Tokyo, 233 (1983).
6. O. Costa de Beauregard, *Ann. Fond. L. de Broglie* **6**, 329 (1981).
7. O. Costa de Beauregard, Causality as identified with conditional probability and quantal nonseparability, *Proc. Int. Conf. on Microphysical Reality and Quantum Formalism, Urbino, Italy, September 1985* (to appear).
8. O. Costa de Beauregard, *Lett. Nuovo Cim.* **29**, 551 (1980).
9. O. Costa de Beauregard, *Found. Phys.* **15**, 871 (1985).
10. O. Costa de Beauregard, *Lett. Nuovo Cim.* **25**, 91 (1979).
11. O. Costa de Beauregard, Comments on a recent article by R. I. Sutherland, preprint, IHP (Dec. 1985).
12. D. Bohm and B. J. Hiley, *Found. Phys.* **14**, 255 (1984).
13. M. Cini, *Nuovo Cim.* B **73**, 27 (1983).
14. C. Møller, *The Theory of Relativity*, Oxford University Press, London (1972).
15. Ph. Droz-Vincent, *Phys. Scr.* **2**, 129 (1970); *Ann. Inst. H. Poincaré* **27**, 407 (1977); **32**, 377 (1980); *Phys. Rev.* D **19**, 702 (1979); L. Bel, *Ann. Inst. H. Poincaré* **3**, 307 (1970); *Phys. Rev.* D **18**, 4770 (1978); A. Komar, *Phys. Rev.* D **18**, 1881 (1978); **18**, 1887 (1978); **18**, 3617 (1978); R. P. Garda, *Sov. J. Part. Nucl.* **13**, 179 (1982).
16. N. Cufaro-Petroni, Ph. Droz-Vincent, and J. P. Vigier, *Lett. Nuov. Cim.* **31**, 415 (1981).
17. D. J. Currie, T. F. Hordan, and E. C. G. Sudarshan, *Rev. Mod. Phys.* **35**, 350 (1963); D. J. Currie, *J. Math. Phys.* **4**, 1470 (1963); *Phys. Rev.* **142**, 817 (1966).
18. L. de Broglie, *La Mécanique Ondulatoire du Photon*, Gautheir-Villars, Paris (1940); L. Bass and E. Schrödinger, *Proc. Roy. Soc. London* A **232**, 1 (1955); S. Deser, *Ann. Inst. H. Poincaré* **16**, 79 (1972); M. Moles and J. P. Vigier, *Compt. Rend. Acad. Sci. paris* B **276**, 697 (1973).
19. C. Dewdney, P. R. Holland, A. Kyprianidis, and J. P. Vigier, *Phys. Rev.* D **31**, 2533 (1985).
20. N. Cufaro-Petroni and J. P. Vigier, *Lett. Nuovo Cim.* **26**, 149 (1979); *Phys. Lett.* A **88**, 272 (1982); Kh. Namsrai, *J. Phys.* A **14**, 1307 (1981).
21. N. Cufaro-Petroni and J. P. Vigier, *Phys. lett.* A **93**, 383 (1983).
22. F. Halbwachs, *Théorie Relativiste des Fluides à Spin*, Gauthier-Villars, Paris (1960).
23. N. Cufaro-Petroni, A. Garuccio, F. Selleri, and J. P. Vigier, *Compt. Rend. Acad. Sci. Paris* B **290**, 111 (1980).

24. R. I. Sutherland, *Nuovo Cim. B* **88**, 114 (1985).
25. C. Dewdney, P. R. Holland, and A. Kyprianidis, *Phys. Lett.* **119A**, 259 (1986); **121A**, 105 (1987).
26. C. Dewdney, P. R. Holland, and A. Kyprianidis, *J. Phys. A* **20**, 4717 (1987).
27. P. A. M. Dirac, *Nature* **168**, 906 (1951).
28. N. Cufaro-Petroni and J. P. Vigier, *Found. Phys.* **13**, 253 (1983).
29. K. P. Sinha, C. Sivaram, and E. C. G. Sudarshan, *Found. Phys.* **6**, 65 (1976); **6**, 717 (1976); **8** 823 (1978).
30. J. P. Vigier, *Lett. Nuovo Cim.* **29**, 467 (1980).
31. K. P. Sinha, E. C. G. Sudarshan, and J. P. Vigier, *Phys. Lett. A* **114**, 298 (1986).
32. D. Bohm and J. P. Vigier, *Phys. Rev.* **96**, 208 (1954); **109**, 1882 (1958).
33. A. Kyprianidis and D. Sardelis, *Lett. Nuovo Cim.* **39**, 337 (1984).
34. E. Nelson, *Phys. Rev.* **150**, 1079 (1966).
35. Z. Maric, M. Bozic, and D. Davidovic, Randomness and determinism in the kinetic equations of Clausius and Boltzmann, Proc. of the Boltzmann Meeting, Vienna (1981).
36. W. Lehr and Park, *J. Math. Phys.* **18**, 1235 (1977); J. P. Vigier, *Lett. Nuovo Cim.* **24**, 258; 265 (1979); N. Cufaro-Petroni and J. P. Vigier, *Int. J. Theor. Phys.* **18**, 807 (1979); Kh. Namsrai, *Found. Phys.* **353**, 731 (1980).
37. F. Guerra and P. Ruggiero, *Lett. Nuovo cim.* **23**, 529 (1979).
38. J. P. Vigier, *Astr. Nachr.* **303**, 55 (1982).
39. A. Kyprianidis, D. Sardelis, and J. P. Vigier, *Phys. Lett. A* **100**, 228 (1984); N. Cufaro-Petroni, A. Kyprianidis, Z. Maric, D. Sardelis, and J. P. Vigier, *Phys. Lett. A* **101**, 4 (1984).
40. N. Cufaro-Petroni, C. Dewdney, P. R. Holland, A. Kyprianidis, and J. P. Vigier, *Phys. Rev. D* **32**, 1378 (1985).

12

Particle Trajectories and Quantum Correlations

C. Dewdney and P. R. Holland

1. Introduction

In this paper we present a series of computer calculations carried out in order to demonstrate exactly how the causal interpretation works for two-particle quantum mechanics. In particular we show how the causal interpretation can account for the essential features of nonrelativistic, two-particle quantum mechanics in terms of well-defined, correlated, individual particle trajectories and spin vectors. We demonstrate exactly how both quantum statistics and the correlations observed in Einstein–Podolsky–Rosen (EPR) experiments can be explained in terms of nonlocal quantum potentials and nonlocal quantum torques which act on the well-defined individual particle coordinates and spin vectors.

Quantum mechanics only presents great difficulties for those who believe that the task of physics is to describe the structure of the material world. For quantum phenomena seem to defy the imagination, and our intuitive notions about how matter behaves, structured by classical physics, do not appear to serve as useful guides when attempting to conceive the structure of the quantum world. The fact that "classical" notions of the world, instead of clarifying our experience only lead to ambiguity when we attempt to conceive what may lie beyond the statistical predictions of the theory, reflects the deep crisis that quantum mechanics has brought about.

C. Dewdney • Department of Applied Physics and Physical Electronics, Portsmouth Polytechnic, Portsmouth PO1 2DZ, England, United Kingdom.   P. R. Holland • Henri Poincaré Institute, 75231 Paris Cedex 05, France.

Of course it is fair to say that many physicists, who have "learned to stop worrying and love the statistics," would deny the existence of such a crisis. If pressed with questions of interpretation these physicists tend to resort to some variation of the Copenhagen interpretation. But how many are really prepared to accept the consequences and to give up any possibility of understanding the statistical predictions of quantum mechanics in terms of some underlying reality? How many are satisfied by Bohr's resolution of the difficulties in terms of a particular restrictive epistemology, that is, a particular opinion about how we come to know and what it is possible to know; or Wigner's idea that consciousness must be introduced in order to make anything definite; or Everett's idea of multiple splitting universes. Is quantum mechanics just an abstract formalism for connecting the statistical results recorded at the presumably unproblematic classical level, or is it indicative of a new order in the structure of the material world, that is, a new ontology?

2. The Causal Interpretation of Quantum Mechanics

One way of exploring a possible underlying structure is through the causal interpretation, proposed originally by de Broglie[1] and rediscovered by Bohm[2] in 1952. In this context the wave function is not held to exhaust the possibilities of description of individual systems, but does, as usual, encompass the limits of prediction. This approach allows a description of quantum phenomena in terms of well-defined individual particle motions; the statistics have no special status and neither does measurement. The disturbance caused by measurement can be analyzed, but not avoided, and the Heisenberg uncertainty relations are interpreted as statistical scatter relations which arise in the repeated measurement, on similarly prepared systems, of well-defined variables. It is assumed that a particle, an electron for example, has a well-defined position, momentum, and spin vector at all times. In addition the particle always has an associated ψ wave. The evolution of the particle coordinates and spin depends on the form, rather than the amplitude, of the associated wave as can be seen from the particle's equations of motion, deduced from the appropriate wave equation. These equations of motion correspond closely to those of classical mechanics but contain additional "quantum" terms. As we shall see these extra terms depend on derivatives of the associated ψ wave rather than its intensity. This means that the particular behavior that will be displayed by an individual particle depends not only on its precise initial coordinates, but also on the initial form and the development in time of its associated ψ wave. Of course the evolution of the ψ wave is determined by the wave equation appropriate to the situation, and the form of the wave will therefore come to reflect the structure of the particle's relevant environment. Through this description it can be seen that a particle, although separable from its environment for the purposes of analysis, in fact forms with its environment one undivided whole, as has been emphasized by Bohm.[3]

In a series of explicit calculations[4-6] for the single-particle case, we have demonstrated how the equations of motion deduced in the causal interpretation imply individual particle motions, determined by quantum potentials and quantum torques, which account completely for the observed quantum phenomena and serve to distinguish classical from quantum behavior.

The specific calculations presented here show clearly the new features that this description entails in the two-particle case. As we shall see the dependence of the individual particle's behavior on the quantum state of the system, $\psi(1, 2)$, implies in the two-particle case that, in addition to the dependence on the environment, an individual particle's behavior will depend on the coordinates of all the other particles constituting the system as well.

3. Nonlocality and the Causal Interpretation of Two-Particle Motion

The two-particle Schrödinger equation

$$i\hbar \frac{\partial \psi}{\partial t} = \left[-\frac{\hbar^2}{2m} (\nabla_1^2 + \nabla_2^2) + V \right] \psi$$

can also be interpreted within the framework of the causal interpretation. Writing

$$\psi(\mathbf{x}_1, \mathbf{x}_2, t) = R(\mathbf{x}_1, \mathbf{x}_2, t) \, e^{iS(\mathbf{x}_1, \mathbf{x}_2, t)/\hbar}$$

we find for two particles of equal mass

$$-\frac{\hbar^2}{2m} \frac{\nabla_1^2 R}{R} - \frac{\hbar^2}{2m} \frac{\nabla_2^2 R}{R} + \frac{(\nabla_1 S)^2}{2m} + \frac{(\nabla_2 S)^2}{2m} + V = -\frac{\partial S}{\partial t}$$

a two-particle Hamilton–Jacobi equation, with

$$\rho = R^2, \qquad \mathbf{V}_1 = \frac{\nabla_1 S}{m}, \qquad \text{and} \qquad \mathbf{V}_2 = \frac{\nabla_2 S}{m}$$

The continuity equation is

$$\frac{\partial \rho}{\partial t} + \nabla_1 \cdot (\rho \mathbf{V}_1) + \nabla_2 \cdot (\rho \mathbf{V}_2) = 0$$

Here V represents external potentials while ∇_1 and ∇_2 operate on coordinates 1 and 2, respectively. Evidently the quantum potential acting on particle one, say, depends not only on the coordinates of particle one but on those of particle two as well. The velocities also show this interrelationship. The additional terms in the Hamilton–Jacobi equation lead, as in the single-particle case, to nonclassical behavior in which even "classically" noninteracting particles can influence each other through the quantum potentials. Since, quantum mechanically, a collection of particles behave in a way that depends also on the form of the associated wave, they may obey a different type of statistics than a collection of noninteracting particles in the classical case.

3.1. Particle Motions and Quantum Statistics

In order to illustrate the conditions under which the correlation between the particles becomes significant, giving rise to quantum statistics, we consider the case of two particles in a harmonic-oscillator potential,[7] $V(x) = 0.5 m\omega^2$. A wave-packet solution may be constructed for the motion of a particle in such a potential which is, incidently, nondispersive:

$$\psi(x, t) = \exp(-i\omega t) \exp[-\tfrac{1}{2}(x - x_0 \cos \omega t)^2]$$

$$\times \exp\left[\frac{i}{2}(\tfrac{1}{2}x_0^2 \sin 2\omega t - 2x x_0 \sin \omega t)\right]$$

Assuming that there are two particles, one in each of the packets ψ_a and ψ_b, centered initially at $x_0 = -x_0 = 1.5$, in our arbitrary units in which $\hbar = 1$, $m = 0.5$, and $\omega = 1$, then there are three possible wave functions that may be written. These are

$$\phi_{MB} = \alpha_{MB}\psi_a(x_1, t)\psi_b(x_2, t)$$

$$\phi_{BE} = \alpha_{BE}[\psi_a(x_1, t)\psi_b(x_2, t) + \psi_b(x_1, t)\psi_a(x_2, t)]$$

$$\phi_{FD} = \alpha_{FD}[\psi_a(x_1, t)\psi_b(x_2, t) - \psi_b(x_1, t)\psi_a(x_2, t)]$$

where the αs are normalization coefficients.

Now in the first case the wave function is a simple product and it is fairly evident that the factorizable wave function enables the two-particle Schrödinger equation to be factored into two separate one-particle equations. The same is true of the equations of motion derived in the causal interpretation and hence each particle exhibits, independently, the motion of a single particle in a harmonic-oscillator potential. The trajectories are shown in Figure 1.

The other two wave functions correspond to those written when the two particles are said to be indistinguishable, being symmetric (the Bose–Einstein case) and antisymmetric (the Fermi–Dirac case). Of course the particles can only be considered to be indistinguishable when the two constituent wave packets overlap. When they do not, the particles are in principle distinguishable by their histories. In the causal interpretation, of course, the particles are always distinguishable, in analysis if not in practice, and hence the different statistics they obey cannot be accounted for by reference to indistinguishability. Instead, as is demonstrated here, the different statistics arise as a result of the development of the two-body quantum potentials which depend on the wave function assumed.

Figure 2 shows a set of correlated pairs of trajectories, for the symmetric wave function ϕ_{BE}, in which the initial position of particle one $x_1(0)$ is

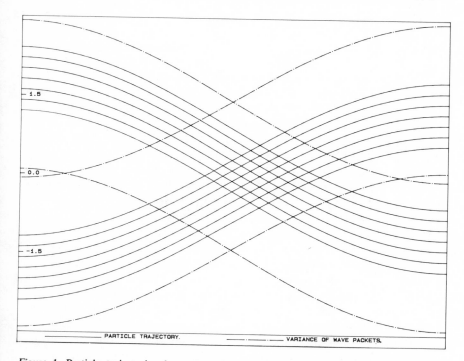

Figure 1. Particle trajectories for two particles in a harmonic-oscillator potential with a factorizable wave function. Maxwell–Boltzmann statistics (solid lines). The interrupted lines plot the variances of the individual wave packets.

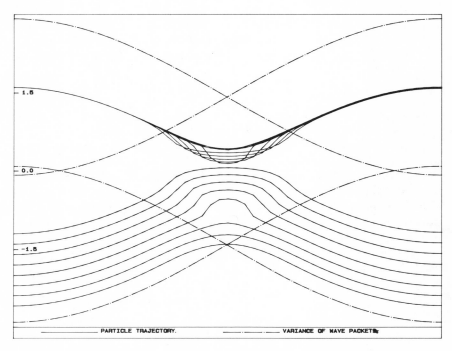

Figure 2. Correlated pairs of particle trajectories for two particles in a harmonic-oscillator potential, symmetric wave function. Initial position of particle one is the same for a range of initial positions of particle two. Bose–Einstein statistics.

chosen to be the same for each pair of trajectories while the initial position of particle two $x_2(0)$ is different for each pair. It is very clearly shown in the form of the correlated trajectories that the motion of each particle depends on the position of the other. To show which of the trajectories are the correlated pairs we plot in Figure 3 the trajectories in the region of overlap. The position of the numbers at the right-hand side of the plot indicates the order of trajectories at the center. In Figure 4 we plot the correlated trajectories that arise from the antisymmetric wave function ϕ_{FD}. The initial positions are chosen as for the symmetric case and clearly the different wave function introduces a different form of correlation in which, on the average, the particles tend to stay further apart. In Figure 5 we plot the quantum potential acting on particle one (Q_1), after one eighth of a cycle, for a fixed position of x_2 ($=-1.0$ in our arbitrary units $h = 1$, $m = 0.5$). The solid line represents the quantum potential arising from the antisymmetric wave function and the interrupted line, the quantum potential arising from the symmetric wave function. Clearly at this stage particle one will not be affected by the presence of particle two, since the quantum potential

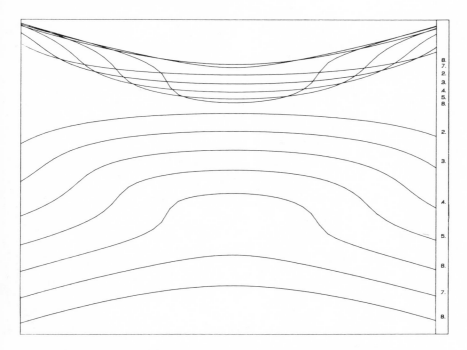

Figure 3. Correlated pairs of trajectories for two particles in a harmonic-oscillator potential, symmetric wave function, in the region of overlap. The numbering of the upper trajectories gives the vertical ordering at the center of the plot.

acting on particle one is a constant in the region where it has a nonnegligible probability density. Figure 6 depicts the same thing at one quarter of a cycle, when the spatial wave functions are superimposed. Figure 7 shows, for the antisymmetric case, the effect of varying the position of particle two $x_2 = -1.5$, -1.0, -0.5 on the quantum potential acting on particle one. Similarly, Figure 8 shows this affect for the symmetric case.

From these illustrations it can be seen that the magnitude of the nonlocal correlation is only appreciable in the region of overlap between the two wave packets ψ_a and ψ_b. This is understandable, since the different statistics arise from the interference terms. In this case then the strength of the nonlocal interaction depends on the amount of overlap between the two wave packets. This is interesting since it tells us that the fate of particle one, for example, will not be appreciably altered by what happens to particle two when we can be sure that they are separated. In the example studied here then we may say that, when the particles can be distinguished, according to the definitions of the usual approach, we can expect no nonlocal correlation of their trajectories, while when they become indistinguishable, by

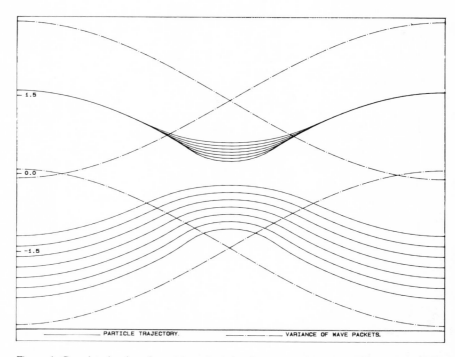

PARTICLE TRAJECTORY.   VARIANCE OF WAVE PACKETS.

Figure 4. Correlated pairs of particle trajectories for two particles in a harmonic-oscillator potential, antisymmetric wave function. Initial position of particle one is chosen to be the same for a range of initial positions of particle two.

virtue of the overlap of the two wave packets, nonlocal correlation will arise. According to this particular example nonlocal interaction between the particles, in the causal interpretation, only exists in those cases in which the particles cannot be said to be definitely separated in space according to the usual interpretation.

3.2. EPR Correlations

In the example of correlated particle motion suggested by Einstein, Podolsky, and Rosen[8] the wave function that they proposed is

$$\psi = \int dk_1 \int dk_2 \, \delta(k_1 + k_2) \, e^{2\pi i(k_1 x_1 + k_2 x_2)} \, e^{-2\pi i k_2 d} = \int dk_1 \, e^{2\pi i k_1 (x_1 - x_2 + d)}$$

This wave function implies that measurement of the momentum of one of the particles allows us to deduce that of the other: $k_1 = -k_2$. Also, measurement of the position of one of the particles allows us to deduce the position of the other: $x_1 = x_2 - d$. The essential point of the EPR argument, at that

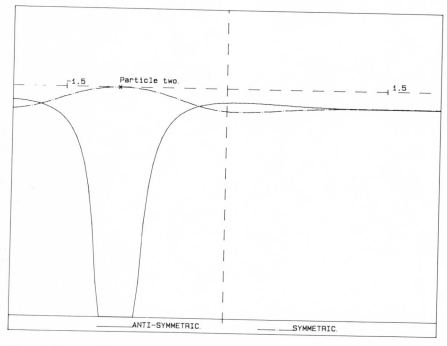

Figure 5. The quantum potential acting on particle one, when $x_2 = -1.0$ after one eighth of a cycle. Solid line for the antisymmetric wave function and interrupted line for the symmetric wave function.

time, was that we may choose whether to measure the momentum or the position of particle one, say, and hence infer the value of the momentum or position of particle two, without interacting with it at all. EPR then argued that if we can, in this way, predict with certainty the values of these quantities of particle two, then they must be preexistent. In other words, the particles must possess a well-defined position and momentum, even before the measurements are carried out. Furthermore, the above reasoning is independent of how far apart the particles actually are situated; the correlation should exist to infinity. Their conclusion was that the quantum-mechanical description of reality through the wave function is incomplete.

From the point of view of our discussion of the two particles in the harmonic-oscillator potential we can see that, in the EPR case, the correlation between the particles exists to infinity, as a result of the particular wave function assumed. In particular the particles, in the state defined by them, cannot be considered to be separated in space, in the usual approach, and hence a nonlocal connection can be expected to exist in the approach of the causal interpretation. It can be seen then that the use of spatially infinite

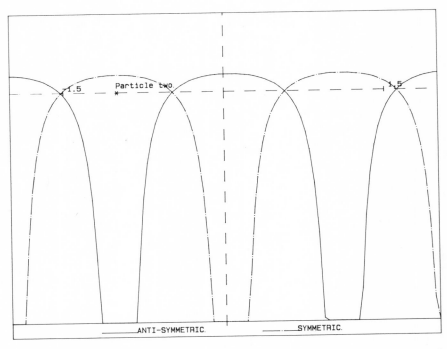

Figure 6. The quantum potential acting on particle one, when $x_2 = -1.0$ after one quarter of a cycle. Solid line for the antisymmetric wave function and interrupted line for the symmetric wave function.

waves by EPR is responsible for the distance independence of the correlation. It was pointed out by de Broglie[1] that it is this feature that enabled Bohr[9] to reply to the problem posed by the EPR example. Further on we examine the same problem from the point of view of the version of the EPR argument proposed by Bohm, dealing with a correlation between the particles' spins. We conclude that even when the particles' spatial packets do not overlap, a nonlocal correlation may still exist between the spins and hence positions. In the next section we discuss the interpretation of spin in the single-particle case to facilitate the discussion in the two-particle EPR example.

4. The Causal Interpretation of Spin

In the usual interpretation spin is simply treated as an empirically required addition to the angular momentum. It is argued that no intuitive model can possibly provide an understanding of phenomena associated

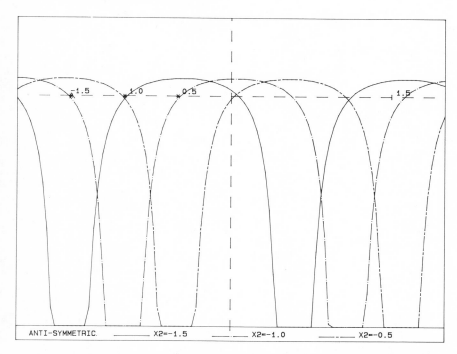

Figure 7. The quantum potential acting on particle one after one quater of a cycle, antisymmetric wave function, plotted for $x_2 = -1.5$, -1.0, and -0.5.

with the spin. (This is the case for all quantum phenomena in the usual approach.) Since the operators for the components of the spin along three mutually perpendicular directions $(\hat{s}_x, \hat{s}_y, \hat{s}_z)$ do not commute, it is argued that they cannot be simultaneously well defined. As is usually the case the value of a quantity, in this case the spin component, does not become definite until a measurement "throws" the system into an eigenstate of the observable being measured, simultaneously throwing the values of noncommuting operators into an indefinite state.

The causal interpretation can be extended to include the description of spin, as has been demonstrated by Bohm *et al.*[10] and also by Takabayasi.[11] More recently we have shown how the causal interpretation of spin actually works in a series of specific calculations, which show explicitly the continuous motion of the well-defined spin vector during a spin-superposition experiment and during the passage through a Stern–Gerlach measuring device.[12-14]

In the causal interpretation, the Pauli spinor is interpreted as defining the state of rotation of a body in terms of the Eulerian rotation angles

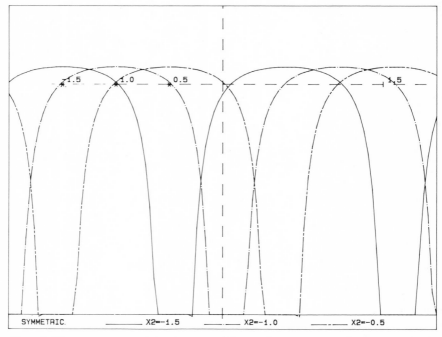

Figure 8. The quantum potential acting on particle one after one quarter of a cycle, symmetric wave function, plotted for $x_2 = -1.5$, -1.0, and -0.5.

θ, ϕ, χ, relative to a standard spinor $\binom{1}{0}$ defining a z direction, according to

$$\psi = R \, e^{i\chi/2} \begin{pmatrix} \cos \dfrac{\theta}{2} \, e^{i\phi/2} \\ i \sin \dfrac{\theta}{2} \, e^{-i\phi/2} \end{pmatrix} \tag{1}$$

The spin vector is defined to be

$$\mathbf{s} = \frac{\hbar}{2} \frac{\psi^\dagger \boldsymbol{\sigma} \psi}{\psi^\dagger \psi}$$

The Pauli spinor evolves according to the equation

$$i\hbar \frac{\partial \psi}{\partial t} = \left[-\frac{\hbar^2}{2m} \left(\nabla - \frac{ie}{\hbar c} \mathbf{A} \right)^2 + \mu \mathbf{B} \cdot \boldsymbol{\sigma} + eA_0 + V \right] \psi \tag{2}$$

where $\mathbf{A}$ is the external vector potential, $\mathbf{B} = \nabla \times \mathbf{A}$, A_0 is the external electric potential, and V any other scalar potential.

The velocity of the particles is given by

$$\mathbf{v} = \frac{\hbar}{m}\left(\frac{\nabla\chi}{2} + \cos\theta\frac{\nabla\phi}{2} - \frac{e}{\hbar c}\mathbf{A}\right)$$

The equations of motion can be derived by direct substitution of equation (1) in (2) and yields

$$\frac{\hbar}{2}\left(\frac{\partial\chi}{\partial t} + \cos\theta\frac{\partial\phi}{\partial t}\right) + \tfrac{1}{2}mv^2 + Q + Q_S + \frac{2\mu}{\hbar}\mathbf{B}\cdot\mathbf{s} + eA_0 + V = 0 \qquad (3)$$

a Hamilton–Jacobi equation, where

$$Q = -\frac{\hbar^2}{2m}\frac{\nabla^2 R}{R}$$

is the quantum potential and

$$Q_S = \frac{\hbar^2}{8m}[(\nabla\theta)^2 + \sin^2\theta(\nabla\phi)^2]$$

is a spin-dependent addition. The total energy is given by

$$\frac{\hbar}{2}\left(\frac{\partial\chi}{\partial t} + \cos\theta\frac{\partial\phi}{\partial t}\right)$$

The equation of motion of the spin vector can be written in the form

$$\frac{d\mathbf{s}}{dt} = \frac{1}{m\rho}\mathbf{s}\times\partial_i(\rho\partial_i\mathbf{s}) + \frac{2\mu}{\hbar}\mathbf{B}\times\mathbf{s} \qquad (4)$$

where the first term on the right-hand side is an additional quantum torque and $\rho = R^2$. The continuity equation is

$$\frac{\partial\rho}{\partial t} + \nabla\cdot(\rho\mathbf{v}) = 0 \qquad (5)$$

From the equations of motion (3), (4), and (5) it can be seen that even in the absence of magnetic fields, free spinning particle trajectories will not be the same as Schrödinger trajectories, nor will the spin vector orientation remain constant if the particle is in a nonstationary spin state.

4.1. Spin Measurement

In the usual approach it is said that when a component of the spin is measured, along z say, the system is "thrown" into an eigenstate of the corresponding operator. However, the effect of such a measurement is to disturb the other x and y components which become indefinite, and hence the spin cannot be considered to be well defined even after the measurement has been carried out. In the following we demonstrate how the causal interpretation can account for these features of the quantum description in terms of a well-defined but continuously variable spin vector. The process has been described in detail elsewhere[13] and here we simply review the results.

We represent the initial state of an atom with angular momentum $h/2$ by the spinor wave function

$$\psi_0(z) = f_0(z)(c_+ u_+ + c_- u_-)$$

where c_+ and c_- are unknown real coefficients and $f_0(z)$ a localized packet, which we assume to be of Gaussian form. On writing

$$c_+ = |c_+| \, e^{iS_+/\hbar} \qquad \text{and} \qquad c_- = |c_-| \, e^{iS_-/\hbar}$$

the initial spin-vector orientation is

$$\theta_0 = \cos^{-1}(|c_+|^2 - |c_-|^2)$$

$$\phi_0 = \frac{S_+ - S_-}{\hbar} + \pi/2$$

$$\chi_0 = \frac{S_+ + S_-}{\hbar} - \pi/2$$

Solution of the Pauli equation with approximate interaction Hamiltonian H_I yields $\psi(z, t)$. The velocities and orientations of a set of representative single-particle motions can then be calculated for various choices of the parameters c_+ and c_-. Figures 9, 10, and 11 show the spin-dependent trajectories, the field of orientations $\theta(z, t)$ and $\phi(z, t)$, respectively, for the choice $|c_+| = |c_-| = (0.5)^{1/2}$. This illustration demonstrates explicitly that it is possible to describe the process in terms of well-defined particle motions. Clearly the quantum torque aligns the particle's spin vector parallel or antiparallel to the field. Which of the two alternatives is in fact realized in a particular case depends on the actual initial values of the hidden variables of both the system (here the spin-vector direction) and the apparatus (here

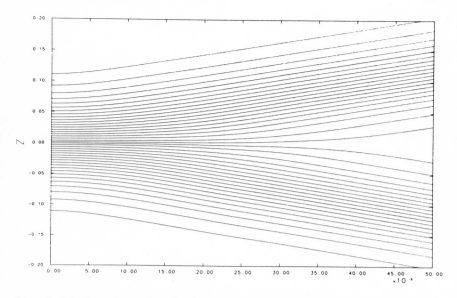

Figure 9. Spin-dependent trajectories, from a Gaussian distribution of initial positions, at the exit of a Stern–Gerlach field oriented in the *z* direction. Initial spin-vector orientation perpendicular to the field.

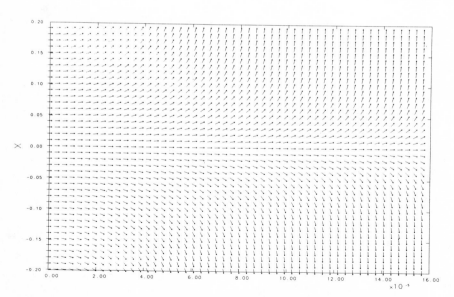

Figure 10. The field of orientation $\theta(z, t)$ corresponding to the trajectories of Figure 5.

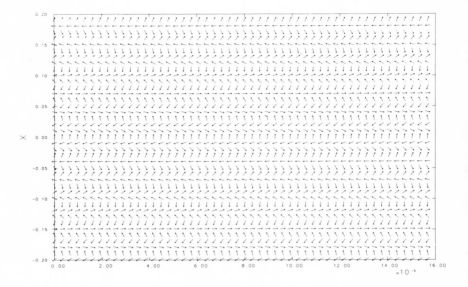

Figure 11. The field of orientations $\phi(x, t)$ corresponding to the trajectories of Figure 5.

the particle position, since by observing this we can deduce the value of the spin). In the particular example described here the spin-dependent quantum potential splits the packet at its center, and as the two packets separate the quantum torque rotates the spin vector to align parallel or antiparallel to the analyzing field. In the causal interpretation, after the measurement in the z direction $s_x = s_y = 0$ and $s_z = \frac{1}{2}$.

Evidently, in this description the outcome of the measurement is related deterministically to the actual (uncontrollable) initial values of the hidden variables, but measurement does not simply reveal them. Rather the evolution of the spin variable is correlated with the evolution of the apparatus variable (the particle position), the correlation being introduced by the inhomogeneous field. It is the existence of well-defined variables in the system and in the apparatus, evolving according to the causal equations of motion, that ensures unique initial conditions lead to unique and well-defined outcomes. In this way it can be seen that wave-packet collapse is a redundant hypothesis.

Clearly there is nothing special or extraordinary about measurement; it is simply a particular case of the correlated evolution of the variables of two systems according to the laws of quantum mechanics. During this evolution the apparatus variable enters the space of one of a series of unambiguously distinguishable states, each of which is correlated with a different state of the system under observation.

5. Nonlocal Spin Correlations in the Two-Particle Case

The description given above of spin-half particles can be extended to the two-particle case. This enables a description to be given of the EPR experiment in the form proposed by Bohm,[15] in which the correlations are between spin measurements carried out on each of two particles in a singlet state. Bohm's version of the experiment has been of great historical importance and the results obtained by Aspect *et al.* can be easily discussed in terms of spin rather than polarization. Although it is still possible to question whether the experiments of Aspect *et al.* have finally demonstrated the existence of nonlocal phenomena in physics, this has become increasingly difficult. Indeed it only seems to be possible by denying the validity of many-body quantum theory.

We have already seen in the foregoing that the particle motions calculated in the causal interpretation of many-body phenomena exhibit nonlocal correlations under certain circumstances. In particular we saw that when the wave function can be written as a sum of products of individual spatial wave functions, nonlocal correlation will only produce observable results when these functions overlap. We now demonstrate that the causal interpretation of the two-body Pauli equation naturally implies that nonlocal correlations will indeed exist between two spin-half particles in the singlet state. We present plots of the correlated trajectories of the particles and the evolution of their spin vectors as a result of the measurement of the spin components of each. Our conclusion is that although the model presented here is an idealized one, it provides an insight into the meaning of nonlocality in terms of an underlying causal process, in a way that no other interpretation of quantum mechanics has managed to do.

5.1. The Two-Body Pauli Equation

We consider a system of two spin-half particles of masses m_1, m_2 and charges e_1, e_2 respectively, which are placed in external electromagnetic fields and possibly interact. The two-body Pauli equation is

$$i\hbar \frac{\partial \psi}{\partial t} = \left\{ -\frac{\hbar^2}{2m_1} \left[\nabla_1 - \frac{ie_1}{\hbar c} \mathbf{A}_1(\mathbf{x}_1, \mathbf{x}_2) \right]^2 \right.$$

$$\left. - \frac{\hbar^2}{2m_2} \left[\nabla_2 - \frac{ie_2}{\hbar c} \mathbf{A}_2(\mathbf{x}_1, \mathbf{x}_2) \right]^2 + W_1 + W_2 + V \right\} \psi$$

where $\mathbf{x}_1$ and $\mathbf{x}_2$ are the coordinates of particles 1 and 2 while

$$\psi = \psi_{ab}(\mathbf{x}_1, \mathbf{x}_2, t)$$

is the wave function of the system; $V = V(\mathbf{x}_1, \mathbf{x}_2, t)$ is the total external plus internal scalar potential,

$$W_1 = W_1(\mathbf{x}_1, \mathbf{x}_2, t) = \mu_1 \mathbf{H}_1(\mathbf{x}_1, \mathbf{x}_2) \cdot \boldsymbol{\sigma}_1$$

$$W_2 = W_2(\mathbf{x}_1, \mathbf{x}_2, t) = \mu_2 \mathbf{H}_2(\mathbf{x}_1, \mathbf{x}_2) \cdot \boldsymbol{\sigma}_2$$

where μ_1, μ_2 are the particles' magnetic moments with

$$\mathbf{H}_1 = \nabla_1 \times \mathbf{A}_1 \qquad \text{and} \qquad \mathbf{H}_2 = \nabla_2 \times \mathbf{A}_2$$

and $\boldsymbol{\sigma}_1, \boldsymbol{\sigma}_2$ are two sets of Pauli matrices which commute and operate independently.

Writing

$$\psi_{ab}(\mathbf{x}_1, \mathbf{x}_2, t) = R(\mathbf{x}_1, \mathbf{x}_2, t) \, e^{iS(\mathbf{x}_1, \mathbf{x}_2, t)/\hbar} \phi_{ab}$$

where R and S are real amplitude and phase functions respectively and $\phi^{\dagger}\phi = 1$, we may deduce a Hamilton–Jacobi equation

$$\frac{\partial S}{\partial t} - i\hbar \phi^{\dagger} \frac{\partial \phi}{\partial t} + \tfrac{1}{2} m_1 \mathbf{v}_1^2 + \tfrac{1}{2} m_2 \mathbf{v}_2^2 + Q_1 + Q_2 + H_{1S} + H_{2S}$$

$$+ \frac{2}{\hbar} \mu_1 \mathbf{H}_1 \cdot \mathbf{S}_1 + \frac{2}{\hbar} \mu_2 \mathbf{H}_2 \cdot \mathbf{S}_2 + V = 0$$

and a continuity equation

$$\frac{\partial \rho}{\partial t} + \nabla_1 \cdot (\rho \mathbf{v}_1) + \nabla_2 \cdot (\rho \mathbf{v}_2) = 0$$

where $\rho = \psi^{\dagger}\psi = R^2$ is the configuration-space probability density,

$$\mathbf{v}_i = \frac{-i\hbar}{2m_i\rho} \psi^{\dagger} \overleftrightarrow{\nabla}_i \psi - \frac{e_i \mathbf{A}_i}{m_i c} = \frac{1}{m_i} \left(\nabla_i S - i\hbar \phi^{\dagger} \nabla_i \phi - \frac{e_i}{c} \mathbf{A}_i \right) \qquad i = 1, 2$$

are the velocities of the particles which contain spin-dependent contributions,

$$Q_i = -\frac{\hbar^2}{2m_i} \frac{\nabla_i^2 R}{R}$$

are the usual quantum potentials which arise in the two-body case,

$$H_{is} = -\frac{\hbar^2}{2m_i}[\nabla_i\phi^\dagger \cdot \nabla_i\phi + (\phi^\dagger\nabla_i\phi)^2] \qquad i = 1, 2$$

are spin-dependent additions to the quantum potentials, and

$$\mathbf{s}_i = \frac{\hbar}{2}\phi^\dagger\boldsymbol{\sigma}_i\phi = \frac{\hbar}{2}(\psi^\dagger\boldsymbol{\sigma}_i\psi)|\rho \qquad i = 1, 2$$

are the vectors which we shall adopt here to describe the local spin orientation of each particle. The total energy of the system

$$\frac{\partial S}{\partial t} - i\hbar\phi^\dagger\frac{\partial\phi}{\partial t}$$

is clearly spin-dependent. Each of the above functions depends on the coordinates of both particles, and it is simple to show that the trajectories and spin vectors of the two particles will only evolve independently when the wave function factorizes:

$$\psi_{ab}(\mathbf{x}_1, x_2, t) = \psi_{1a}(\mathbf{x}_1, t)\psi_{2b}(x_2, t)$$

5.2. EPR Spin Correlations

The basic setup is as shown in Figure 12. A pair of spin-half particles of mass m and magnetic moment μ are formed at 0 in a simultaneous eigenstate of the spin operator in the z direction $\frac{1}{2}\hbar(\sigma_{z_1} + \sigma_{z_2})$ and the total spin operator $\frac{1}{4}\hbar^2(\boldsymbol{\sigma}_1 + \boldsymbol{\sigma}_2)^2$ of eigenvalue zero. The particles separate in the y direction and pass through Gaussian slits oriented so as to produce packets in the directions of the analyzing fields of two identical Stern-Gerlach devices. The magnet 2 is set to measure spin in the z direction, and magnet 1 has been rotated counterclockwise through an angle δ about the y axis so that it has a gradient in the z' direction.

At the entrance to the fields the wave function is

$$\psi_0 = \frac{1}{\sqrt{2}}f_1(z_1')f_2(z_2)(u_+v_- - u_-v_+) \tag{6}$$

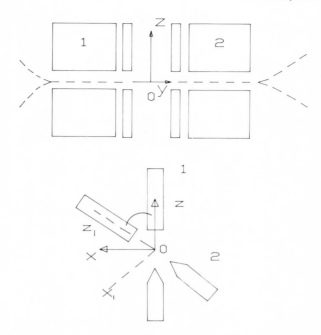

Figure 12. The experimental arrangement for the Einstein–Podolsky–Rosen experiment.

where $f_1(z_1')$ and $f_2(z_2)$ are normalized packets, z_1 and z_2 are the coordinates of particles 1 and 2 in the z' and z directions, respectively, and $\sigma_{z_1} u_\pm = \pm u_\pm$, $\sigma_{z_2} v_\pm = \pm v_\pm$. The state (6) predicts the following expectation value for the correlations of the spins measured in the z, z' directions:

$$\langle \sigma_{z_1'}, \sigma_{z_2} \rangle = -\cos \delta$$

In treating this problem we have suppressed the motion in the y direction, since this is not relevant to the measuring process. We only assume that the particles are sufficiently far apart on the y axis that they do not interact and the measuring devices cannot influence one another. As we saw in the single-particle case discussed above, the state before the measurement takes place is one in which the spin is independent of position. The Stern–Gerlach devices introduce couplings between the spins (the variables measured) and the particle positions (the apparatus coordinates). As in the single-particle case at the exit of each magnet, two superposed packets are formed which separate with time along the direction of the analyzing fields. The calculation of the motions when the fields are aligned along different directions are given in our paper.[16] Here we demonstrate explicitly the correlated particle motions which arise in the causal interpretation by presenting the results of these calculations in terms of correlated particle motions and spin-vector orientations.

If we write

$$f_1(z_1') = R_1(z_1') \, e^{iS_1(z_1')/\hbar} \qquad \text{and} \qquad f_2(z_2) = R_2(z_2) \, e^{iS_2(z_2)/\hbar}$$

then the velocities are simply given by

$$\mathbf{v}_1 = \frac{\nabla_1' S_1}{m} \qquad \text{and} \qquad \mathbf{v}_2 = \frac{\nabla_2 S_2}{m}$$

The spin vectors are $s_1 = s_2 = 0$. This clearly demonstrates the context-dependence of particle properties in the causal interpretation, since the individual spin vectors are here zero in the singlet state, something that cannot occur in the single-particle case.

In order to understand what happens during the measuring process, we consider first the simplest case, namely, a measurement on one side only. The calculation yields the result that at the exit of the field the particle undergoing measurement enters one of the separating packets, depending on its initial position on the z axis, under the influence of the total spin-dependent quantum potential, and its spin-vector component in the direction of the analyzing field changes continuously from 0 to $+\hbar/2$ or $-\hbar/2$ as the packets become separated in the z direction. Simultaneously, the spin-vector component in the z direction of the second particle not undergoing measurement changes in the opposite sense from 0 to $-\hbar/2$ or $+\hbar/2$ as a result of the operation of the nonlocal quantum torque. That is, the spin of particle two depends on the position and hence spin of particle one. The velocities however in this case remain independent:

$$\mathbf{v}_2 = \frac{\nabla_2 S_2}{m} \qquad \text{and} \qquad \mathbf{v}_1 = \frac{|R_{1+}|^2 \nabla_1 S_{1+} + |R_{1-}|^2 \nabla_1 S_{1-}}{m(|R_{+1}|^2 + |R_{-1}|^2)}$$

The trajectory of particle two is unaffected by the measurement on particle one and the trajectory of particle one depends simply on local factors. If we were subsequently to measure the spin of particle two in the z direction, we would of course find the opposite result to that found for particle one. The trajectories and spin-vector magnitudes (indicated by the length of the arrow which always lies in the z direction) are shown in Figure 13. Consider now the case in which both Stern–Gerlach devices are operational and set to measure the spin component in the z direction on both particles simultaneously. The motion of each particle for any pair of trajectories depends sensitively on the choice of both initial positions at the entrance slits to the Stern–Gerlach devices. The calculation in this case yields the results plotted in Figure 14. These results were calculated by taking the initial position of particle one to be the same in each case and then calculating the correlated trajectories which develop for a representative range of initial positions of particle two. The situation when the initial position of particle one $z_{1,0}$ is chosen to be equal to that of particle two $z_{2,0}$ represents a bifurcation point (actually, a bifurcation line in configuration space). If $z_{2,0} < z_{1,0}$, then particle two has a negative velocity and its z spin component decreases form 0 to $-\frac{1}{2}$ while the correlated particle one has a positive velocity and its z spin component increases from 0 to $+\frac{1}{2}$, with a corresponding result if $z_{2,0} > z_{1,0}$. Clearly the fate of each particle depends sensitively on what happens to the other one. In Figure 15 we illustrate the same phenomenon with a different choice of the constant $z_{2,0}$.

The description given above demonstrates the manner in which the causal interpretation can account for EPR correlations. The description is based on that proposed by Pauli, however it must be born in mind that Pauli simply introduced the minimal extra mathematical structure necessary to deal with the statistics of phenomena dependent on spin. A more natural way to treat the spin is through an extension of the dependency of the

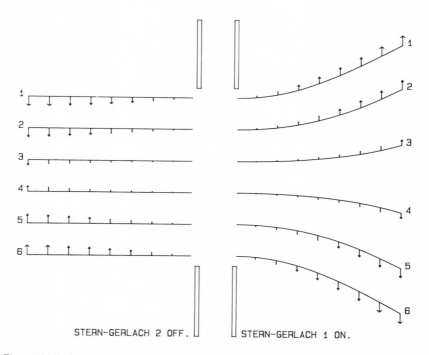

STERN–GERLACH 2 OFF.   STERN–GERLACH 1 ON.

Figure 13. Trajectories and correlated spin-vector orientations for two particles initially in a singlet state after the impulsive measurement of the z component of the spin of particle two only.

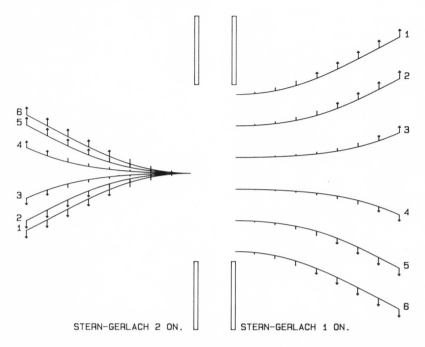

Figure 14. Correlated pairs of trajectories and correlated spin-vector orientations for two particles initially in a singlet state after the impulsive measurement of the spin in the z direction on both particles: $z_{1,0}$ = constant, $z_{2,0}$ variable.

Schrödinger wave function to encompass internal orientation coordinates. The Schrödinger equation must then include the contribution to the energy from the rotational motion of the particle and appropriate operators must be defined. A full description of this approach is beyond the scope of the present chapter, but we might mention here that in this more general approach the results of the Pauli theory are recovered as averages over the internal coordinates. The trajectories and spin-vector orientations plotted here then represent these averages over the internal coordinates and this leads to a more natural interpretation of the zero spin vector of each particle in the singlet state. In fact, in the more general theory although the individual spin vectors will not be zero in the singlet state, the average values however will be.

6. Conclusion

The illustrations presented here demonstrate clearly how the results of experiments in quantum mechanics, including EPR-type experiments, can

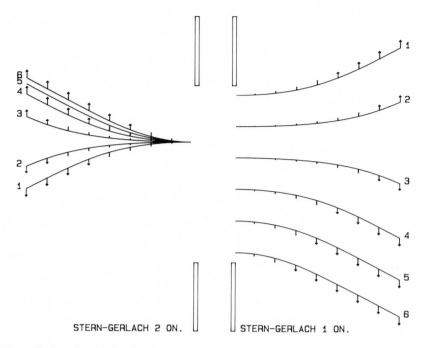

STERN–GERLACH 2 ON.   STERN–GERLACH 1 ON.

Figure 15. Correlated pairs of trajectories and correlated spin vectors for two particles initially in a singlet state after the impulsive measurement of the spin in the z direction on both particles: $z_{1,0}$ = a different constant to that of Figure 12, $z_{2,0}$ variable.

be accounted for in terms of a reality in which well-defined and continuously variable quantities evolve in a deterministic manner according to the equations of motion of the causal interpretation. They also illustrate that the fundamentally new feature of matter introduced in the quantum theory is a kind of wholeness in which the behavior of an individual particle is irreducibly connected with its context (expressed through the wave function), evidenced in the two-particle case by the existence of nonlocal connection. Einstein's elements of reality exist, they are not described by quantum theory which deals with eigenvalues of operators, and in general they are not simply revealed in measurement interactions. As we have seen the interactions regarded as measurements are, in fact, those in which a particular variable of a "measured" system becomes correlated with a particular apparatus coordinate according to deterministic laws of evolution of the whole undivided system-plus-apparatus. In this sense the elements of reality in quantum theory are essentially different to the elements of reality of Newtonian physics. Although we use the same terms, position, momentum, kinetic and potential energies, to describe the particle's motion, the individual and its relation with these attributes is not similar to that which exists in classical mechanics. In the history of science each new epoch-making discovery has indicated a new feature matter, one of which was the introduction of the idea of the field. In our opinion, assuming reality has this fundamentally new feature of wholeness is preferable to assuming that it does not exist except when we are looking at it!

References

1. L. de Broglie, Non-Linear Wave Mechanics, Elsevier, Amsterdam (1969).
2. D. Bohm, *Phys. Rev.* **85**, 166, 180 (1952).
3. D. Bohm, *Wholeness and the Implicate Order*, Routledge and Kegan Paul, London (1980).
4. C. Philipidis, C. Dewdney, and B. J. Hiley, *Nuovo Cim.* **52B**, 15 (1979).
5. C. Dewdney and B. J. Hiley, *Found. Phys.* **12**, 27 (1982).
6. C. Dewdney, *Phys. Lett.* **109A**, 377 (1985).
7. C. Dewdney, A. Kyprianidis, and J. P. Vigier, *J. Phys. A* **17**, L741 (1984).
8. A. Einstein, N. Rosen, and B. Podolsky, *Phys. Rev.* **48**, 777 (1935).
9. N. Bohr, in: *Albert Einstein: Philosopher-Scientist* (P. A. Schlipp, ed.), pp. 200–241, Tudor, New York (1949).
10. D. Bohm, R. Schiller, and J. Tiomno, *Suppl. Nuovo Cim.* 1.
11. T. Takabayasi, *Progr. Theor. Phys.* **14**, 283 (1955).
12. C. Dewdney, P. R. Holland, A. Kyprianidis, and J-P. Vigier, Trajectories and spin vector orientations in the causal interpretation of the Pauli equation, preprint, Institut Henri Poincaré (1985).
13. C. Dewdney, P. R. Holland, A. Kyprianidis, and J. P. Vigier, What happens in a spin measurement, *Phys. Lett.* (to appear)
14. C. Dewdney, P. R. Holland, A. Kyprianidis, and J. P. Vigier, Spin superposition in neutron interferometry, *Phys. Lett. A* (to appear).
15. D. Bohm, *Quantum Theory*, Prentice Hall, New York (1951).
16. C. Dewdney, P. R. Holland, A. Kyprianidis, and J. P. Vigier, Particle trajectories, spin vectors and Einstein–Podolsky–Rosen correlations, preprint, Institut Henri Poincaré (1987).

13

Bell's Inequality and the Nonergodic Interpretation of Quantum Mechanics

Vincent Buonomano

1. Introduction

The nonergodic interpretation has been described and discussed in various works.[1-3] Here we very briefly review it in Section 2 and make some observations. In Section 3 we clarify the various types of averages that one actually deals with in a laboratory experiment and also establish some notation. Section 4 discusses theories which involve only a flow of information from the source to the polarizers for time averages. In Section 5, theories in which there is also a flow of information from the polarizers to the source are considered for time averages. Some miscellaneous comments are made in Section 6. Appendix 1 describes an experimental test in a low-intensity interference experiment and Appendix 2 examines Nelson's stochastic mechanics in relation to our view. Parts of this work follow Buonomano.[1,2]

2. The Physical Viewpoint

2.1. The Physical Idea

The nonergodic interpretation is a local realistic view which assumes a certain physical viewpoint of how particles behave in the microworld that

VINCENT BUONOMANO ● Institute of Mathematics, State University of Campinas, Campinas, São Paulo, Brazil.

is distinct from the usual interpretations of quantum mechanics (e.g., Copenhagen, statistical, causal, and De Broglie's). It assumes that a sequence of particles, which consecutively pass through an apparatus separated by large times, is not independent in general. That is, it imagines that particles may indirectly interact with each other via memory effects in an hypothesized medium. Particles which first pass through a region will affect the medium, which then affects the particles, which later pass through that same region. By large times we mean times sufficiently great to guarantee that there is almost never more than one particle in the apparatus at a time. One might make the logical analogy of two professors who are never simultaneously in a room together, but communicate with each other only indirectly by leaving messages on the blackboard.

For example, consider the double-slit experiment which motivated this interpretation. The above type of indirect interaction permits one to say that a particle passing through one slit knows if the other slit is open (closed) from this information being recorded in the medium in their common path by particles which previously passed (didn't pass) the other slit. Here interference can only happen after a sufficiently large number of particles have traveled through the apparatus and conditioned our imagined medium. Particles interfere with other particles, but only indirectly via the medium. There can be no interference whatsoever for an ensemble average. See Buonomano[1] for more details. Appendix 1 describes an experimental test.

In other words, one might describe the nonergodic interpretation by saying that it questions the ergodic-type assumption that it is currently necessary to make in interpreting both the existing low-intensity interference experiments and the polarization-correlation experiments as giving the true quantum-mechanical *ensemble* averages.*

2.2. The Formal Definition

The nonergodic interpretation is a well-defined alternative interpretation of quantum mechanics. It assumes the same Hilbert-space formalism used in both the Copenhagen and the statistical interpretations except for the following. In these usual interpretations one associates the mathematical object, $\langle A \rangle = \langle \psi | A | \psi \rangle$, representing the average of the observable A in the state ψ, with the laboratory procedure of taking an ensemble average. Instead, the nonergodic interpretation identifies this same mathematical

---

* We recall that, for example, in the double-slit experiment the ideal quantum-mechanical ensemble average for a one-particle system should be made over many identical independent apparatus with exactly one photon in each apparatus, all of which are prepared in the same quantum-mechanical state. Real laboratory averages are time averages and, of course, necessarily involve an ergodic-type assumption to interpret the averages as ensemble averages. See discussions of this in Glauber[4] and Margenau.[5]

object with the laboratory procedure of taking a time average.* This is the *only* difference in the formalism. Here we are making a different association between *mathematical objects* of the Hilbert-space formalism and *laboratory procedures* than either the Copenhagen or the statistical interpretations. The nonergodic interpretation always makes the same numeric predictions as the usual interpretations, but it makes them only for time averages and not for ensemble averages.

2.3. The Medium

We would like to emphasize that we are not presenting a concrete physical theory. We are only talking about a physical idea of a medium with some sort of memory effects which permits us to justify an *indirect* interaction between particles. One may try to justify this medium as a stochastic medium or as a fluid medium. One may wish to remain abstract and refer to it as a field, whose properties depend on what passed through it. One could think about vacuum states. One might talk about an index of refraction of the medium or field which depends on its past history. Concepts from cooperative phenomena look particularly promising. Any future physical theory would have to somehow justify a medium of some type which can be forced into stable modes by many similarly prepared particles which consecutively pass through it. Lorentzian invariance of this memory would also have to be dealt with. Despite this lack of concrete physical basis, it is as well defined an interpretation as the others, in addition to being completely falsifiable (Appendix 1).

2.4. Memory Decay Time

The existing interpretations of quantum mechanics predict that interference must exist independent of the light intensity. This forces us to consider a more specific type of memory. We must imagine that we can obtain the same memory buildup in the case of intense light, using relatively few photons, as in the case of weak light, using many photons. That is, in this view, as one reduces the light intensity one may compensate for this by using many more photons (i.e., much longer time averages) in order to obtain the same level of memory development. In the limit of infinitely-low-intensity light one must use infinitely many photons to condition the medium.[1] Another way of saying this is that the memory decay time (after an equilibrium has been established between the particles and the medium) depends on the intensity; the lower the intensity the larger the decay time. On the other hand, it will take more particles to establish the equilibrium

---

* Actually we should say a grand-time average. This will be defined in the next section.

in the first place when the intensity is low. In any case, this concept of memory decay time does not appear to be a very satisfactory way of trying to describe the kind of equilibrium phenomena that we are imagining.

3. Time Averages

Here we discuss the various averages that must be considered in laboratory experiments and establish some notation. Consider an ideal laboratory experiment, namely, there is no noise, we have perfect counters and polarizers, and there is no drift or warm-up effects (see Figure 1). In particular, this means we may turn on our apparatus and immediately begin to collect meaningful data in any of the individual experimental runs. Let R represent the number of experimental runs which we assume to be strictly independent,* and N be the number of photon pairs in each of the runs. Also, λ_{rn} represents the characteristics of the nth particle pair in the rth run, and s_{rn} (s'_{rn}) is the state of polarizer A (B) of Figure 1 prior to particle λ_{rn} interacting with it. When we say Polarizer A (B) we mean to include the counters and any other instrumentation on side A (B) of the apparatus. We let $c_{rn} \equiv c_{rn}(a, b)$ be the correlation measurement of the particle pair λ_{rn} and define the following averages:

$$C \equiv C(a, b) = \left(\sum_{r,n} c_{rn} \right) \Big/ RN \qquad (1)$$

$$C_r \equiv C_r(a, b) \equiv \left(\sum_{n} c_{rn} \right) \Big/ R \qquad (2)$$

$$C_n t \equiv C_n(a, b) \equiv \left(\sum_{r} c_{rn} \right) \Big/ N \qquad (3)$$

---

* For an ensemble average it is essential to speak of *independent* individual measurements. In the nonergodic interpretation it is essential to speak of independent experimental runs. It is difficult to define what independence means in both cases in terms of concrete laboratory procedures. It seems obvious to almost all the research community that as long as the particles are not in the apparatus together, then they must be independent. This work is calling attention to this assumption. On the other hand, we will assume that runs that are separated by the procedure of turning the apparatus off for a short period of time and on again will guarantee independence of the runs.

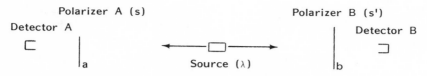

Figure 1. This is just the canonical polarization correlation experiment with light: λ represents the state of the photon pair; s (s') represent the state of the polarizer-detector system A (B respectively); a and b are the angles of polarizers A and B, respectively.

The indexes r and n go from 1 to R and N, respectively. Quantity C is sometimes called a *grand-time average*, C_r is the *run-time average* in run r, and C_n is the *ensemble average* at "time" n. The indexes on the character C identify the average being referred to. We will frequently say time average to mean either a run-time average or a grand-time average. It is well known that in general these averages (1–3) are unrelated.

Unless one makes some type of ergodic assumption it is only $C_{m=1}$ that represents rigorously the conceptually correct quantum-mechanical average (assuming that the experimental runs are independent). It is almost always implicitly or explicitly assumed by the research community that

$$C_n(a, b) = C_r(a, b) = C(a, b) \tag{4}$$

for all n and r in an ideal experiment. This involves an ergodic-type assumption and it is exactly what the nonergodic interpretation is questioning. There is no experimental evidence for the validity of equation (4) that we are aware of. For example, in the double-slit experiment, one might say that this assumption is based on the belief that two particles which are not in the same place at the same time cannot affect each other, or, in other words, particles must be "touching" to affect each other. This is a very local assumption considering the kind of global behavior that one is trying to explain.

The nonergodic interpretation says that $C_n(a, b)$ must agree with Bell's inequality for small n, while for large n it will agree with (converge to) the quantum-mechanical predictions. No criteria for small or large n are known.

4. One-Directional Information Flow

Here we will introduce local realistic theories for time averages which involve only a flow of information from the source to the polarizers (via the photons) with no flow of information from the polarizers to the source. We call such theories *one-directional theories*. These are, of course, the theories for which Bell's inequality is usually derived for ensemble averages.

For time averages they are much more general and complicated. For example, in the notation of the previous section, we may imagine a theory where

$$s_{rn} = s_{rn}(s_{r,n-1}, \lambda_{r,n-1}) \tag{5}$$

In other words, the state of a polarizer depends on its previous state and the state of the previous photon which passed through it. The states of the distant polarizers are not factorizable in this case. We note that here it is not necessary to imagine a medium to provide for the type of communication between particles, since the memory may exist in the polarizers.

The following is a more concrete formal example of this. Let our particle source produce orange and white photons under a conservation-of-color law. That is, both photons are either orange or white. We also assume that the colors are random with a probability of one-half each. Let polarizer A be such that it has the following properties. It is a filtering device with only two states, o and w. It will permit an orange photon to pass only if it is in state o, and will permit a white photon to pass only if it is in state w. In addition, it is assumed that the polarizer's state becomes o (w) after an orange (white respectively) photon impinges on it regardless of whether or not the photon passes and regardless of the polarizer's previous state. We assume polarizer B is identical to A, and we make R experimental runs with N measurements (i.e., photon pairs) in each of the runs. Assume that the initial states of polarizers, $s_{r,n=0}$, are o and w randomly with a probability of one-half. Now it is easy to see the following. The ensemble average $C_{n=1}$ is zero. After the first measurement the polarizers are always in the same state, o or w, since the photons are both the same color. Therefore both will pass or both will not pass their respective polarizers. The run-time averages $C_r = 1$ for all r, the ensemble averages $C_n = 1$ for $n > 1$, and the grand-time average $C = 1$. One may modify this example to produce any desired correlation and also make it so that C_n converges to the correlation 1. The point of this example is that it clearly shows that for time averages, the states of the distant polarizers are not independent. They cannot be factorized and therefore the usual proof of Bell's inequality does not apply.

In Buonomano[6] it was claimed that even these theories must obey Bell's inequality if one averages over all initial states of the apparatus and all sequences of particle pairs (i.e., for a grand-time average). More recently, a series of works[7-14] involving time, but not permitting information flow from a polarizer to the source or to the other polarizer, have been published. In some cases they reported a violation of Bell's inequality. We are currently analyzing these works in light of these seemingly conflicting claims and hope to publish a detailed comparison of all these works. Therefore we must leave the analysis of all these one-directional theories incomplete here.

5. Two-Directional Information Flow

By a *two-directional theory* we mean a local realistic theory in which there is also a flow of information from a polarizer to the source or to the other polarizer. It is, of course, trivial that such theories may violate Bell's inequality for either ensemble or time averages. In this section, we simply want to argue that for time averages, such theories would appear to be physically much more reasonable than for ensemble averages. First we show a formal example of a two-directional theory which violates Bell's inequality for a time average (either a run-time or a grand-time average).

5.1. A Formal Example

We consider the left-hand side of a polarization-correlation experiment as shown in Figure 2. The space between the polarizer and source is divided into M cells, numbered from left to right. Hence the polarizer is in cell 1 and the source in cell M. Each photon then passes through each of these cells. Let s_{rn}^m be the state of cell m after the nth photon has passed through it in run r, while λ_{rn}, s_{rn}, and c_{rn} are as defined in Section 3. It is assumed that

$$s_{rn}^m = s_{rn}^m(s_{r,n-1}^{m-1}, S_{r,n-1}^{m+1}, \lambda_{r,n-1}) \qquad (6)$$

that is, the state of cell m depends on the states of the two neighboring cells as well as the state of the previous photon that passed. Relation (6) is valid every time a photon passes through a cell. Therefore, after one photon has passed, cell 2 depends on the state of the polarizer. After two photons have passed cell 3 depends on the state of the polarizer, and so on. Then after $N > M$ photons, cell M, and therefore the source, depends on the state of the polarizer. The other side of the polarization correlation is treated in the same manner. This is, of course, sufficient to violate Bell's inequality, since the state of the source depends on the states of the polarizers.

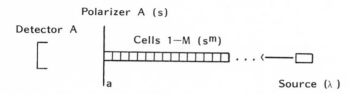

Figure 2. This is the left-hand side of Figure 1. The little boxes represent the cells in the formal example of the text; s^m is the state of cell m.

5.2. Discussion

As stated above, constructing a formal example for a two-directional theory that violates Bell's inequality is completely trivial. The important question is: "Does there exist a physically reasonable theory with such properties?" This is a much more difficult question being partially subjective. We want to argue that for time averages, two-directional theories are more physically reasonable than for ensemble averages.*

Consider our formal example, where the cells are arbitrary divisions in a medium of some type. After a sufficient number of photons have passed through the system, we have imagined that the state of the photons (i.e., the state preparation of the photon) is affected by the polarizer. If we change the polarizer angle, then the state preparation is affected differently, but only after a sufficient number of photons have again passed to establish a new equilibrium between the polarizer and the state preparation. Is it so unreasonable to think in terms of some sort of cooperative phenomena here, perhaps something analogous to the Bénard phenomena (e.g. Haken[15]) where water molecules organize themselves into hexagonal shapes when slowly heated? It is important to point out that there are very large numbers of photons in these experiments. For example, in those of Aspect *et al.*[16] there were millions at each given configuration. With these numbers it is not farfetched to speak of some basically probabilistic phenomenon that orders the medium into an equilibrium which depends on the polarizer angle. If the distance between the polarizer and source is increased, then one may still imagine the same development except that many more photons would have to be used in order to obtain the same average.

The nonergodic interpretation imagines that quantum mechanics is basically an equilibrium phenomenon with feedback between the state preparation and measurement parts of apparatus. That is, the strong quantum-mechanical correlations do not initially exist, but develop after many

---

* We think that there may sometimes be a double standard applied here in terms of what type of explanation of information flow may be considered physically reasonable. For example, there is a gravitational force between the polarizer and the source. For certain types of polarizer, this force depends on the angle. What is the mechanism for transmission of this force or information between the polarizer and source when we suddenly change the angle of the polarizer? Neither Newton's nor Einstein's theory gives us an entirely satisfying answer. In the former, one speaks of the propagation of a gravitational field. In the latter, one speaks of the propagation of a geometry (i.e., the metric tensor). These are more mathematical than physical explanations, and furnish us with little understanding of how information flows from one place to the other or what is the information (i.e., the mechanism of gravity). Copenhagen has been telling that such questioning is naive. They may be right, but this is what is at issue.

particles pass through the system. In this view, if for example you do things, like increase the distance or lower the particle intensity, then the convergence to the quantum-mechanical predictions is slower (i.e., you need more particles).

5.3. Aspect's Experiment

We now discuss Aspect's last experiment.[16] If in this experiment the polarizers were randomly switched between parameter values, then the nonergodic interpretation could not agree with quantum mechanics in this experiment. This is because, even if there is a mechanism for information flow between a polarizer and the source, the information must arrive at the source at a delayed time and therefore must be uncorrelated with the actual state of the polarizer. In Aspect's experiment, the switching between parameter values is not random but periodic. One might imagine theories in which the polarization information about the parameter value of a polarizer is modulated by the commutator. In other words, one may justify saying that the source accumulates information such that it knows when the photons leave the source what the parameter settings of the polarizers will be. Such theories are more complicated, but are not eliminated by Aspect's experiment, as he has pointed out. In order to have the equivalent of random commutator switching, the autocorrelation time of the switching should be small in comparison to the time of flight of the photons. Aspect intended to have this condition satisfied in his experimental proposal. In the actual experiment this condition was not satisfied.

6. Some Comments

We make some miscellaneous comments in this section.

6.1. Interference Is Independent of Intensity, but . . .

It is commonly stated that interference is independent of the intensity. This could be misleading in the following sense. Quantum mechanics does predict that you obtain interference independent of the intensity, and this is confirmed in a wide range of experiments with various types of particles. But if Buonomano[17] (see also Buonomano and Bartmann[18]) it is argued that the number of particles, M, needed to measure the visibility of an interference pattern with some specified precision and probability depends on the intensity, in general. For laser light, M is independent of the intensity, but not for thermal light. In other words, as you reduce the intensity, the number of particles that you need to measure the visibility within that same specified precision and probability changes. The number M depends on the intensity, that is, the statistics of interference is not independent of the intensity, in general. Even though you consider the consecutive particles to be independent, in a sense each particle must know the intensity of its preparation.

6.2. A Comment on the Joint Probability Question

In the nonergodic interpretation, there is a logical mechanism for feedback between the measuring part of the apparatus and the state preparation part. They are not independent. The measurement you choose to make affects the state preparation. The logical mechanism is simply our time average. (We have offered no concrete physical mechanism for this feedback other than for some speculations about cooperative phenomena.) For a true ensemble average this would not make sense. In our view the phase distribution of the properties of our particles will, in general, depend on the observable (but only for a sufficiently long time average). This means that the marginality requirement (Cohen) is not valid here. Therefore the conclusions about the nonexistence of classical joint probability distributions do not apply to the nonergodic interpretation (Mugur-Schachter[19] and Buonomano[3]).

This same comment may be applied to the question of understanding the object–apparatus interface in quantum mechanics. In the nonergodic interpretation, the properties of the object are not independent of the properties of measuring apparatus and vice versa for sufficiently long time averages. It says that when you first begin an experimental run, they (the states of the object and apparatus) will be independent, but after a sufficient number of particles pass through the apparatus they will not. One may easily think of examples of this type of behavior from both the physical and social sciences.

Appendix 1. An Experimental Test

It is clear that the nonergodic interpretation is experimentally distinguishable from all the usual views in interference and polarization-correlation experiments, since it predicts no interference effects and no strong correlations whatsoever for a real ensemble average. Of course, it is not practical to take a true ensemble average. We now sketch a limited low-intensity interference experimental test which avoids the necessity of having to take a genuine ensemble average. Various other, perhaps more difficult, experiments have been described in Buonomano.[2,3] Although the basic idea of some of these proposals may be easily applied to the polarization-correlation experiments, they would appear to be much more difficult to perform.

As in Section 3 we will again assume that our experiment is performed under absolutely ideal conditions with perfect instrumentation. Consider a neutron or photon interferometer as in Figure 3. We have shown it in the form of a Mach–Zehnder interferometer. There are two counters, C1 and C2, placed at the maximum and minimum of the interference pattern, respectively. There is a chopper in arm A which is half solid and half transparent. If the chopper frequency is f, then the chopper lets no particles pass for the first $1/(2f)$ seconds and lets all of them pass for the next $1/(2f)$ seconds, and so on and so on. Quantity f is assumed to be small, for example, it might be chosen to be of the order of 1 cycle per second for a photon experiment. Also, it is assumed we have a timing circuit that permits us to unambiguously know if it was a two- or one-arm configuration for each of the detected particles. We let n_{max} and n_{min} be the counting rates at C1 and C2 with both arms open, with $n \equiv n_{max} + n_{min}$, and so $n/4$ is the counting rate at C1 and C2 when the chopper is blocking arm A.

What we want to examine are the average counting rates at the two counters when we initially change the configuration from a two-arm to a one-arm experiment, and vice versa. (We will only discuss the former case, as the latter case is the same but with the counting rates changing in the opposite sense.) The nonergocdic interpretation makes dramatically different predictions than the usual interpretations in how the counting rates must initially change when the configuration is first altered.

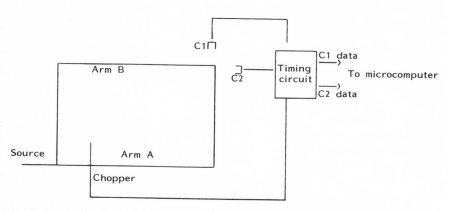

Figure 3. A Mach–Zehnder interferometer; C1 and C2 are two counters which are positioned at the maximum and minimum of the interference pattern. The chopper rotates slowly and is half solid and half transparent. The timing permits us to unambiguously know if arm A was open or closed for each of the detected particles.

The usual views clearly say that when you first close arm A, then the counting rate at C1 will go from n_{max} to $n/4$ *abruptly*. That is, in the very next counting interval after the arm is closed C1 will give $n/4$ on the average. This is because we definitely know that all the particles that are detected must have come from path B and therefore there can be no interference. Our view says that the counting rate at C1 will *gradually* change from n_{max} to $n/4$. More accurately, it will change from n_{max} to $n_{max}/2$ in the first counting interval after arm A is closed, and then it will change gradually to $n/4$. That is, the initial particles (which all must have passed arm B) are fooled, so to speak. They will act as if the other arm was still open, since there was not yet time for the medium to become reconditioned to a one-arm experiment. After a sufficient number of particles have passed the medium becomes reconditioned (i.e., a new one-arm equilibrium has been reached). At this point C1 will give the rate $n/4$. No criteria for how many particles are needed for this reconditioning to date are known, but it certainly must depend on the intensity among other factors. The counting rates at C2 may be described in an analogous way in the two views.

We will now be somewhat more precise about the above. Imagine that we perform one experimental run in which our chopper has made M revolutions, where M is large. We let T be a small counting time interval ($T \ll 1/f$) during which we measure the number of particles, $N_1(i, j)$ and $N_2(i, j)$, detected at the counters C1 and C2, respectively. The indexes i and j refer to the following: $N_k(1, 1)$, $k = 1$ or 2, is the number of counts at C_k in the time interval $[0, T]$ during the first revolution of the chopper; $N_k(2, 1)$ is the number of counts in the second time interval $[T, 2T]$, again in the first revolution of the chopper; $N_k(i, j)$ is then the number of counts at C_k in the time interval $[iT, (i + 1)T]$ in the jth revolution of the chopper. Now the average over the various revolutions, $N_k(i)$, is given by

$$N_k(i) = \left[\sum_{j=1}^{M} N_k(i, j) \right] \Big/ M$$

The quantities $N_k(i)$ are the basic objects of interest. They permit us to see the (statistical) behavior of the counting rates when we first change the configuration (i.e. before, during, and after we change from a two-arm to a one-arm configuration).

Figure 4 shows the nonergodic predictions and the usual predictions for counter C1. The predictions are dramatically different. In particular $N_1(1)$ represents the number of counts just after arm A was closed. Again the usual views say that since you know with certainty which path these photons traveled, then there can be no interference. So $N_1(1) = N_2(1) = n/4$, that is, the rates must be the same for both counters. Our view says

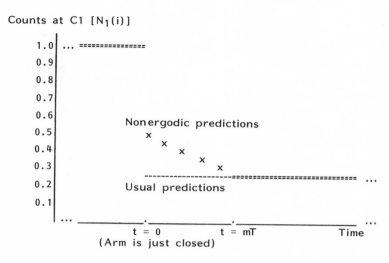

Figure 4. Average normalized counts at C1 in the time intervals $[iT, (i + 1)T]$ as predicted by the nonergodic and the usual interpretations. Curve $= = =$ is where the interpretations agree, curve $- - -$ gives the predictions of the usual interpretations alone, and curve $\times \times \times$ gives the nonergodic predictions alone. The entire vertical axis has the value n, where we have taken $n = n_{max}$ and $n_{min} = 0$. The horizontal axis is in units of T seconds; m is the number of intervals of time T for the medium to become reconditioned. No criteria for m are known. Curve $\times \times \times$ is shown as a straight line, but its shape is not known.

the medium in the common region is still conditioned to a two-arm configuration and therefore mostly all the photons will still register in C1 when the arm is first closed [i.e., $N_1(1)$ is approximately equal to $n_{max}/2$] but will converge to $n_{max}/4$. The counting rate $N_2(1)$ will be approximately $n_{min}/2$ and $N_2(i)$ will converge to $n/4$. In the figure we have taken n_{min} to be zero and normalized for simplicity.

In the above we have not mentioned concrete values for T, f, n, and so on. We have also used freely undefined terms like "small," "large," and "the initial particles." The experimental parameters will, of course, depend on whether one is conducting photon or neutron experiments. Works are in preparation that will give concrete criteria and more details.*

---

* *Note added in proof.* Buonomano[20] contains some of this information. It also corrects the above in the sense that a chopper cannot be considered as giving a test of all conceivable nonergodic theories. A random shutter in place of the chopper would test all nonergodic theories. Also J. Summhammer has performed a rough test of the nonergodic interpretation with a neutron interferometer. It shows that if memory exists, then it must develop in under 200 neutrons at the most (at the intensity and coherence conditions of the experiment). A more sensitive experiment is underway.

Appendix 2. Nelson's Stochastic Mechanics

Given the fact that the nonergodic interpretation conceptually depends on a medium, it is natural to modify Nelson's stochastic quantum mechanics[21] in the manner for time averages (see Prado[22] for details). A review of the stochastic interpretation is given in Ghirardi *et al.*[23] In the stochastic interpretation one derives a stochastic process, X, from Schrödinger's equation and $\psi(x, 0)$. The first-order probability density $p(xt)$ of X then equals $|\psi(x, t)|^2$; X is a stochastic process like Brownian motion and is defined over an ensemble of particles. That is, the average is defined over a set of strictly independent particles, each of which experiences a statistically identical medium. By definition there can be no question of an interaction (direct or indirect) between the particles.

The physical situation we must imagine in our view is the following. In each of the experimental runs, a first particle passes through the medium, affecting the medium and being affected by it. A second particle then passes through the medium, again affecting the medium and being affected by it, and so on. (We assume for simplicity that there is rarely more than one particle in the apparatus at any given instant of time.) We want to define an average that will represent the motion after very many particles have passed through the medium, i.e., after an equilibrium between the particles (state preparation) and the medium has had a chance to set in.

One way of doing this is the following. We must average over all independent experimental runs; each experimental run will have many particles consecutively passing through the apparatus. We define a stochastic process, X_1, as the stochastic representation of the first particle in each of the experimental runs. We let X_2 represent the motion of the second particle in each of the experimental runs in our ensemble of experimental runs, and X_n represent the stochastic motion of the nth particle in each of our experimental runs, etc. Then, in the nonergodic interpretation, we associate ψ with the limit (in a sense we now define) of the X_n as n goes to infinity. If $p_n(xt)$ is the first-order probability density of X_n, and $p_n(x_1t_1, \ldots, x_mt_m)$ the mth-order probability density, then X is defined to be the stochastic process defined by the probability densities

$$p(xt) = \lim_{n \to \infty} = p_n(xt)$$

$$p(x_1t_1, \ldots, x_mt_m) = \lim_{n \to \infty} p_n(x_1t_1, \ldots, x_mt_m)$$

X itself does not necessarily represent stochastic motion of real particles, but instead the limit of such motion. In the nonergodic interpretation associate ψ with X and not any of the X_n.

The Criticism against the Stochastic Interpretation

The stochastic interpretation has been criticized in a number of recent articles. The review by Ghirardi *et al.*[23] describes most of these criticisms and contains the references. Here we argue that many of these criticisms do not apply to the nonergodic interpretation.[22]

The first criticism that we consider relates to the fact that the transition probability density P of the stochastic process, associated with the Schrödinger equation, depends also on the initial value of ψ. Another way of saying this is that the properties of the medium are affected by $\psi(x, 0)$, i.e., the state preparation affects the properties of the medium[23] For example, in an interferometer experiment the medium in the common path region must be imagined to have different properties in the three cases of one or other arm being open and both arms being open. Here we have all the well-known difficulties in understanding how the opening or closing of an interferometer arm can affect the medium on which the stochastic interpretation implicitly depends. We observe that we may block an arm with a piece of paper or a large piece of lead. We may block the arm at any of various positions. Also we may leave the arm open and put that same piece of paper or lead near the arm but not blocking it. The medium must only be sensitive to whether the arm is blocked or not, and not the distribution of matter. In a Michelson interferometer the medium in the common path region is sensitive to a half-wavelength change in the position of a mirror in one of the arms. But the medium is totally insensitive to any other distribution of matter in the interferometer as long as a path is not blocked. What is relevant is whether a path is open or closed and not the distribution of matter. In the independent laser interference experiments, the medium in the common path region must be sensitive to a path being blocked as well as one of the lasers being turned on or off. Further, in a many-slit interference experiment (i.e., with a grading) the medium would have to be sensitive to the width of the light beam, as the diffraction pattern is different depending on the number of slits through which the beam can pass. Such causal relationships are difficult to understand in the stochastic interpretation. Existing stochastic mechanics has shed no new insight on these *physical* problems despite its very elegant mathematical nature.

The nonergodic interpretation hypothesizes a structure to understand these types of causal relations. That the state preparation affects the medium is trivially true, by assumption in our viewpoint. (The transition probability density of any stochastic process which represents particles that may interact, directly or indirectly, will in general depend on the state preparation.)

A second criticism is related to the fact that in the presence of nodal surfaces, such as in the hydrogen atom, the stochastic interpretation has more solutions than Schrödinger's equation (see Ghirardi *et al.*[23] for references). Such extra solutions can only be ruled out by using quantum mechanics. These extra solutions in the case of the hydrogen atom correspond to an electron being trapped permanently between two nodal surfaces. The demonstration of the above does not apply to our view, since we do not associate ψ with a diffusion process but with the limit of a sequence of diffusion processes. In our view we may imagine, again in the example of the hydrogen atom, that for long time averages we obtain the quantum-mechanical averages. However, there is a tendency for an electron between two nodal surfaces to remain there for disproportionate times, but eventually to escape. Here it is only in the unrealized physical limit of the X_n that the electron becomes trapped. The above is a modification of Nelson's[21] own attempt to confront this difficulty, which we consider less *ad hoc.* Here we have only one electron, which arrives at an equilibrium with itself and the medium as it revolves.

The last criticism that we consider is the lack of distinction between a pure and mixed state in the stochastic interpretation. In the nonergodic interpretation one may make a conceptually clear distinction between the ways one may superimpose state preparations. If one does it in such a way as to physically permit indirect interference between the particles via our medium, then we have a pure superposition.* If, on the other hand, one superimposes the preparations in such a way that the particles cannot interact with each other via the medium, then the superposition is a mixed one. For example, for us a true ensemble average in an interference experiment is always a mixed state superposition, since there is no possibility of indirect interference between the particles in this case.

References

1. V. Buonomano, in: *The Proceedings of the International Conference on Quantum Violations: Recent and Future Experiments and Interpretations, University of Bridgeport, USA, 1986* (W. M. Honig, D. W. Kraft, and E. Panarella, eds.), Academic Press, New York (to appear).
2. V. Buonomano, in: *The Proceedings of the International Conference on Microphysical Reality and Quantum Formalism, Urbino, Italy, 1985* (G. Tarozzi and A. van der Merwe, eds.), D. Reidel, Dordrecht (to appear).
3. V. Buonomano, *Nuovo Cim.* **57B**, 146 (1980).
4. R. J. Glauber, in: *Quantum Optics and Electronics* (C. de Witt, A. Blandin, and C. Cohen-Tannoudji, eds.), p. 63, Gordon and Breach, New York (1965).
5. H. Margenau, *Philos. Sci.* **30**, 1 (1963).
6. V. Buonomano, *Ann. Inst. Henri Poincaré, Sect. A* **29**, 379 (1978).

---

* The fact that particles, physically, may pass through the same region of space consecutively does not mean necessarily that they may interfere. The particles must be sufficiently coherent to affect the medium in a consistent way to arrive at an appropriate equilibrium which causes the interference.

7. G. C. Scalera, *Lett. Nuovo Cim.* **38**, 16 (1983).
8. G. C. Scalera, *Lett. Nuovo Cim.* **40**, 353 (1984).
9. S. Notarrigo, *Nuovo Cim.* **83B**, 173 (1984).
10. S. Pascazio, *Phys. Lett.* **111**, 339 (1985).
11. S. Pascazio, Time and Bell-type inequalities, preprint (1986).
12. E. Santos, *Phys. Lett.* **101A**, 379 (1984).
13. S. Caser, On a local model that violates Bell's inequality, preprint (1985).
14. J. D. Franson, *Phys. Rev. D* **31**, 2529 (1985).
15. H. Haken, *Synergetics, An Introduction*, Springer-Verlag, Berlin (1978).
16. A. Aspect, J. Dalibard, and G. Roger, *Phys. Rev. Lett.* **49**, 1804 (1982).
17. V. Buonomano, *Lett. Nuovo Cim.* **43**, 69 (1985).
18. V. Buonomano and F. Bartmann, *Nuovo Cim.* **95B(2)**, 99 (1986).
19. M. Mugur-Schachter, *Found. Phys.* **13**, 419 (1983).
20. V. Buonomano, Neutron interferometry and the nonergodic interpretation of quantum mechanics, *Phys. Lett. A* (submitted). Also *Proceedings of the International Workshop on Matter-Wave Interferometry*, Vienna 1987 (to appear).
21. E. Nelson, *Phys. Rev.* **150**, 1079 (1966).
22. F. Prado, 1984, Particulas que Interagim Indiretamente e a Interpretação Estocástica de Mecânica Quântica, PhD Thesis, Instituto de Matemática, Universidade Estadual de Campinas, Campinas, S.P., Brasil. Also V. Buonomano and F. Prado, *Found. Phys.* April, 1988.
23. G. C. Ghirardi, C. Omero, A. Rimini, and T. Weber, *Riv. Nuovo Cim.* **1**, 1 (1978).

14

An Extended-Probability Response to the Einstein-Podolsky-Rosen Argument

W. MÜCKENHEIM

1. Introduction

A variety of answers to the Einstein–Podolsky–Rosen (EPR) argument have been proposed. Why not let us consider a different one—as unattractive as (but not more unattractive than) all the others—which does not obey Kolmogorov's axiom according to which probabilities p are restricted to the range $0 \le p \le 1$? Instead of starting with a conventional introduction, this topic appears strange enough to allow us to dispense with the usual form taken by scientific expositions and to take the liberty of beginning by eavesdropping on a fictitious conversation *which could have happened but (probably) did not*. Other than that this mode of occurrence is frequently involved in EPR considerations, the content of the discussion is not in any way related to the main part of this chapter.

Let us imagine a dark and frosty winter's evening, say December 31, 1733, to give it a date. Two scholars, Speculati and Orthocredo, are sitting comfortably in a chimney corner and, while imbibing some exquisite wine, are discussing metaphysical and philosophical problems.

Spec. I'm occupied with this strange idea. Could it be that the geometry of nature does not exactly follow the rules which Euclid has taught us?

Ortho. What a silly question! What makes you wonder about something of that kind?

W. MÜCKENHEIM ● Südring 10, D-3400 Göttingen, Federal Republic of Germany.

Spec. There's no real reason. I'm just considering whether it might be possible.

Ortho. Oh, do stop it! You're wasting your time. It's well known that Euclid's authority has never failed and, besides, it's obvious that the world must obey his laws.

Spec. But you know, his fifth postulate, in particular, is rather difficult and lengthy.

Ortho. That's right. But its validity has been proven so recently by Saccheri. The title of his work is just to the point: *Euclides ab omni naevo vindicatus.*

Spec. Are you quite sure that Saccheri's proof isn't based on invalid conclusions?

Ortho. As the result is correct, his conclusions must be accepted. But if you're not open to logical argument, then be convinced by your own eyes: could you even imagine two parallel lines that cross each other or that vary in their distance apart by even the hundredth part of a foot?

Spec. Consider a sphere. The pair of lines which could reasonably be regarded as the straightest lines possible between two given pairs of points, even though they are parallel at one of each of their points, will cross each other at two other points.

Ortho. That example doesn't prove anything. Euclid's geometry deals with our world and not with a sphere. According to Plato straight lines are represented by light rays, and they could never twist around a sphere: Straight lines which cross each other aren't parallel, and straight lines which are parallel don't cross each other. If you're really that insistent on disputing this point, we could settle it once and for all using a couple of well-made rulers that I happened to buy a few days ago. But that's just too silly. Any fool could tell you the truth of the matter.

Spec. I was really thinking in terms of extremely long distances, rather than in terms of rulers.

Ortho. What's the difference? Take thousands of rulers! Lay them end-to-end precisely, along two parallel lines and you'll find that the distance between them always remains the same. You'd only be testing the accuracy of your own work by the very definition of parallelism.

Spec. Perhaps one would have to replace this definition by a different one.

Ortho. I can't see any advantage in complicating the wonderful system of geometry by so doing. Lines that are subject to other specifications are simply not parallel or not straight!

Spec. I can't see any advantage either. However, that doesn't rule out the possibility; and that's just what keeps going around in my head.

Ortho. There are things which are possible and others which simply aren't. Your idea is of the latter sort. But in order to finish this discussion, let me tell you that all scholars I know are convinced that Euclid is right. They haven't even thought along different lines. What more proof can you ask?

Meanwhile the fire is waning, the flask of wine is empty, and New Year's Eve has turned into New Year's Day. Speculati and Orthocredo bid each other goodnight and go to sleep. Thereafter, owing to the quantity of wine they consumed, they are unable to remember their discussion, other than as a confused dream. How lucky for Speculati!

Any reader who cannot sympathize with Speculati's attitude (as it relates to extended probabilities, of course; nowadays nobody will deny the usefulness of Riemannian geometry) is advised to skip the rest of this chapter.

2. Axioms

In order to set the stage for our discussion, we will first delineate our point of departure. The following propositions will be considered axioms throughout this chapter:

1. There is something outside ourselves that exists whether or not a living being observes it. We call it "objective reality."
2. Reality is "local" or "separable" for it is simply impossible even to define a unique time relation between space-like separated events.*
3. The *results* of quantum theory, as far as they are utilized in this chapter, are correct.

Owing to their logical status, these axioms can be neither proved nor disproved. This in no way precludes a world description, based on different axioms, which may be more successful. But our choice has been fixed, assuming the impossibility of the instantaneous collapse of any nonsingular wave packet (because even nature would not "know" what "instantaneous" meant in this case).

Starting off with the EPR argument,[1] Bohm,[2] Bell,[3] Clauser,[4] Aspect,[5] and many others have contributed to showing that the world described by our three axioms is not the one that we live in.

---

* "I take Relativity seriously *as a theory of space-time structure*—not just as a prohibition on sending *detectable* signals faster than light" (H. Putnam).

It might, however, be possible to maintain these axioms and to remove the resulting conflict (with our world) by sacrificing one or another self-evident truth, e.g., Kolmogorov's axiom according to which probabilities p are restricted to the range $0 \leq p \leq 1$. In spite of its apparent inconsistency, this idea has been applied more frequently than is commonly perceived.[6,7] Its formal consistency has been shown by Bartlett,[8] and Dirac[9] has furnished a plausible interpretation, at least of small negative or complex probability numbers, simply saying that the corresponding events are unlikely.

3. Extended Probabilities

The usual interpretation of probabilities is based upon the relative frequency of observed events exhibiting a certain property A (so-called "successful" trials) in a number N_0 of observed events (or trials). We let

$$N_0 = N(A) + N(\neg A) \tag{1}$$

with $\neg A$ denoting the absence of A. Then the probability of observing A is given by

$$p(A) = \lim_{N_0 \to \infty} N(A)/N_0 \tag{2}$$

This event-related definition serves very well in all practical applications, but it entails a degree of uncertainty because N_0 can never reach infinity. (In any case, most physical quantities can be verified only approximately.)

In order to avoid the uncertainty of the *event-related* definition, the *fact-related* density definition can be applied. The chance, for instance, that the reader has turned to this page during the last quarter of any minute is 0.25, independently of how precisely this result is verified by observation (of course, only one trial per reader is allowed), because the density of "last quarters" is known to be 0.25. The problem with this definition is that, in physics, not only are densities not always known, but sometimes even their objective existence is doubted.

Both definitions given above obviously lead to probabilities which obey Kolmogorov's axiom strictly. Extended probabilities cannot emerge from them, just as negative energies cannot emerge from the classical definition of energy: $E = \frac{1}{2}mv^2$. In order to exchange or extend the meaning of a notion the replacement of its definition is usually required. The notion of probability may presently be subject to such a process—as is suggested by Wigner's carefully worded statement[10]:

> I fully agree that the concept of a negative probability is in contradiction to the usual definition of the probability concept. However, other quantities from which an actual probability can be calculated are often called "probabilities."

If reality includes features which can appropriately be described by extended probabilities, we have to look for them at the microscopic level: Extended probabilities are not involved in everyday life. We know that quantum-mechanical phase-space distributions (usually termed "Wigner functions") necessarily entail negative values,[11] which have to be interpreted as negative probabilities if considered in the light of axiom (1). This feature disappears if the phase-space functions are smeared out over phase-space regions of order $\hbar$.[12] Also, the time-resolved probability distribution for photon emission from an excited state, as calculated by the Weisskopf-Wigner formalism[13] as well as from the model of the classical electromagnetic oscillator, leads to probabilities of both less than zero and larger than unity.[6] The latter case has even been verified by experiment.[14,15] Not surprising this behavior is also found when analyzing the Glauber–Sudarshan P-functions,[16] which are applied in quantum optics in order to describe modes of an electromagnetic field, and which are closely related to Wigner functions.

These and some other[6] observations suggest that we revise our concept of probability at the microscopic level. How this can best be accomplished is not yet known. One very simple approach will be presented below, but even without a suitable definition, it may be possible to calculate probability values by applying only symmetry and invariance considerations, as Jaynes has shown in his beautiful resolution of Betrand's paradox.[17]

One concrete approach to obtaining a definition for extended probabilities makes use of two kinds of events (in close analogy to the theory of matter):

Let us consider a number N_0 of particles prepared in a pure quantum state but, deviating from quantum theory, populating different "microstates" $\lambda_i, i = 1, \ldots, n$. For the sake of simplicity we assume that the different states λ_i are taken with equal positive probability p_i. As the result of an interaction, the particles change their microstate from λ_i to $\lambda_j', j = 1, \ldots, m$, the transition probability being denoted by p_{ij}. We assume, further, that the primed microstates belong to different quantum states, one of them being called A. Those values of λ_j' which form state A are denoted by $\lambda_j'(A)$. The result of a measurement shows $N(A)$ particles in state A and $N(\neg A)$ particles in different states $(\neg A)$. Conservation of particle number necessitates the validity of equation (1). The combined probability for transition to state A follows, by experiment, in the case of $N_0 \to \infty$:

$$p(A) = n^{-1} \sum_i \sum_{j(A)} p_{ij} = \lim_{N_0 \to \infty} N(A)/N_0 \tag{3}$$

If single transitions could interfere with each other, it would be possible to introduce, in addition to "normal" events, a second kind of event by which normal events are eliminated. Then a negative probability could be interpreted as a positive probability for the occurrence of such an eliminating event; and a probability exceeding unity would imply that sometimes more than one transition may take place from a microstate populated by only one particle.

While in this picture the usual macroscopically observable probabilities are restricted to $0 \le p(A) \le 1$, the individual probabilities p_{ij} need only satisfy the normalization condition

$$n^{-1} \sum_i \sum_j p_{ij} = 1 \tag{4}$$

and, for distinguishable results, the left-hand side of equation (3).

Of course, negative probabilities are not directly observable, just as electrons with negative energy are not directly observable. Only the absence of an electron in the Dirac sea, or its transition from and to a negative-energy state, can be observed. Similarly, evidence for negative probabilities would be obtained when a probability distribution is known to be normalized, and the experimental integration over a part of this distribution yields a value larger than unity.

The application of extended probabilities to problems of quantum electrodynamics (QED) has been discussed by Dirac,[18,19] and an appropriate formalism has led to Gupta's "Indefinite Metric"[20] which has served as a paradigm for Heisenberg's unified field theory of elementary particles.[21] The following application to QED has been discussed most recently by Feynman.[22]

In order to account for the Coulomb interaction, a photon is assumed to have *four* directions of polarization x, y, z, and t, no matter which way it is going. The time component is coupled with ie instead of e. Then for real photons, the probability of a t-photon emission is negative, proportional to $-|\langle f|j_t|i\rangle|^2$ (j denotes the current with components j_x, j_y, j_z, and j_t). The probability of emission of an x-photon is $+|\langle f|j_x|i\rangle|^2$. The total probability of emission of any sort of photon is

$$|\langle f|j_x|i\rangle|^2 + |\langle f|j_y|i\rangle|^2 + |\langle f|j_z|i\rangle|^2 - |\langle f|j_t|i\rangle|^2 \tag{5}$$

We paraphrase Feynman's own explanation[22] of this expression as follows: It is always positive for, by conservation of current, there is a relationship between j_t and the space components of j: $k_\mu j_\mu = 0$ if k_μ is the four-vector of the photon. For example, if k is in the z direction, then $k_z = \omega$ and $k_x = k_y = 0$ so that $j_t = j_z$ and we see that equation (5) is equal to the usual result in which we sum only the transverse emissions. The probability of emission of a photon of definite polarization e_μ is (assuming e_μ is not a null vector):

$$-|\langle f|j_\mu e_\mu|i\rangle|^2/(e_\mu e_\mu) \qquad (6)$$

This has the danger of producing negative probabilities. The rule whereby to avoid them is to allow only photons whose polarization vector satisfies $k_\mu e_\mu = 0$ and $e_\mu e_\mu = -1$ to be observed asymptotically in the final or initial states. But this restriction is not to be applied to virtual photons: intermediary negative probabilities are not to be avoided. Only in this way is the Coulomb interaction truly understandable as the interchange of virtual photons, i.e., photons with time-like polarization which are radiated as real photons with a negative probability.

One problem with an interpretation of extended probabilities in terms of normal and eliminating events is to understand how and why both sorts of events always interfere sufficiently closely to cancel each other out and thus prevent negative probabilities from being detected. This applies in particular to the EPR experiment in which it is possible to detect single events in correlation measurements. Another problem connected with the application of any probabilistic hidden-variables theory to the EPR problem is that strict correlations do not ever apply to single events [see equation (7) in Section 4] but only to the average of several events. If and how these obstacles can be overcome is open to question.

Notwithstanding this disadvantage we will now discuss some approaches to preserve the physics of our world while maintaining our three axioms, in spite of Bell's inequalities.

4. Bell's Inequalities Circumvented

Bohm's popular version[2] of the EPR argument deals with a pair of spatially separated spin-$\frac{1}{2}$ particles in the "singlet" state. That is, if the spin component s of one particle (arbitrarily labeled "1") is measured in direction $\hat{\mathbf{a}}$ yielding the result $s_1(\hat{\mathbf{a}})$, the same measurement performed on the second particle will yield the result

$$s_2(\hat{\mathbf{a}}) = -s_1(\hat{\mathbf{a}}) \qquad (7)$$

As equation (7) is valid for any direction $\hat{\mathbf{a}}$ whatsoever, it must be concluded that information about the result for every direction of measurement is contained in each particle (unless nonlocal interactions are permitted), although equation (7) can be verified in only one direction without one's possibly destroying the information about all the other directions. It has been argued that measurements which could have been performed, but were not, are meaningless. With respect to our axiom (1) we cannot but follow Schrödinger, who disproves this line of argument using the plain picture: *From so many experiments performed in advance, I know that the pupil always answers my first question correctly. Hence it follows, that in every case he <u>knows</u> the answers to <u>both</u> questions.*[23] According to quantum theory the pupil does not know more than one answer. And Bell[3] has shown that this claim is unconditionally entailed by the predictions of quantum mechanics with respect to correlation experiments (subsequently verified on several separate occasions by means of correlated photon pairs) thus turning the EPR argument into a paradox.*

Nevertheless, it is easily seen that Bell used Kolmogorov's axiom implicitly, in the step between inequalities (14) and (15) of his famous treatise.[3] By rejecting this axiom his result can be circumvented. There may be different ways of so doing, but applying some plausible restrictions we arrive at the simplest one dealing with spin-$\frac{1}{2}$ particles[26] (which, to the knowledge of the present writer, is the only one that adheres to the technical spin properties).

4.1. Correlated Pairs of Spin-$\frac{1}{2}$ Particles

In Bell's version[3] of the EPR experiment the spin component of particle 1 is measured in direction $\hat{\mathbf{a}}$ and the spin component of particle 2 in direction $\hat{\mathbf{b}}$. The quantum-mechanical expectation for the product of both measurements is (if $\hat{\mathbf{a}}$ and $\hat{\mathbf{b}}$ are unit vectors, and $\hbar = 1$):

$$\langle s_1(\hat{\mathbf{a}}) \cdot s_2(\hat{\mathbf{b}}) \rangle \equiv \langle \eta | \sigma_1 \hat{\mathbf{a}} \otimes \sigma_2 \hat{\mathbf{b}} | \eta \rangle = -\tfrac{1}{4} \hat{\mathbf{a}} \cdot \hat{\mathbf{b}} \qquad (8)$$

Here η is the (nonlocal) wave function of the particle pair in the singlet state. This result cannot be reproduced by any local hidden-variables theory obeying Kolmogorov's axiom.

In order to find a formalism which satisfies equation (8) while leaving the measurement results independent of each other, we assume the total spin of each particle to be represented by a "spin vector" $\mathbf{S}$ of length $\frac{1}{2}\sqrt{3}$. After interaction with a magnetic field the spin component in the field direction is either $s_+ = +\frac{1}{2}$ (spin-up) or $s_- = -\frac{1}{2}$ (spin-down) although, beforehand, the spin vector may have pointed in any random direction.

The probability $w_+(\hat{\mathbf{a}}, \mathbf{S})$ for the result spin-up depends on the orientation of the incident spin vector $\mathbf{S}$ with respect to the field direction $\hat{\mathbf{a}}$. For

---

* A comprehensive review of the first experiments of this kind is given by Clauser and Shimony.[24] A table including the more recent results can be found elsewhere.[25]

conservation of particle number, the probability $w_-(\hat{\mathbf{a}}, \mathbf{S})$ for the result spin-down is then related to $w_+(\hat{\mathbf{a}}, \mathbf{S})$ by

$$w_+(\hat{\mathbf{a}}, \mathbf{S}) + w_-(\hat{\mathbf{a}}, \mathbf{S}) = 1 \qquad (9)$$

In order to correctly describe the result for particles which are already polarized in direction $\hat{\mathbf{a}}$, the condition

$$w_+(\hat{\mathbf{a}}, \mathbf{S}) = 1 \qquad \text{for } \hat{\mathbf{a}} \cdot \mathbf{S} = \tfrac{1}{2} \qquad (10)$$

must hold.

Symmetry considerations supply the condition

$$w_+(\hat{\mathbf{a}}, \mathbf{S}) = \tfrac{1}{2} \qquad \text{for } \hat{\mathbf{a}} \cdot \mathbf{S} = 0 \qquad (11)$$

In order to determine a simple expression for $w_+(\hat{\mathbf{a}}, \mathbf{S})$ we require that it varies only linearly with $\hat{\mathbf{a}} \cdot \mathbf{S}$

$$w_+(\hat{\mathbf{a}}, \mathbf{S}) = C_1 + C_2 \hat{\mathbf{a}} \cdot \mathbf{S} \qquad (12)$$

From equations (9)–(12) we obtain the probability functions

$$w_+(\hat{\mathbf{a}}, \mathbf{S}) = \tfrac{1}{2} + \hat{\mathbf{a}} \cdot \mathbf{S} \qquad \text{and} \quad w_-(\hat{\mathbf{a}}, \mathbf{S}) = \tfrac{1}{2} - \hat{\mathbf{a}} \cdot \mathbf{S} \qquad (13)$$

or, if the angle between $\hat{\mathbf{a}}$ and $\mathbf{S}$ is denoted by θ,

$$w_+(\theta) = \tfrac{1}{2} + \tfrac{1}{2}\sqrt{3} \cos \theta \qquad \text{and} \qquad w_-(\theta) = \tfrac{1}{2} - \tfrac{1}{2}\sqrt{3} \cos \theta \qquad (14)$$

In the interval $(0, \pi)$ w_- is the mirror image of w_+, reflected about $\pi/2$. Thus the discussion will be restricted to w_+ (see Figure 1). Between 0 and $\theta_{1/2}$, $w_+(\theta)$ exceeds unity where

$$\theta_{1/2} = \arccos(1/\sqrt{3}) \approx 54.7° \qquad (15)$$

while it takes negative values between $\theta_{-1/2}$ and π, where

$$\theta_{-1/2} = \arccos(-1/\sqrt{3}) \approx 125.3° \qquad (16)$$

If a beam of unpolarized particles is measured, the probability for the result spin-up is given by

$$P_+ = \frac{1}{4\pi} \int_0^{2\pi} d\tau \int_0^{\pi} d\theta \, \sin \theta \, w_+(\theta) = \tfrac{1}{2} \qquad (17)$$

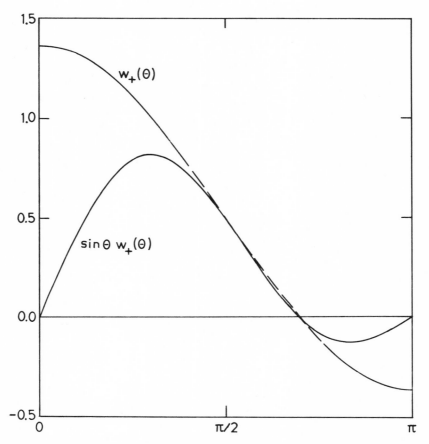

Figure 1. Probability function $w_+(\theta)$ and weighted probability function $\sin\theta\, w_+(\theta)$. $w_+(\theta)$ is the probability for the result spin-up if the spin vector of an incident spin-$\frac{1}{2}$ particle forms an angle θ with the direction of measurement.

as it should for a random $(4\pi)^{-1}$ distribution of spin vectors. The average of the angles θ_+ the original angles of those of the incident spin vectors which have yielded the result spin-up, is

$$\langle\theta_+\rangle = \int_0^{2\pi} d\tau \int_0^\pi \theta\, d\theta \sin\theta\, w_+(\theta) \Big/ \int_0^{2\pi} d\tau \int_0^\pi d\theta \sin\theta\, w_+(\theta)$$

$$= (4-\sqrt{3})\pi/8 \approx 51.0° \tag{18}$$

Accordingly the mean square can be calculated, yielding

$$\sqrt{\langle\theta_+^2\rangle} = \sqrt{[(4-\sqrt{3})\pi^2/8]-2} \approx 51.2° \tag{19}$$

and, from equations (18) and (19), we obtain the standard deviation:

$$\Delta\theta_+ = \sqrt{\langle\theta_+^2\rangle - \langle\theta_+\rangle^2} \approx 4.0° \tag{20}$$

Of course, this low value is generated by the negative parts of $w_+(\theta)$ and has only formal character, because the shape of the probability distribution $w_+(\theta)$ does not actually look like that. But it is remarkable that those spin vectors which, after measurement, form an angle $\theta_{1/2}$ with the field direction arise from an incident angular distribution with average angle $\langle\theta_+\rangle$ differing by less than one standard deviation from $\theta_{1/2}$.*

The probability function $w_+(\theta)$ and the weighted probability function $\sin\theta\, w_+(\theta)$ are shown in Figure 1. The latter takes its maximum value $\sqrt{2/3}$ at

$$\theta_{max} = \theta_{1/2} \tag{21}$$

precisely, and its minimum value $-1/8$ at

$$\theta_{min} = 5\pi/6 \tag{22}$$

It goes to zero for

$$\theta_{01} = 0, \qquad \theta_{02} = \theta_{-1/2}, \qquad \theta_{03} = \pi \tag{23}$$

One further property of the probability function $w_+(\theta)$ should be noted from the point of view of practical calculations: Its average $w_+(\overline{\theta_1, \theta_2})$ over any interval (θ_1, θ_2) on the sphere is always equal to its average value at the borders of the interval:

$$w_+(\overline{\theta_1, \theta_2}) = \int_{\theta_1}^{\theta_2} 2\pi\, d\theta\, \sin\theta\, w_+(\theta) \Big/ \int_{\theta_1}^{\theta_2} 2\pi\, d\theta\, \sin\theta$$

$$= \tfrac{1}{2}[w_+(\theta_1) + w_+(\theta_2)] \tag{24}$$

In the case of Bell's version of the EPR experiment, with detectors oriented in directions $\hat{\mathbf{a}}$ and $\hat{\mathbf{b}}$, and correlated spin vectors $\mathbf{S} = \mathbf{S}_1 = -\mathbf{S}_2$, we obtain for the expectation value of the product of both measurements:

$$\langle s_1(\hat{\mathbf{a}}, \mathbf{S}) \cdot s_2(\hat{\mathbf{b}}, -\mathbf{S})\rangle = (4\pi)^{-1} \int_\Omega [w_+(\hat{\mathbf{a}}, \mathbf{S})s_+ + w_-(\hat{\mathbf{a}}, \mathbf{S})s_-]$$

$$\times [w_+(\hat{\mathbf{b}}, -\mathbf{S})s_+ + w_-(\hat{\mathbf{b}}, -\mathbf{S})s_-]\, d\Omega \tag{25}$$

---

* It is easy to construct probability functions $w_+(\theta)$ which lead to a zero or even imaginary standard deviation $\Delta\theta_+$ while satisfying equations (9), (10), and (11). Unfortunately they fail to reproduce the correct quantum correlations.

Denoting the angles as follows: $\sphericalangle(\hat{\mathbf{a}}, \mathbf{S}) = \theta$, $\sphericalangle(\hat{\mathbf{a}}, \hat{\mathbf{b}}) = \phi$, $\sphericalangle(\hat{\mathbf{b}}, -\mathbf{S}) = \psi = \pi - \sphericalangle(\hat{\mathbf{b}}, \mathbf{S})$, and $d\Omega = d\tau\, d\theta \sin\theta$ where τ is the angle between ϕ and θ (see Figure 2), we can utilize the relation

$$\cos(\pi - \psi) = \cos\phi \cos\theta + \sin\phi \sin\theta \cos\tau \qquad (26)$$

to arrive at the desired result:

$$\langle s_1(\hat{\mathbf{a}}, \mathbf{S}) \cdot s_2(\hat{\mathbf{b}}, -\mathbf{S}) \rangle = -\tfrac{1}{4}\hat{\mathbf{a}} \cdot \hat{\mathbf{b}} \qquad (8')$$

for a random distribution of incident spin-vector pairs in the singlet state. Similarly, all predictions of quantum theory that concern averages of measurements on spin-$\frac{1}{2}$ particles are reproduced.

If, for instance, a beam of spin-$\frac{1}{2}$ particles is polarized in direction $\hat{\mathbf{o}}$ (i.e., the particles have passed a magnetic field oriented in the direction $\hat{\mathbf{o}}$, and those which have shown spin-down have been removed), the expectation for a spin measurement in the direction $\hat{\mathbf{a}}$ is (see Figure 3):

$$\langle s(\hat{\mathbf{a}}, \mathbf{S}(\hat{\mathbf{o}})) \rangle = (2\pi)^{-1} \int_0^{2\pi} d\tau [w_+(\hat{\mathbf{a}}, \mathbf{S})s_+ + w_-(\hat{\mathbf{a}}, \mathbf{S})s_-] \qquad (27)$$

A straightforward calculation, like the one above, results in

$$\langle s(\hat{\mathbf{a}}, \mathbf{S}(\hat{\mathbf{o}})) \rangle = \tfrac{1}{2}\hat{\mathbf{a}} \cdot \hat{\mathbf{o}} \qquad (28)$$

These examples show that every version of Bell's inequality which has been derived hitherto, or will be derived in the future, is invalid if extended probabilities are permitted. Also Wigner's refutation,[27] obviously not based on probabilities, does not apply because it involves determined sets which, in a probabilistic theory, simply do not exist.

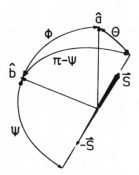

Figure 2. Angular configuration for a certain direction of the incident spin vector $\mathbf{S}$ in Bell's version of the EPR experiment. τ (not indicated) is the angle between ϕ and θ.

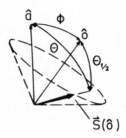

Figure 3. According to the adopted model, the spin vectors $S(\hat{o})$ of a beam of spin-$\frac{1}{2}$ particles polarized in direction $\hat{o}$ lie on a cone of angle $\theta_{1/2}$ about $\hat{o}$.

The probability functions (13) for spin-$\frac{1}{2}$ particles have also been derived by Barut and Meystre.[28] Their approach is based on a decomposition of the operator $\sigma_1\hat{\mathbf{a}} \otimes \sigma_2\hat{\mathbf{b}}$ similar to the joint probabilities: P_{++} for both particles showing spin-up, P_{--} for both particles showing spin-down, and P_{+-} and P_{-+} for opposite results. Equation (8) can then be written

$$-\tfrac{1}{4}\hat{\mathbf{a}} \cdot \hat{\mathbf{b}} = -\tfrac{1}{4}\cos\phi$$

$$= \tfrac{1}{8}[\sin^2(\phi/2) + \sin^2(\phi/2) - \cos^2(\phi/2) - \cos^2(\phi/2)]$$

$$= \tfrac{1}{4}[P_{++}(\hat{\mathbf{a}}, \hat{\mathbf{b}}) + P_{--}(\hat{\mathbf{a}}, \hat{\mathbf{b}}) - P_{+-}(\hat{\mathbf{a}}, \hat{\mathbf{b}}) - P_{-+}(\hat{\mathbf{a}}, \hat{\mathbf{b}})] \qquad (29)$$

The operator can be given an analogous form by decomposing it in terms of the projection operators as follows:

$$\sigma_1\hat{\mathbf{a}} \otimes \sigma_2\hat{\mathbf{b}} = \tfrac{1}{4}[(1 + \sigma_1\hat{\mathbf{a}})(1 + \sigma_2\hat{\mathbf{b}}) + (1 - \sigma_1\hat{\mathbf{a}})(1 - \sigma_2\hat{\mathbf{b}})$$

$$- (1 - \sigma_1\hat{\mathbf{a}})(1 + \sigma_2\hat{\mathbf{b}}) - (1 + \sigma_1\hat{\mathbf{a}})(1 - \sigma_2\hat{\mathbf{b}})] \qquad (30)$$

Denoting $\hat{\mathbf{a}} \cdot \mathbf{S}$ by x and $\hat{\mathbf{b}} \cdot (-\mathbf{S})$ by y, one can use the identity

$$xy \equiv \tfrac{1}{4}[(\alpha + x)(\beta + y) + (\alpha - x)(\beta - y)$$

$$- (\alpha - x)(\beta + y) - (\alpha + x)(\beta - y)] \qquad (31)$$

valid for arbitrary constants α and β, in order to obtain a decomposition similar to equation (29). With an appropriate choice: $\alpha = \beta = \frac{1}{2}$, the bracketed terms become equal to the probability functions (13) which, if averaged over the sphere, exactly reproduce the quantum-mechanical joint probabilities for Bell's version of the EPR experiment:

$$P_{++}(\hat{\mathbf{a}}, \hat{\mathbf{b}}) = \langle w_+(\hat{\mathbf{a}}, \mathbf{S}) \cdot w_+(\hat{\mathbf{b}}, -\mathbf{S}) \rangle = \tfrac{1}{4}(1 - \cos\phi) = \tfrac{1}{2}\sin^2(\phi/2) \qquad (32)$$

and similarly for P_{+-}, P_{-+}, and P_{--}.

Barut and Meystre[28] note that these probability functions are not positive definite and, therefore, cannot be interpreted as probabilities. This feature can be avoided by rejecting some events[29] or recording each event with a weight factor equal to or less than unity.[28] A full discussion of this type of approach is beyond the scope of this chapter.

To complete our own discussion of this type of approach, we note that a two-dimensional probability function (using the angle* θ between a hidden variable and a polarizer direction within a plane) has been given by Scully.[30] This "passage probability" (for the result spin-up) is given by

$$w_+(\theta) = \tfrac{1}{2} + \cos\theta/\sqrt{2} \tag{33}$$

The distribution of spins which have passed a polarizer set up in the $\hat{x}$ direction ($\phi = 0$), showing spin-up, consists of two components

$$\rho_+(\theta) = \tfrac{1}{2}[\delta(\theta - \pi/4) + \delta(\theta + \pi/4)] \tag{34}$$

For this distribution we find that the probability $P_+(\phi)$ of showing spin-up after passing a second polarizer oriented at an angle ϕ with respect to the $\hat{x}$ direction is

$$P_+(\phi) = \int_0^{2\pi} d\theta\, \rho_+(\theta) w_+(\theta - \phi) = \cos^2(\phi/2) \tag{35}$$

This result is correct, but obviously the probability function $w_+(\theta)$ takes values between $\tfrac{1}{2}(1 - \sqrt{2})$ and $\tfrac{1}{2}(1 + \sqrt{2})$.

In a closely-related hidden-variables treatment, Scully arrives at a nonnegative passage probability. This theory, however, violates our axiom (2) in that it is nonlocal.

4.2. Correlated Photon Pairs

The most convincing evidence against the existence of local hidden variables has been gathered by employing photon pairs stemming from an atomic cascade.[4,5,24] Such experiments measure the transmission or deflection of both photons of a pair incident on polarizers which are oriented along axes $\hat{a}$ and $\hat{b}$, respectively. Of course, any description of the axis of polarization in terms of hidden variables has to obey Malus' $\cos^2$ law.

---

* Throughout this chapter ϕ is the angle between the directions of measurement while θ and ψ are reserved to denote the angles between analyzers and hidden variables. Thus the couple denoted elsewhere[30] by (α, θ) is here written (θ, ϕ).

Hence, if the polarization of the first photon was known to be along θ, and that of the second photon was known to be along ψ, the probability for simultaneous transmission of both photons is given by

$$p_{++} = \cos^2(\theta - \hat{\mathbf{a}}) \cos^2(\psi - \hat{\mathbf{b}}) \tag{36}$$

In order to verify this, one could prepare pairs of suitably polarized photons by means of auxiliary polarizers.

In Bell's version of the photon-correlation experiment the axes of polarization are not known but are assumed to exist. In order to predict the expectation value, one has to calculate the average over the right-hand side of equation (36) by inserting the initial distribution $\rho(\theta, \psi)$ of hidden polarizations. In the case of photon pairs emitted from a 0–1–0 cascade, $\theta = \psi$ and θ is randomly distributed. This means that

$$\rho(\theta, \psi) = \delta(\theta - \psi)/2\pi \tag{37}$$

Thus, if two photons are emitted from a 0–1–0 cascade, the averaged probability for simultaneously passing the polarizers is at most[31]

$$P_{++}(\hat{\mathbf{a}}, \hat{\mathbf{b}}) = \int_0^{2\pi} d\psi \int_0^{2\pi} d\theta\, \rho(\theta, \psi) \cos^2(\theta - \hat{\mathbf{a}}) \cos^2(\psi - \hat{\mathbf{b}}) \tag{38}$$

No local hidden-variables theory can do better, i.e., can produce stronger correlations. But integration of equation (38) yields (with ϕ the angle between $\hat{\mathbf{a}}$ and $\hat{\mathbf{b}}$):

$$P_{++}(\phi) = \tfrac{1}{8} + \tfrac{1}{4}\cos^2 \phi \tag{39}$$

violating Malus' law and experimental observation for every choice of ϕ, except $\pi/4$, $3\pi/4$, and related angles.

The way to improve this result, considered by Meystre,[32] leads us into the realm of extended probabilities. As the δ-function is apparently not sharp enough to reproduce the strong quantum correlations, he adds a "quantum correction" such that the corrected distribution is then:

$$\rho_c(\theta, \psi) = \delta(\theta - \psi)/2\pi - \delta''(\theta - \psi)/8\pi \tag{37'}$$

and, by inserting this into equation (38), he arrives at the correct result:

$$P_{++}(\phi) = \tfrac{1}{2}\cos^2 \phi \tag{39'}$$

In order to show explicitly that extended probabilities occur, we choose $\delta''(\theta - \psi) = \partial^2[\delta(\theta - \psi)]/\partial\theta^2$. Then equations (37') and (38) give:

$$P_{++}(\hat{\mathbf{a}}, \hat{\mathbf{b}}) = (2\pi)^{-1} \int_0^{2\pi} d\theta \cos^2(\theta - \hat{\mathbf{a}})[2\cos^2(\theta - \hat{\mathbf{b}}) - \tfrac{1}{2}] \tag{40}$$

and, within the interval $(0, \pi)$, the integrand becomes negative for $\pi/3 < (\theta - \hat{\mathbf{b}}) < 2\pi/3$, no matter what the value of $(\theta - \hat{\mathbf{a}})$. Owing to our choice of δ'', this asymmetry is not surprising, but it is unphysical. Equation (40) can be written in a symmetrical form:

$$P_{++}(\hat{\mathbf{a}}, \hat{\mathbf{b}}) = (2\pi)^{-1} \int_0^{2\pi} d\theta[2\cos^2(\theta - \hat{\mathbf{a}})\cos^2(\theta - \hat{\mathbf{b}})$$

$$- \tfrac{1}{4}\cos^2(\theta - \hat{\mathbf{a}}) - \tfrac{1}{4}\cos^2(\theta - \hat{\mathbf{b}})] \tag{40'}$$

The integrand of equation (40') can be considered to be a probability function that assumes values between $-\tfrac{1}{4}$ and $+\tfrac{3}{2}$, and reproduces precisely the quantum-mechanical expectation for Bell's version of the EPR experiment employing photons. But the present writer has been unable to find a decomposition of the integrand in the form of a product of two probability functions, each of which depends on the orientation of one polarizer only, as would be required by a local theory.

In order to describe the interaction of photons with polarizers in a completely symmetrical and local way, we begin by stipulating the correct result.

The simultaneous transmission probability of two photons, emitted from a 0–1–0 cascade, through polarizers which are set up at a relative angle ϕ is thus to be calculated by means of two independent probability functions:

$$P_{++}(\phi) = (2\pi)^{-1} \int_0^{2\pi} d\theta\, w_+(\theta)w_+(\theta - \phi) = \tfrac{1}{2}\cos^2\phi \tag{41}$$

This condition is satisfied by[33]

$$w_+(\theta) = \tfrac{1}{2} + \cos(2\theta)/\sqrt{2} \tag{42}$$

Accordingly, the probability function for deflection is

$$w_-(\theta) = \tfrac{1}{2} - \cos(2\theta)/\sqrt{2} \qquad (42')$$

As required, these probability functions supply the correct results for the outcome of Bell's version of the EPR experiment, if inserted in equation (41) and in the analogous equations:

$$P_{++}(\phi) = P_{--}(\phi) = \tfrac{1}{2}\cos^2\phi, \; P_{+-}(\phi) = P_{-+}(\phi) = \tfrac{1}{2}\sin^2\phi$$

However, the angle θ cannot be regarded as being between polarizer axis and polarization axis, because $w_+(\theta = 0)$ exceeds unity. Instead, we have to introduce a hidden variable, which we call $\hat{\mathbf{u}}$, that is situated in the plane perpendicular to the motion of the photon.

For an unpolarized beam, the hidden variables $\hat{\mathbf{u}}$ assume a homogeneous $(2\pi)^{-1}$ distribution. After having passed through a polarizer, they form an angle of $\theta_p = \pm\pi/8$ with the polarizer axis, while those of the deflected photons form an angle of $\theta_d = \pm 3\pi/8$ with the polarizer axis. These directions are indicated in Figure 4 for a beam of photons which has interacted with a polarizer oriented along the x-axis. These conditions are necessary because the passage probability for the respective angles is unity or zero, and repeated measurements in the $\hat{\mathbf{x}}$ or $\hat{\mathbf{y}}$ directions have to be taken into account.

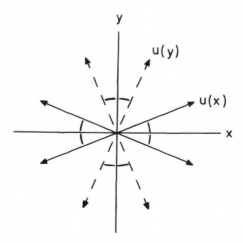

Figure 4. Four directions to be assumed by the hidden variables $\hat{\mathbf{u}}$ of a photon beam polarized along the x-axis (solid arrows) and the y-axis (dashed arrows), respectively. The angles indicated are always $\pi/8$.

The distribution of the $\hat{u}$ in a beam which is polarized under $\phi = 0$ is given by (normalized to unity):

$$\rho_p(\theta) = \tfrac{1}{4} \sum_{i=1}^{4} \delta(\theta - \theta_i) \tag{43}$$

where θ_i represents the angles $\pi/8$, $\pi - \pi/8$, $\pi + \pi/8$, and $-\pi/8$. The transmission through a polarizer for a beam polarized under the angle ϕ is then

$$P_+(\phi) = \int_0^{2\pi} d\theta \, \rho_p(\theta) w_+(\theta - \phi) = \tfrac{1}{4} \sum_{i=1}^{4} w_+(\theta_i - \phi) \tag{44}$$

The result is $P_+(\phi) = \cos^2 \phi$, as desired (because the polarized beam had been normalized to unity).

Consequently, we find the correct results for both cases: Bell's version of the EPR experiment and polarizers set up consecutively, if we accept probabilities between $(1 - \sqrt{2})/2$ and $(1 + \sqrt{2})/2$, and the existence of microstates which, after interaction with a polarizer, are populated with equal probability (otherwise it would be possible to enhance one of them, and another polarizer, oriented suitably, would show a transmission exceeding unity).

It is not known whether the $\hat{u}$ have a vector-like character or whether they are to be treated as axes only; there is no experimental evidence for them at all, other than it being precisely that angle of $\pi/8$ that lies on the threshold where the difficulties with most local hidden-variables theories of photons start and the most significant violations of Bell's inequality appear. In the above treatment the $\hat{u}$ were assumed to be vectors, thus resulting in four different microstates. If they had been treated as axes, we would have had to deal with only two different microstates. This is not an unfamiliar situation because every linearly polarized beam can be decomposed into two beams with different circular polarization. However, it should be emphasized that this treatment has only been given in order to show that extended probabilities are capable of supplying a formal resolution of the EPR problem in the case of photon correlations as well. The very existence of the hidden variables $\hat{u}$ can quite reasonably be doubted and a more plausible theory is to be developed.

5. Conclusion

From the examples discussed one might conclude that extended probabilities are required for elementary particles only, i.e., where the quantum behavior manifests itself most significantly. This is correct insofar as extended probabilities disappear as soon as the corresponding phase space is smeared out over regions of order $\hbar$, but it has been proved by Mermin[34,35] that without such a procedure a description of composed systems in terms of our three axioms involves extended probabilities even for macroscopic objects. Thus the EPR problem has become more pressing then ever.

It is sometimes advocated that the locality requirement, i.e., our axiom (2), should be relinquished in order to leave quantum theory as it is. But it can be shown explicitly that even nonlocal deterministic or probabilistic interactions do not permit us to maintain the quantum-theoretical notion according to which the total value of the spin is $\sqrt{3}/2$ and the measured component is $\frac{1}{2}$.[36] Thus if locality is abolished, which is in fact suggested by a variety of recent experiments, then this quantum-theoretic definition will also have to be modified.

It has been shown in this chapter that by introducing extended probabilities a formalism can be constructed which, if applied to spin-$\frac{1}{2}$ particles or photons, always yields the correct results by utilizing local interactions only. But it is really inconceivable that probabilities should always interfere sufficiently precisely to prevent the negative probabilities from being detected. It is also inconceivable that they interfere at all when the measurements are performed at distant places. Their so doing necessarily entails some kind of nonlocality—not on the physical level but on a level at which the experimental results are evaluated and compared.

This theory is ugly enough to restrain the author from any further dealings with it—if someone would only show him a more plausible resolution of the EPR paradox.

References

1. A. Einstein, B. Podolsky, and N. Rosen, *Phys. Rev.* **47**, 777 (1935).
2. D. Bohm, *Quantum Theory*, pp. 611–622, Prentice-Hall, New York (1951).
3. J. S. Bell, *Physics* **1**, 195 (1964).
4. J. F. Clauser, *Phys. Rev. Lett.* **36**, 1223 (1976).
5. A. Aspect, J. Dalibard, and G. Roger, *Phys. Rev. Lett.* **49**, 1804 (1982).
6. W. Mückenheim, *Phys. Rep.* **133**, 337 (1986).
7. W. Mückenheim (ed.) *Erweiterte Wahrscheinlichkeiten*, A collection of papers on extended probabilities, in preparation.
8. M. S. Bartlett, *Proc. Cambridge Phil. Soc.* **41**, 71 (1945).
9. P. A. M. Dirac, *Rev. Mod. Phys.* **17**, 195 (1945).
10. E. P. Wigner, private communication.
11. E. P. Wigner, in: *Perspectives in Quantum Theory* (W. Yourgrau and A. van der Merwe, eds.), pp. 25–36, MIT Press, Cambridge, MA (1971).
12. D. Iagolnitzer, *J. Math. Phys.* **10**, 1241 (1969).
13. V. Weisskopf and E. Wigner, *Z. Phys.* **63**, 54 (1930).
14. F. J. Lynch, R. E. Holland, and M. Hamermesh, *Phys. Rev.* **120**, 513 (1960).

15. C. S. Wu, Y. K. Lee, N. Benczer-Koller, and P. Simms, *Phys. Rev. Lett.* **5**, 432 (1960).
16. M. D. Reid and D. F. Walls, *Phys. Rev. Lett.* **53**, 955 (1984).
17. E. T. Jaynes, *Found. Phys.* **3**, 477 (1973).
18. P. A. M. Dirac, *Proc. Roy. Soc. London Ser. A* **180**, 1 (1942).
19. P. A. M. Dirac, *Comm. Dublin Inst. Adv. Stud. A* **1**, 1 (1943).
20. S. N. Gupta, *Can. J. Phys.* **35**, 961 (1957).
21. W. Heisenberg, *Introduction to the Unified Field Theory of Elementary Particles*, Interscience, London (1966).
22. R. P. Feynman, in: *Quantum Implications* (B. J. Hiley and F. D. Peat, eds.), pp. 235–248, Routledge and Kegan Paul, London (1987).
23. E. Schrödinger, *Naturwissenschaften* **23**, 844 (1935).
24. J. F. Clauser and A. Shimony, *Rep. Prog. Phys.* **41**, 1881 (1978).
25. W. Mückenheim, *Phys. Blätter* **39**, 331 (1983).
26. W. Mückenheim, *Lett. Nuovo Cim.* **35**, 300 (1982).
27. E. P. Wigner, *Am. J. Phys.* **38**, 1005 (1970).
28. A. O. Barut and P. Meystre, *Phys. Lett. A* **105**, 458 (1984).
29. P. M. Pearle, *Phys. Rev. D* **2**, 1418 (1970).
30. M. O. Scully, *Phys. Rev. D* **28**, 2477 (1983).
31. F. J. Belinfante, *A Survey of Hidden-Variables Theories*, p. 282, Pergamon Press, Oxford (1973).
32. P. Meystre, in: *Quantum Electrodynamics and Quantum Optics* (A. O. Barut, ed.), pp. 443–458, Plenum, London (1984).
33. W. Mückenheim, Extended probabilities and the EPR-paradox, unpublished.
34. N. D. Mermin, *Phys. Rev. D* **22**, 356 (1980).
35. N. D. Mermin, Generalizations of Bell's theorem to higher spins and higher correlations, preprint to appear in the proceedings of the Symposium on Fundamental Questions in Quantum Mechanics, SUNY Albany, April 1984.
36. W. Mückenheim, *Ann. Fond. Louis de Broglie* **11**, 173 (1986).

15

The Search for Hidden Variables in Quantum Mechanics

EMILIO SANTOS

1. Can Quantum-Mechanical Description of Physical Reality Be Considered Complete?

The main task of the Einstein–Podolsky–Rosen (EPR) paper,[1] as posed in the above title, was to question the completeness of quantum mechanics, completeness which has been assumed by the majority of the scientific community since 1927. Theories attempting to complete quantum mechanics are called "theories with supplementary parameters" or "hidden-variables theories." Although the first name is more correct, I shall use the second, which is more popular. The conclusion of the EPR paper was that quantum mechanics is not complete, so that paper is regarded as one of the main supports of hidden-variables theories. The EPR theorem is illustrated in Figure 1.

The reason quantum mechanics is not complete can be seen most clearly from the example presented by Einstein at the Solvay Conference of Physics in 1928.[2] It is slightly altered here. A spin-zero radioactive atom is placed at the center of a sphere of photographic emulsion. After a while a dark spot appears at some point in the photographic plate. According to quantum mechanics, the emitted particle is represented by a spherical wave that travels from the atom to the emulsion, where it arrives with equal amplitude at all points. However, the particle is detected at a single point and, at this moment, the wave vanishes at all other points, this process being the

EMILIO SANTOS ● Department of Theoretical Physics, University of Cantabria, Santander 39005, Spain.

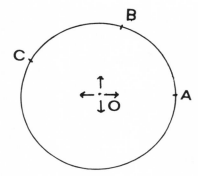

Figure 1. Illustration of the EPR theorem: quantum mechanics is either nonlocal or incomplete. An atom placed at 0 decays with emission of a particle, which is detected at A. Just before the detection, the particle is represented by a spherical quantum-mechanical wave (having the same intensity at A, at B, etc.). If the wave is a physical field possessing energy and momentum, then the wave-packet collapse implies an instantaneous energy transfer from B, C, etc. to A (nonlocal); if the wave is just a formal representation of our knowledge, then more detailed information should be possible (incomplete).

"collapse of the wave function." Now, either the quantum-mechanical wave is just a representation of our knowledge about the particle's position, or it represents a real field propagating in space. In the first case, quantum mechanics is not complete because it does not describe completely the real world. In the second case, quantum mechanics is not local because the wave collapse takes place at all points of the screen simultaneously, i.e., the detection event produces action at a distance (propagating with infinite velocity). The EPR argument, which is reproduced in detail in other chapters of this book, is just a refinement of the above example.

The trouble with quantum theory is that both alternatives—complete and incomplete—are unsatisfactory. The first alternative, which is assumed by the Copenhagen interpretation, gives rise to many paradoxical situations some of which will be commented on briefly below: de Broglie's box, Schrödinger's cat, Wigner's friend, quantum Zeno's paradox, and so on. The second alternative—sometimes called the statistical interpretation—is quite reasonable at a superficial glance, but it involves the following difficulties:

1. Sixty years after the discovery of quantum mechanics, nobody has been able to develop it in a coherent form.
2. There are numerous proofs, of various degrees of generality, that it is really not compatible with the quantum formalism (proofs of impossibility of hidden variables).
3. It is considered disproved empirically by the results of the recent experimental tests of Bell's inequalities (but see below).

The purpose of the present paper is to show that the second alternative is still possible, with some small changes in the accepted formalism of quantum theory, maybe not implying a change in the predictions of actual experiments.

We shall now comment on the paradoxes which arise from the completeness assumption. These paradoxes are very well known but, strangely enough, the scientific community seems unaware of them, to the extent that it is at present more respectable to speculate about what happened 10^{-43} second after the "birth of the universe" than to search for completions or alternatives to quantum mechanics. It is certainly to the credit of Niels Bohr[3] and his followers that they offered the scientific community an extremely subtle set of conventions able to hide, if not destroy, the problems of the completeness assumption. This subtle—and, in my opinion, confused—way of thinking is called the Copenhagen interpretation. Although confused, this interpretation has survived for sixty years because no real alternative has been found.

All paradoxes derive from the fact that, according to quantum mechanics, there are two types of evolution. Isolated systems evolve according to unitary transformations of the Hilbert space (the Schrödinger equation), but systems interacting with measuring apparatus evolve according to nonunitary transformations consisting of projections onto a subspace related to the observables measured. This type of evolution is usually called the *reduction* or *collapse* of the wave packet.

A typical example is an initially excited atom, which decays to the ground state with emission of an α particle, i.e.,

$$A \rightarrow B + \alpha \tag{1}$$

The Schrödinger equation for this problem is cumbersome but, for our purposes, the solution can be represented schematically by the wave function

$$\psi(t) = (1 - e^{-\gamma t})^{1/2}\psi_A + e^{-\gamma t/2}\psi_B \tag{2}$$

where ψ_A represents atom A with lifetime $1/\gamma$ and ψ_B represents the system consisting of atom B plus the α particle. If a measurement is made at time $t_1 > 0$, then the system is found either as atom A (with probability $1 - e^{-\gamma t_1}$) or as atom B plus the α particle (with probability $e^{-\gamma t_1}$). The problem is how to interpret the wave function (2) prior to the measurement. Two possibilities exist. If we assume that $\psi(t)$ fully describes the real state of the system, then it is very difficult to understand what such a state is. In order to illustrate the problem more clearly, Schrödinger[4] considered an imaginary example where, besides the radioactive atom, there is some device that kills a cat when the atom decays. Then, the corresponding $\psi(t)$ for the full system (atom + device + cat) is similar to equation (2) and corresponds to a state in which the cat is partially dead and partially alive. The obvious alternative is to think that $\psi(t)$ does not describe the real system, but rather our knowledge about the system. In this case, however, the quantum description is not complete because it does not tell us whether the cat is actually dead or alive at time t_1. We note that the interpretation of ψ as describing knowledge does not necessarily give it a subjective character. In fact, the wave function can be objectively related to the available data about the system in a similar way, as each possible result of tossing a (nonbiased) dice has an objective probability of $1/6$.

There is another argument, known already in 1927, that prevents the possibility of a naive interpretation of the wave function as a real field propagating in space. It derives from the fact that the wave function of a system of N particles depends on $3N$ coordinates—plus time—and so is a wave in an abstract space of $3N$ dimensions (configuration space) and not in real three-dimensional space. This fact has been confirmed by detailed calculations in the helium atom,[6] the simplest nontrivial many-body problem. As a consequence we can consider as firmly established that the wave function *cannot* represent a real field. (Surprisingly enough, recent papers in respected review journals[7] seem unaware of this sixty-year-old knowledge.) It should be stressed that this fact is established for electron systems, but certainly not for other entities like the quantized electromagnetic field, where a "many-photon configuration space" simply does not exist. Much confusion has arisen by the attempt to put all material entities on the same footing, and derives from the belief that all of them possess both a particle and a wave character. Actually, photons are sharply different from electrons even in quantum theory, a distinction extendible to all elementary bosons and fermions, respectively.

The remaining possibility is to consider the (many-electron) wave function as just a mathematical construction able to represent the available information. This leads to the so-called statistical interpretation.[8] It is a fact that this is the interpretation supported—unconsciously—by the majority of working physicists. The most aware supporters, however, realize that this interpretation leads naturally to the idea that quantum mechanics is incomplete, and therefore one should search for hidden-variables theories which, on the other hand, are forbidden by the impossibility theorems. As a consequence, supporters of the statistical interpretation try to avoid the problem by detaching the "knowledge interpretation" of the wave function from the incompleteness of quantum mechanics and/or this incompleteness from the need for hidden-variables theories. In my opinion, this detachment is artificial and transforms the statistical interpretation into a close relative of the Copenhagen interpretation.

Finally, we comment on the Copenhagen interpretation. It starts with the assumption that quantum mechanics is complete and therefore rejects emphatically hidden-variables theories. Indeed, the impossibility proofs or the empirical disproofs of hidden variables have been celebrated as triumphs of the Copenhagen interpretation. Then, what about the paradoxes of Schrödinger, EPR, and so on, or the configuration-space domain of the wave function? These problems are solved by a change in the concept of science. In the prequantal era or in the view of most present-day "non-orthodox" people, the purpose of science is to describe as closely as possible a real external world, "which is independent of any theory," as EPR[1] put it. Following also EPR, the description can be complete only if there is some element in the theory corresponding to each element of the physical reality. In contrast, for Bohr[3] and the followers of the Copenhagen interpretation, the purpose of science is to predict (to be able to calculate) the results of experiments or observations. The question of the existence of a real world is rejected as metaphysical, i.e., outside the realm of science. Then, completeness is not the adequacy between the theory and the real world (whose existence is not necessarily assumed), but the adequacy between what can be predicted and what can be measured. In this sense, quantum mechanics is considered complete. For instance, the Heisenberg uncertainty relations—a consequence of the quantum formalism—correspond fairly well to the precision bounds in the measurements as shown, for instance, by the Heisenberg microscope.

The criticism of the Copenhagen interpretation is that it hides rather than solves the problems. In particular, the question of completeness involves a circular reasoning, because nothing can be measured if there does not exist a previous theory—even tentative—to define what is to be measured. For instance, prior to Einstein's theory of Brownian motion[9] many workers tried to measure the average velocity of Brownian particles, without success, because the different measurements gave apparently quite diverse results. Einstein's theory showed that the relevant quantity is not the ratio $\Delta x/\Delta t$ (velocity), but the ratio $(\Delta x)^2/\Delta t$ (diffusion coefficient), and thereafter it was measured quite accurately. In a similar way, quantum mechanics is considered complete (in the Copenhagen sense) simply because there is no other more detailed theory able to guide us toward measurements at the subquantum level. To summarize, as Schrödinger put it, the Copenhagen interpretation is a desperate attempt at confronting a serious crisis.[10]

2. The Various Kinds of Hidden Variables

As discussed in the previous section, the debate about the interpretation of the quantum formalism is closely related to the controversy about the possibility or not of hidden-variables theories. The controversy began in the founding years of quantum mechanics, 1925–7, and seemed closed with the 1932 book of von Neumann,[5] who established rigorously the mathematical structure of the theory and showed that it is not compatible with hidden variables. Due to von Neumann's authority the search for hidden variables was blocked for twenty years. The criticisms of the current interpretation of quantum mechanics were not absent, however, the EPR argument being the outstanding example. In 1952 Bohm[11] was able to construct a particular hidden-variables theory, which showed the incorrectness of the usual interpretation of von Neumann's theorem. Bohm's work reactivated the field and a large number of papers[12,13] were devoted to the subject in the following decade (such as Mackey,[14] Gleason,[15] Kochen, Specker, Iauch,[16] Piron[17]; see Hooker[13]) culminating in the two celebrated papers of John S. Bell in 1965-6.[18,19] In the last twenty years Bell's inequalities and their empirical tests have been the subject of most discussions in relation to the foundations of quantum mechanics.[20,21] The strange fact that there are both proofs of impossibility and particular examples of hidden-variables theories shows that there are several different concepts included under that heading. The purpose of this section is to clarify the subject by discussing the various types of hidden variables used.

Hidden-variables theories attempt to do for quantum mechanics what statistical mechanics achieved for thermodynamics, namely, to provide a more detailed theory which agrees with the former when suitable averages are considered. Quantum mechanics considers two kinds of states for physical systems: pure states and mixtures. A pure state is represented by a wave function or, more generally, a vector in a Hilbert space. Actually, all vectors in a ray (one-dimensional subspace) are assumed to represent the same state, so that the most appropriate representation for a pure state is by means of a projection operator

$$p_\psi = |\psi\rangle\langle\psi| \tag{3}$$

where $|\psi\rangle$ (or $\langle\psi|$) is a ket (or bra) vector in Dirac's notation. Mixed states are represented by density operators, which can be obtained by convex (i.e., with positive coefficients) linear combinations of projection operators. On the other hand, any hidden-variables theory associates with a physical system a set of hidden parameters λ with domain Λ. The quantum states, either pure or mixed, correspond to probability distributions on Λ and are therefore mixed states in the theory. Pure states are now associated with a given value of λ, and will be called "microstates." Observables correspond to self-adjoint operators in quantum mechanics and to real functions on Λ in the hidden-variables theory. The condition that this theory reproduces quantum predictions leads to the assumption that all expectation values agree, i.e.,

$$\operatorname{tr}\hat{\rho}\hat{A} = \int_\Lambda A(\lambda)\,d\mu(\lambda) \equiv \langle A\rangle_\rho \tag{4}$$

where $\hat{A}$ and $\hat{\rho}$ [$A(\lambda)$ and $\mu(\lambda)$] represent the quantum operator (hidden-variables observable) and state, respectively. In the case of a pure quantum state, equation (4) reduces to

$$\langle\psi|\hat{A}|\psi\rangle = \int A(\lambda)\,d\mu(\lambda) \tag{5}$$

Similarly, correlations between compatible observables should also agree, i.e.,

$$\operatorname{tr}\hat{\rho}\hat{A}\cdots\hat{B} = \int_{\Lambda} A(\lambda)\cdots B(\lambda)\,d\mu_{\rho}(\lambda) \equiv \langle A\cdots B\rangle_{\rho} \tag{6}$$

If no other restriction than equation (6) is imposed, it is rather obvious that a large degree of freedom exists for the construction of hidden-variables theories. However, a number of additional conditions could be imposed on physical grounds. von Neumann[5] considered the following:

Q1. All self-adjoint operators correspond to observables.
Q2. All density operators correspond to possible states.
Q3. If $\hat{A}$ is the operator associated with observable A, then $\hat{A}^n$ is the operator associated with A^n.
 N. The linear (vector space) structure of the set of operators is isomorphic with the corresponding structure of the set of hidden-variables observables (functions of λ).

From these postulates von Neumann proved that hidden-variables theories are not possible. It must be stressed that assumptions Q1, Q2, and Q3 are considered a part of the quantum formalism and are usually not questioned in the search for hidden-variables theories. On the other hand, assumption N (for von Neumann) defines a kind of hidden-variables theory and can be certainly questioned.

The theorem can be proved by showing the impossibility in a simple example: the ground state of the one-dimensional harmonic oscillator. According to postulates Q1 and Q2 above, plus the general condition (4), we have

$$\langle x^n\rangle = \langle\psi|\hat{x}^n|\psi\rangle \tag{7}$$

whence the probability distribution for the position is a Gaussian, a well-known result. Now, quantum mechanics as well as classical mechanics assumes the following relation between kinetic energy T, potential energy $V = \frac{1}{2}m\omega^2 x^2$, and total energy E:

$$E = T + V \geqslant \frac{1}{2}m\omega^2 x^2 \tag{8}$$

where the inequality originates from the assumption $T \geq 0$. From condition N above, a similar inequality must hold for the observables of the hidden-variables theory, i.e.,

$$E(\lambda) \geq \tfrac{1}{2}m\omega^2 x(\lambda)^2 \tag{9}$$

Multiplying this inequality by $d\mu(\lambda)$ and integrating over the subset of Λ associated with a given position y, we get

$$\langle E \rangle_y \geq \tfrac{1}{2}m\omega^2 y^2 \tag{10}$$

where $\langle E \rangle_y$ is the average energy of those microstates whose position is y. But in quantum mechanics the total energy of the ground state is dispersionless and has the value $\tfrac{1}{2}\hbar\omega$, so that relation (10) leads to

$$y^2 \leq \hbar/m\omega \tag{11}$$

However, the set of values of λ violating relation (11) has a nonzero probability, thus proving that at least one of the assumptions is false. In consequence, there are no hidden-variables theories fulfilling the four conditions stated above. von Neumann's theorem is illustrated in Figure 2.

The most frequent criticism of von Neumann's theorem is that the assumption N is unreasonable physically. In particular, an implication like

$$\alpha\hat{A} + \beta\hat{B} = \hat{C} \Rightarrow \alpha A(\lambda) + \beta B(\lambda) = C(\lambda), \qquad \alpha, \beta \in \mathbf{R} \tag{12}$$

postulated by N, should not be imposed if $\hat{A}$ and $\hat{B}$ do not commute, because then the measurements of A, B, and C require different experimental arrangements. This argument against the relevance of von Neumann's

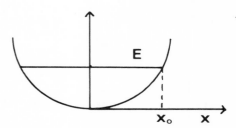

Figure 2. A particle in a parabolic potential well (a harmonic oscillator) has a sharp energy $E = \tfrac{1}{2}\hbar\omega$ in the ground state, but a measurement of position may give $x > x_0 = (\hbar/m\omega)^{1/2}$ with finite probability. These predictions of quantum mechanics are incompatible with the relation $E - \tfrac{1}{2}m\omega^2 x^2 = p^2/2m \geq 0$. Consequently, the quantum-mechanical state cannot be considered as a probability distribution over dispersion-free states, each having well-defined position and momentum. This illustrates von Neumann's theorem of the impossibility of hidden variables in quantum mechanics. The possibility exists, however, if the above relation is modified to $E = \tfrac{1}{2}m\omega^2 x^2 + p^2/2m + E_{\text{hidden}}$.

theorem is today widely accepted, but it should be noted that the rejection of postulate N has dramatic consequences. For instance, in the above-considered example of the harmonic oscillator, the conservation of energy equation (7) will no longer be valid. In order to restore this conservation law, it would be necessary to assume, besides potential and kinetic energy, another "hidden energy," maybe originating from some subquantum medium. Then, hidden-variables theories compatible with assumptions Q1, Q2, Q3 but not N are nontrivial.

The problem, however, is even worse because an impossibility theorem can also be proved on replacing von Neumann's assumption N by the following weaker one.

G. The linear structure of any subset of *commuting* self-adjoint operators is isomorphic with the structure of the corresponding subset of hidden-variables observables.

This is equivalent to assuming the validity of implication (12) only if $\hat{A}$ and $\hat{B}$ commute, which is quite reasonable because then A and B can be measured simultaneously, so that the probability distribution of C can be easily obtained from the joint distribution of A and B by just using the second equation (12) (without any need of additional measurements). This important theorem was first shown by Gleason[15] in 1957. The proof of Gleason's theorem is more involved than that of von Neumann's and is not given here. However, a similar theorem will be proved in Section 4.

In 1966 Bell criticized Gleason's theorem, as well as von Neumann's and similar ones, on the basis that there is an implicit assumption in equation (6) which is not physically reasonable. It is the hypothesis that the numerical value $A(\lambda)$ depends only on the "microstate" λ of the system, besides the measuring apparatus which defines the observable A. According to Bell, that value will also depend on all other things which may possibly influence the process of measurement, such as other measuring apparatus acting simultaneously. All these things define the "context" of the measurement and we should replace equation (6) by

$$\operatorname{tr} \hat{\rho}\hat{A} \cdots \hat{B} = \int A(\lambda, C) \cdots B(\lambda, C)\, d\mu_\rho(\lambda) \equiv \langle A \cdots B \rangle_{\rho, C} \quad (13)$$

It is noteworthy that the context includes, in particular, all observables being measured at the same time. Contextual theories, resting on equation (13) instead of equation (6), are always possible, as shown by Gudder.[22] In other words Gleason's—and similar—theorems do not exclude hidden-variables theories of the contextual kind.

In his 1965 paper (written after the 1966 one) Bell[19] considered local theories, which are somewhat intermediate between contextual and noncontextual. In these theories, it is assumed that the context of an observable contains only those things able to influence the measurement of this observable by means of actions propagating not faster than light. The typical situations are EPR-type experiments, where two measurements are performed on two subsystems (e.g., two particles of a correlated pair) in spatially separated regions (i.e., the measurement events are conducted within a time interval smaller than the distance between them divided by the velocity of light). In this case, equation (13) must be replaced by

$$\text{tr } \hat{\rho}\hat{A}\hat{B} = \int A(\lambda, C_1)B(\lambda, C_2) \, d\mu_\rho(\lambda) = \langle AB \rangle_\rho \tag{14}$$

The important point is that the context C_1 (C_2) does not contain any reference to B (A). Hence if one measures A and B' or A' and B or A' and B', the correlations could be obtained from

$$\langle RS \rangle = \int R(\lambda, C_1)S(\lambda, C_2) \, d\mu_\rho(\lambda) \tag{15}$$

where R (S) stands for A or A' (B or B'). The factorization property of relation (15) was exploited by Bell to derive some inequalities that should be fulfilled by every local hidden-variables theories. Bell was able to show that these inequalities are violated by the quantum-mechanical prediction in some particular examples. In summary, local theories are contextual within each separate region, but noncontextual for different regions. They have been shown to be incompatible with quantum mechanics. Thus, Bell's theorem is stronger than the impossibility proofs of noncontextual hidden-variables theories (by Gleason and others) in that it excludes the contextual theories which are local besides all noncontextual theories.

As the distinction between contextual, noncontextual, and local theories is very important, a new but equivalent characterization of them is given by introducing the concept of "formal joint probability distribution" for several observables. If $A(\lambda)$ and $B(\lambda)$ are the functions associated with two observables, in a noncontextual hidden-variables theory, it is possible to define a formal joint probability density, $f_\rho(A, B)$, related to the quantum state ρ by means of

$$f_\rho(A, B) \equiv (2\pi)^{-2} \int e^{-i\alpha A - i\beta B} \, d\alpha \, d\beta \int_\Lambda e^{i\alpha A(\lambda) + i\beta B(\lambda)} \, d\mu_\rho(\lambda) \tag{16}$$

We note that, if the operators $\hat{A}$ and $\hat{B}$ commute, we can also write

$$f_\rho(A, B) = (2\pi)^{-2} \int e^{-i\alpha A - i\beta B} \, d\alpha \, d\beta \, \mathrm{tr}[\hat{\rho} \, e^{i\alpha\hat{A} + i\beta\hat{B}}] \tag{17}$$

in view of our assumption G. However, if $\hat{A}$ and $\hat{B}$ do not commute, the operator associated with $\alpha A(\lambda) + \beta B(\lambda)$ is *not* $\alpha\hat{A} + \beta\hat{B}$ in general, and equation (17) may not agree with equation (16). This can be generalized to more than two observables. After that, we see that a necessary condition for noncontextual hidden-variables theories is the existence of a formal joint probability distribution for all observables associated with each quantum state. A proof that this probability does not exist will therefore prove Gleason's theorem. The word "formal" has been included in order to stress that the distribution cannot always be determined empirically. In particular, it cannot be determined whenever the operators involved are not jointly measurable. This is the reason why we distinguish carefully between "empirical joint distributions," that require joint measurability, and "formal joint distributions," which are purely mathematical constructions not requiring that condition. However, the mere existence of the mathematical function called "formal joint probability distribution" gives rise to empirically testable consequences (Bell's inequalities).

If we try to apply the previous construction to contextual hidden-variables theories, we realize that we cannot get a joint distribution for all observables, but only for those which are jointly measurable in a given context. Then, the corresponding quantum operators commute and we get a result similar to equation (17), namely

$$f_{\rho,C}(A, B) = (2\pi)^{-1} \int e^{-i\alpha A - i\beta B} \, \mathrm{tr}[\hat{\rho} \, e^{i\alpha\hat{A} + i\beta\hat{B}}] \, d\alpha \, d\beta \tag{18}$$

It should be noted, however, that now the probability density corresponds to a given *state and* a given *context*. Nevertheless, if the theory is to agree with quantum mechanics, the marginal probability distribution of a given observable depends only on the state of the measured system and *not* on the context, a truly remarkable fact showing that contextual hidden-variables theories are somewhat artificial, contrary to a widespread opinion.[19]

The lack of a clear definition of contextual, local, and noncontextual theories is currently the origin of much misunderstanding. For instance, there are people[23-25] who insist very strongly that Bell's inequalities are irrelevant as regards the question of locality, because they can be derived just from the existence of a joint probability distribution for observables which are not simultaneously measurable. There is some truth in the second part of the assertion, but the first part simply means that these people are not worried by the nonexistence of local (in Bell's sense) hidden-variables theories, an opinion they share with the supporters of the Copenhagen interpretation.

3. Are Noncontextual Hidden-Variables Theories Still Possible?

A frequently expressed opinion in recent years is that local hidden-variables theories have been disproved by the recent experiments of Aspect and collaborators.[26] This opinion is wrong, as we shall see in the next section. Even more widespread is the opinion that noncontextual theories are impossible. For instance, Shimony has recently stated that "Gleason's theorem doomed the program of noncontextual hidden-variables theories." It is a fact, however, that no empirical disproof of these theories has yet been exhibited. On the other hand, noncontextual theories were those implicitly supported by EPR[1] when they concluded: "The wave function does not provide a complete description of physical reality. . . . We believe that such a theory is possible." In fact, they "arrived at the conclusion that two physical quantities, with noncommuting operators, can have simultaneous reality." This simultaneous reality, which implies the existence of a (formal) joint probability distribution, is the characteristic of noncontextual theories, as discussed in the previous section. We are thus obliged to search for either noncontextual hidden-variables theories or *empirical* disproofs of them.

Gleason's theorem shows that noncontextual hidden-variables theories are not compatible with *all* currently accepted postulates of quantum mechanics. Therefore, if these theories exist, at least one of these postulates is incorrect. Quantum mechanics is so firmly established at present that just suggesting a change in any of its postulates sounds like a scandal. It is clear, however, that some of the postulates are extrapolations that could never possibly be tested. For instance, the postulate Q1 of the previous section states that *all* self-adjoint operators correspond to observables, but there are infinitely many such operators and certainly only a finite number of them will ever be measured. Similarly, according to Q2, *all* vectors represent states, but only a finite subset of them could ever be prepared. Finally, in Q3 *all* powers of an operator are assumed to correspond to the powers of the associated observable. Again, the axiom could possibly be tested only for a finite number of such powers. The point is that if we replace the word "*all*" in the three axioms by "a suitable subset of the set of," then Gleason's theorem cannot be proved, thus opening the door for noncontextual hidden-variables theories. The important question is: Will the agreement with experiments be destroyed by this change? The answer to this question consists in finding empirical tests of noncontextual hidden-variables theories. One of the two great lessons of Bell's work[19] has been to show us that we should change from searching for general impossibility theorems to looking for particular empirical tests. The other lesson, actually taken from EPR, is the relevance of locality for the hidden-variables problem, a subject discussed at length in other chapters of this book.

In order to find empirical tests of noncontextual theories, it is convenient to analyze situations where the conflict is more acute. One such case is provided by the Kochen–Specker paradox.[27] These authors considered a system in a state with total spin (or angular momentum in general) equal to one. In this case, it is easy to realize that the three operators J_x, J_y, and J_z fulfill the conditions

$$J_x^2 J_y^2 = J_y^2 J_x^2, \qquad J_x^2 J_z^2 = J_z^2 J_x^2, \qquad J_y^2 J_z^2 = J_z^2 J_y^2, \qquad J_x^2 + J_y^2 + J_z^2 = 2 \quad (19)$$

(in units of $\hbar^2$). Also, the operators J_x^2, J_y^2, and J_z^2 have eigenvalues 0 and 1. In consequence, a joint measurement of the observables corresponding to these operators is possible and, furthermore, if the value of one of them, say J_z^2, is 0, then the value of each one of the other two is 1. If we want to construct a noncontextual hidden-variables model for the system, the possibility must exist of associating either the number zero or one to every spin component squared, with the condition that if the number is 0 for a given direction, it is 1 for all perpendicular directions. Kochen and Specker proved that this is impossible, i.e., one cannot divide the sphere into two regions, the first (second) including the points where $(\mathbf{J} \cdot \mathbf{u})^2 = 1$ ($=0$) and such that if a point, taken as the pole, lies in the second region, all points in the corresponding equator lie in the first region (see Figure 3).

The Kochen–Specker theorem is actually a proof that noncontextual theories are not compatible with quantum mechanics for systems whose

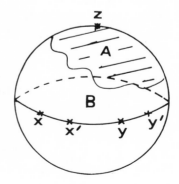

Figure 3. The spin components of a spin-1 system fulfill $J_x^2 + J_y^2 + J_z^2 = 2\hbar^2$ and $J_u^2 = 0$ or $\hbar^2$, so that $J_z^2 = 0$ implies $J_x^2 = J_y^2 = \hbar^2$, and vice versa. A dispersion-free state with these properties should correspond to a division of the sphere into two regions A and B, such that if a point, taken as a pole, lies in A, all points of the equator lie in B, and reciprocally. Kochen and Specker showed that such a division does not exist, so proving the impossibility of hidden variables under weaker assumptions than von Neumann. However, according to Bell, it is still possible to have contextual theories, where the result of measuring J_z^2 depends on whether we measure at the same time J_x^2 and J_y^2 or $J_{x'}^2$ and $J_{y'}^2$.

associated Hilbert spaces have dimension three or more. In fact, in such Hilbert spaces, the construction of operators with the formal properties of $(\mathbf{J} \cdot \mathbf{u})^2$ is always possible and the proof can follow similar lines. (In contrast, general noncontextual models have been found[28] for any two-dimensional Hilbert space.) In consequence, this theorem is similar to Gleason's theorem.

It should be clear that the proof of the Kochen–Specker theorem involves the postulates Q1, Q2, and Q3 of the previous section. For instance, we may assume that pure states with $J = 1$ are not realized in nature, but only some linear combinations of $J = 1$ and $J = 0$, which amounts to modifying postulate Q2. For such states the Kochen–Specker theorem cannot be proved. On the other hand, if the weight of $J = 0$ is small enough it might not be detected empirically. The point is that only actual experiments could show whether or not noncontextual hidden variables are possible.

The Kochen–Specker paradox demonstrates the difficulty of finding probability distributions for the angular-momentum vector that are compatible with the quantum predictions. Similar difficulties arise, for instance, when one tries to find probability distributions in phase space such that their marginals reproduce the quantum prediction for the position and the momentum of one or several particles.[29] Again, a weakening of the postulates Q1, Q2, and Q3 might perhaps solve the problem. In this paper we will not be concerned with the question of how the postulates should be changed without destroying the truly spectacular success of quantum mechanics, but rather with the previous problem of whether this program is possible.

4. Elements of Physical Axiomatics

A rigorous analysis of the possible empirical tests of hidden-variables theories requires an axiomatic approach to quantum mechanics. Consequently, the last two sections of this chapter will necessarily be more formal than the previous one. Several related axiomatics[30] have been used for quantum mechanics, starting from that of Hilbert space, initiated by von Neumann.[5] The most convenient for our purposes is a logicoalgebraic one.[13-17]

We must begin by formalizing the concept of "experiment." This is dangerous because any experiment consists of a sophisticated set of manipulations, especially if the experiment deals with microsystems. Then, any formalization implies an oversimplification (the same could be said, for instance, about the quantum theory of measurement). In spite of this, we shall state that a *simple experiment* consists of the *preparation* of a system followed by a *measurement* on it. The measurement gives the values of some set of compatible observables whose joint probability distribution is conditioned by the previous preparation. Note that we shall call "one experiment" not the actual performance of the manipulations involved once, but the performance of them a very large number of times, so that the joint probability distribution of all the (simultaneously measurable or "compatible") observables can be obtained with enough precision.

A comment is necessary at this stage. Many authors of articles or books dealing with the logicoalgebraic approach to quantum mechanics start writing about "yes–no" experiments in order to introduce the concept of "experimental proposition" or "question" (see, e.g., Gleason,[15] Jauch,[16] and many of the papers reproduced in Hooker[13]). This practice appears to be misleading and gives rise to much confusion. In fact, this approach forces one to define as "compatible propositions" those corresponding to two "yes–no" experiments which can be performed one after the other in either temporal ordering, a definition trying to imitate the commutativity of the quantum (projection) operators. However, commutativity is a mathematical property having no relation with time ordering, contrary to what is sometimes suggested. In practice, "yes–no" experiments are rather exceptional, and nobody says, for instance, that the propositions "this table is longer than one meter" and "this table is shorter than five meters" are compatible just because the answer to each question is independent of the order in which the questions are investigated. They are compatible because a single measurement of the length of the table gives the answer to both. I think that it is essential for a correct understanding of the hidden-variables problem to realize that any experiment involves a whole set of "experimental propositions" (this set has the mathematical structure of a Boolean lattice, as will be shown subsequently).

Besides simple experiments, we define *compound experiments*. These consist of several different experimental runs such that the *same preparation* is always involved, but several *different sets* of compatible observables are measured in different measurements. In other words, a compound experiment is defined as a set of (a few) simple experiments, all with the same preparation for the system. Compound experiments are relevant when they involve noncompatible observables (measured in different experimental runs) because this condition is necessary for the test of Bell's inequalities (see below).

Now, we must investigate the mathematical structure of the sets of observables. It proves convenient, and it is a common feature of the logicoalgebraic approach, to use dichotomic observables or "experimental propositions" as the elementary concept. It is seen that any measurement can be reduced to finding the truth value of a set of such propositions. For instance, measuring the length of a table with a precision of one millimeter is equivalent to finding what propositions in the class "the length is greater than N millimeters" ($N = 0, 1, 2, \ldots$) are true. After that, we may formally

define a *physical system* by a *set of propositions* and a *state* of the system
by a *probability for each proposition* being true.

In the following, a very brief sketch is given of the most relevant
mathematical concepts needed for the next section. The reader is referred
to the literature[13-17] for details. The set of propositions of a system is a
partially ordered set or "poset," $\mathscr{P}$. The partial ordering relation is given
by the *implication*. Proposition a implies proposition b, written $a < b$, if
whenever a is true, then b is also true. As mentioned above, it is assumed
that the implication is an ordering relation, i.e., it fulfills the axioms:

$$(1) \ a < a; \qquad (2) \ \text{if } a < b \text{ and } b < c, \text{then } a < c \qquad (20)$$

If $a < b$ and $b < a$, then a and b are called *equivalent*, denoted $a = b$.
Another important property of the poset of propositions of a physical system
is that for each proposition $a \in \mathscr{P}$ there exists also the negation $a' \in \mathscr{P}$ with
the conditions

$$(1) \ (a')' = a; \qquad (2) \ \text{if } a < b \text{ then } b' < a' \qquad (21)$$

A poset with this property is called *orthocomplemented*. For mathematical
convenience we include in $\mathscr{P}$ the "absurd proposition" ϕ (always false)
and the "obvious proposition" I (always true) with the properties

$$\phi' = I, \qquad \phi < a < I \text{ for all } a \in \mathscr{P} \qquad (22)$$

A state, attaching a probability $p(a)$ to every proposition $a \in \mathscr{P}$, is a mapping
of $\mathscr{P}$ into $[0, 1]$ such that

$$p(\phi) = 0, \qquad p(I) = 1, \qquad p(a) + p(a') = 1,$$

$$\text{if } a < b \text{ then } p(a) \leqslant p(b) \qquad (23)$$

The above properties are rather limited and additional postulates are
introduced in most practical cases. In particular, if we consider not the full
poset of a system, but the subset $\mathscr{L}$ of a simple experiment, it is assumed
that, for any two propositions $a, b \in \mathscr{L}$, there is a greatest lower bound or
"meet," $a \wedge b$, and a least upper bound or "joint," $a \vee b$, such that

$$a \wedge b < a, \qquad a \wedge b < b, \qquad \text{if } c < a \text{ and } c < b \text{ then } c < a \wedge b \quad (24)$$

$$a < a \vee b, \qquad b < a \vee b, \qquad \text{if } a < c \text{ and } b < c \text{ then } a \vee b < c \quad (25)$$

A poset fulfilling these conditions is called a *lattice*. Furthermore, the lattice of a simple experiment is Boolean, i.e., the following distributive identities are valid:

$$a \vee (b \wedge c) = (a \vee b) \wedge (a \vee c), \qquad a \wedge (b \vee c) = (a \wedge b) \vee (a \wedge c) \quad (26)$$

A state p on a Boolean lattice fulfills, in addition to properties (23),

$$\text{if } a < b', \text{ then } p(a \vee b) = p(a) + p(b) \qquad (27)$$

For mathematical convenience, this property is extended to infinite (denumerable) sets (a property called sigma additivity[13-18]), so that the state p becomes a probability distribution with the usual (Kolmogoroff) axioms (a σ-additive Boolean lattice is a σ-algebra, this being the standard mathematical structure used in the axiomatics of probability). To summarize, the poset of a compound experiment has subsets which are Boolean lattices, each associated with a possible single experiment. For this reason, the poset is sometimes called a "partial Boolean algebra."

The crucial question for the problem that we are studying is whether the poset of any compound experiment is itself a Boolean lattice. It turns out that the answer is yes for all phenomena in the classical domain, but it is no in the quantum domain, this being the essential difference between classical and quantum theory.[31] It is precisely the different mathematical structure which is at the root of all impossibility proofs of noncontextual hidden-variables theories. In fact, in the logicoalgebraic language, we may say that any hidden-variables theory tries to associate a Boolean lattice $\mathscr{L}$ with any compound experiment in the quantum domain in such a way that the poset (or non-Boolean lattice) $\mathscr{P}$ of the quantum projection operators is a subset of $\mathscr{L}$. If no other condition is imposed, this can be always carried out, and this is why contextual hidden-variables theories are always possible. However, it seems natural to demand that the (non-Boolean) lattice structure of $\mathscr{P}$ is the restriction of the Boolean lattice structure of $\mathscr{L}$. But this is impossible because any sublattice of a Boolean lattice is also Boolean. This is why noncontextual theories are incompatible with quantum mechanics (if all its axioms are maintained). Hence, our suggestion is to modify postulates Q1, Q2, and Q3 of the previous section in agreement with the following principle:

> H. Only those compound experiments are physically possible which can be interpreted by a noncontextual hidden-variables theory.

The possible modification of Q1–Q3, in order to make the quantum formalism compatible with H, does not contradict the present formalism in the sense of predicting different empirical results. It just prevents the possibility of some special experiments. In the last section it will be shown that no *performed* experiment contradicts H, so showing that the question of noncontextual hidden variables in quantum mechanics is still open, in sharp contrast with the widely held opinion. We note that some restrictions of type H have already been introduced into the quantum formalism in the form of superselection rules. They forbid one, for instance, to consider as physical states those represented by vectors in the Hilbert space that are linear combinations of two vectors representing states with different electric (or barionic, leptonic, etc.) charge.

To end this section, it is convenient to analyze in more depth what is the relevance of associating a Boolean lattice of propositions with a physical system. It derives from the property that equations (24) and (25) are always fulfilled if the propositions a, b, and c have a simultaneous truth value (the proof is trivial if it is taken into account that $a \vee b$ is true whenever a is true or b is true or both, and $a \wedge b$ is true only when both a and b are true). In other words, the simultaneous truth value of all propositions of a lattice is a sufficient condition for the lattice to be Boolean. Then, a non-Boolean lattice implies that not all propositions can have a truth value simultaneously.

After that, it is possible to identify the realistic position (expressed, e.g., by EPR) with the belief that the set of propositions of any physical system is a Boolean lattice. Let us consider, for instance, the conclusion of EPR that a particle has two elements of reality associated with the position and the momentum. In our logicoalgebraic language, this is equivalent to assuming that the propositions "the position is q" and "the momentum is p" have simultaneous truth values. In consequence, it should be possible to define the meet and the join of these propositions. As this can be made for any two propositions, the set is a lattice, and the lattice is Boolean by the property of having simultaneous truth values. Then, the realistic statement "the properties of the systems exist independently of whether or not they are measured" can be formalized by assuming that the set of propositions is a Boolean lattice. (It has been proposed elsewhere[32] to call BEL theories those describing nature by means of "Boolean Extended Lattices" of propositions, where "extended" emphasizes the fact that the full set of propositions is considered, in contradistinction to the set associated with a simple experiment, which is always Boolean.)

The reader who has followed this brief introduction to the logicoalgebraic language will realize that noncontextual hidden-variables theories have a Boolean lattice. The question, whether or not contextual theories have a Boolean lattice, depends on how one defines a proposition. This question, however, is not very relevant for us so this study will no longer be pursued.

5. Tests of Noncontextual Hidden-Variables Theories

The possibility of empirical tests of noncontextual theories rests upon the following theorem:

If, on a Boolean lattice $\mathscr{L}$, there is defined a probability distribution [i.e., a mapping of $\mathscr{L}$ into $[0, 1]$ fulfilling properties (23) and (27)], then the following inequality holds for every three elements $a, b, c \in \mathscr{L}$:

$$d(a, b) + d(b, c) \geq d(a, c) \qquad (28)$$

where

$$d(a, b) \equiv p(a) + p(b) - 2p(a \wedge b), \qquad a, b \in \mathscr{L} \qquad (29)$$

The proof is easy[33] but will not be reproduced here. We are interested in applying the theorem to orthocomplemented lattices of propositions. In this case, the distance function (29) fulfills

$$d(a, a) = 0, \qquad d(a, a') = 1, \qquad d(a, b) + d(a, b') = 1 \qquad (30)$$

where a' is the negation of a. If the changes $a \to a'$, $b \to b$, and $c \to c'$ are introduced into inequality (28) and relations (30) employed, then

$$d(a, b) + d(b, c) + d(c, a) \leq 2 \qquad (31)$$

for any three propositions in $\mathscr{L}$. We may use either inequality (28) or (31) in the applications.

The incompatibility of noncontextual theories with the *full* formalism of quantum mechanics can be illustrated by inequalities (28) and (31) in the following manner (here "full" means that the restriction H of the previous section is not incorporated). The propositions or dichotomic observables are represented in quantum mechanics by projectors, i.e. operators, which are Hermitian ($\hat{A} = \hat{A}^+$) and idempotent ($\hat{A}^2 = \hat{A}$). It is easy to see that, if the projectors $\hat{A}$ and $\hat{B}$ associated with two propositions a and b commute, then $\hat{A}\hat{B}$ is the projector associated with $a \wedge b$. In consequence, the "distance" between commuting projectors, corresponding to relation (29), is given by

$$d(a, b) = \text{tr}[\hat{\rho}(\hat{A} - \hat{B})^2] \equiv \langle (A - B)^2 \rangle \qquad (32)$$

It is easy to show that equations (30) and inequalities (28) and (31) are fulfilled for any set of commuting projectors. This shows that the construction of a (contextual) hidden-variables theory for each (simple) experiment is possible.

The problem appears with noncommuting operators. In fact, if $\hat{A}$ and $\hat{B}$ do not commute, quantum mechanics does not make any prediction for the proposition $a \wedge b$. (Workers in quantum "logic" have made different assumptions, like associating $\lim(\hat{A}\hat{B})^n$ with $a \wedge b$ as Birhoff and von Neumann[31] or saying that $a \wedge b$ is not defined as Reichenbach.[35]) The problem is that, if we attempt to generalize relation (32), several nonequivalent definitions appear, like the family

$$d(a, b) = \langle (A - B)^{2n} \rangle, \qquad n = 1, 2, 3, \ldots \tag{33}$$

[The equivalence of all the relations (33) for commuting projectors follows from the fact that $(\hat{A} - \hat{B})^2$ is itself a projector when $\hat{A}$ and $\hat{B}$ commute.] Now, a proof of the impossibility of noncontextual theories could be obtained if it is shown that *no* definition of distance can be found so that inequality (28) is fulfilled for any triple of projectors. Such a proof is achieved by Bell's theorem, as shown below. We may illustrate the problem by choosing the simplest definition (32). Then, the definition equations (30) remain valid but neither relation (28) nor (31) is fulfilled. Instead, weaker inequalities hold, such as, for instance,

$$d(a, b) + d(b, c) + d(c, a) \leqslant \tfrac{9}{4} \tag{34}$$

which follows easily from the obvious inequality

$$3\langle A + B + C \rangle - \langle (A + B + C)^2 \rangle \leqslant \tfrac{9}{4}$$

Hence, it follows also that

$$d(a, b) + d(b, c) + \tfrac{1}{4} \geqslant d(a, c) \tag{35}$$

As an example, we may consider a mixed quantum state with angular momentum $J = 1$ represented by the density operator

$$\hat{\rho} = \tfrac{1}{3}(|10\rangle\langle 10| + |11\rangle\langle 11| + |1-1\rangle\langle 1-1|)$$

(which arises in connection with the Kochen–Specker paradox;[27] see Section 3). It is straightforward to calculate the distance between two projectors of the form $(\mathbf{J} \cdot \mathbf{u})^2$, $\mathbf{u}$ being a unit vector. It depends only on the angle θ between the unit vectors, the function being

$$d(\theta) = \tfrac{2}{3}\sin^2\theta \tag{36}$$

It is enough to choose $\theta_1 = \theta_2 = \frac{1}{2}\theta_3 \ll 1$ in order to exhibit a violation of inequality (28), but if the projectors commute namely $\theta_1 = \theta_2 = \theta_3 = \pi/2$, no violation is found of either inequality (28) or (31). Also, no violation of relation (34) or (35) is possible.

After this discussion, it is clear that empirical tests of noncontextual hidden-variables theories can be found if there are particular instances of inequality (28) where a violation is predicted by quantum mechanics. A difficulty appears as follows. If the three projectors involved in inequality (28) commute, then the inequality is certainly fulfilled and no test exists. Alternatively, if two of the projectors do not commute, then the corresponding observables cannot be measured simultaneously and, again, the test is not possible. (An exception arises when the observables can be measured at different times and tests of this kind have been proposed.[36]) The way out of this difficulty is to use quadrilateral inequalities, easily derivable from relation (28), instead of triangle inequalities. For instance,

$$d(a, b) + d(b, c) + d(c, e) \geqslant d(e, a) \qquad (37)$$

Tests using inequality (37) are possible by choosing four projectors such that not all pairs commute (e.g., $\hat{A}\hat{C} \neq \hat{C}\hat{A}$ or $\hat{B}\hat{E} \neq \hat{E}\hat{B}$ or both) but only commuting pairs appear in the distances involved. If this inequality is expressed in terms of probabilities, then relation (29) enables us to obtain

$$p(c) + p(b) \geqslant p(a \wedge b) + p(b \wedge c) + p(c \wedge e) - p(a \wedge e) \qquad (38)$$

which is the Clauser and Horne[34] form of Bell's inequality. (See Figure 4.)

Inequality (38) and similar ones can be tested in compound experiments consisting of the preparation of a microsystem a large number of times (e.g., a pair of correlated photons from an atomic cascade) followed by four different measurements (one at a time). In one measurement $p(a \wedge b)$

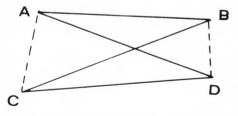

Figure 4. Given two "propositions" (or dichotomic observables) A and B, it is possible to define their "distance" by $d(A, B) = p(A) + p(B) - 2p(A, B)$, where $p(A)$, $p(B)$, and $p(A, B)$ denote the probabilities that A, B, or both are true. Distances range between 0 (if B is equivalent to A) and 1 (if B is the negation of A). In any classical theory, triangle and quadrilateral inequalities are fulfilled, such as $d(A, B) + d(B, C) + d(C, D) > d(A, D)$. Quantum mechanics predicts violations of these (Bell's) inequalities, so allowing for crucial tests. The tests are more dramatic in EPR-type experiments where A and C are measured in a region and B and D in another region, spatially separated from the first. In this case, contextuality cannot be invoked without violating relativity theory.

is determined (e.g., the probability of joint detection after the photons have crossed two polarizes at angles a and b, respectively), in another $p(b \wedge c)$, and so on. The probability $p(b)$ can be measured either in the first or second measurement. In any case, the inequalities can be used with two quite different scopes according to the experimenter's desire of testing either noncontextual hidden-variables theories or the wider family of local ones. The latter are tested in EPR-type experiments. (We have proposed elsewhere[32] using the name BEL inequalities—for Boolean extended lattice—for tests on noncontextual theories, reserving the name BELL for those testing locality.)

This section is brought to a close by stressing an extremely important distinction which, however, is not usually made, namely, that between *homogeneous* and *inhomogeneous* inequalities. Inequality (38), as all genuine inequalities derived from noncontextual (or local) hidden-variables theories, are inhomogeneous in the sense of comparing probabilities of single events, like $p(c)$ or $p(b)$, with probabilities of coincidence events, like $p(a \wedge b)$. A single event is, for instance, the detection of a particle, and a coincidence event corresponds to a coincidence detection on two detectors placed at two regions R_1 and R_2. The test of an inhomogeneous inequality like (38) is difficult, due to the fact that only a fraction of the signals produced in the source can be analyzed with the apparatus placed at R_1 and R_2. First, the signals may travel in the wrong direction not arriving at regions R_1 and R_2. Second, the measuring apparatus have a finite efficiency, so that not all signals arriving are detected. If f is the fraction of the signals emitted which are actually detected, then $p(c)$ and $p(b)$ are of order f while $p(a \wedge b)$, and so on, are of order f^2. Therefore, inequality (38) cannot be violated unless f is close to unity. More or less plausible estimates could be made about the fraction of all signals which arrive at the measuring apparatus, hence estimating the different probabilities involved in inequality (38) from the counting rates, but the procedure is uncertain. It is even more dangerous to extrapolate the results actually measured with low-efficiency detectors in order to estimate the results with ideal (100% efficient) detectors. This point will be treated further in the next section.

A suggested procedure to avoid these problems is to test *homogeneous* inequalities, involving only coincidence probabilities. Such inequalities cannot be derived from the above-stated theorem, but the derivation involves additional assumptions. For instance, it has been postulated, on the basis that the apparatus placed at R_1 (R_2) usually involves a selector (polarizer, Stern–Gerlach, etc.) and a detector, that the detection probability with the selector in place is less than the probability with the selector removed. Therefore, some homogeneous inequalities have been derived that have been tested in actual experiments. To summarize, homogeneous inequalities do not derive from local realism and inhomogeneous inequalities, like relation (38), could only be tested with very efficient selector–detector systems.

6. Experimental Tests of the Inequalities

Many empirical tests of Bell's inequalities have been proposed so far and several have been performed. Practically all tests involve measuring the correlation between the spin projections (or polarizations) of pairs of particles (or photons) prepared in a pure quantum state (such as a spin singlet). We do not attempt to review all proposed tests but only discuss two most important kinds: atomic cascade (or related ones) and molecular tests. (A review of the empirical tests previous to 1978 was given by Clauser and Shimony[20] and only the Orsay[26] and Stirling[37] groups have performed later experiments.) A common feature of these two kinds of tests is that they involve low energies (of the order of electron-volts), while energies one thousand times larger are involved in most other tests (e.g., pairs of gamma rays produced in the decay of positronium or proton–proton scattering[20]). Low-energy tests have the advantage that the selector used in the EPR experiments (polarizers in the atomic-cascade tests or Stern–Gerlach analyzers in the molecular tests) can be described in classical terms. In contrast, the spin projection (or polarization) of a high-energy particle (or photon) can only be measured by the interaction with another particle, and this process must be analyzed using quantum concepts. Recourse to quantum theory for the analysis of the experiment invalidates the test of local realistic theories in most high-energy tests. (The word photon is used only as a shorthand for light signal, without necessarily attaching quantum properties to it.)

In atomic-cascade experiments, the inhomogeneous inequality (38) is very well fulfilled, the left-hand side being about one thousand times greater than the right-hand side. This is due to the combination of two facts. In the first place the pair of photons to be analyzed appears in a three-body decay, as a result of which the two photons only just propagate in opposite directions. The probability that a photon enters the system of lenses (covering an angle not greater than 60°) is about 10%, and the probability that both photons of a pair enter is of order 1%. It is possible, in principle, to derive the inhomogeneous inequality (38) only for pairs of photons such that both enter the system of lenses. It is not so easy, however, to estimate the probabilities involved in relation (38) from the measured rates, because the ratio between coincidence and single counting rates is not the same as the ratio between coincidence and single probabilities. But even if the fraction of single counts corresponding to pairs both entering the measuring apparatus could be accurately estimated, a problem arises due to the low efficiencies of photon detectors. In fact, the efficiencies are of order 15%, so that the left-hand side of inequality (38) is about one hundred times higher than the right-hand side even for the restricted subensemble of pairs just discussed (see Figure 5).

The standard procedure to circumvent the problem is to assume that the ensemble of pairs actually detected is a representative sample of the pairs arriving at the selectors. But this implicitly uses an assumption of indistinguishability for the photons which is typically quantal. It is obvious that this assumption means nothing in the case of, say, classical wave packets. It is not necessary to repeat here the arguments given in the previous section showing that homogeneous inequalities are not genuine Bell's inequalities, derived only from local realism. In conclusion, *atomic-cascade experiments are inadequate for the test of local realism,* a fact already recognized 12 years ago,[34] but not seriously taken into account until recently.[38]

In view of the impossibility of testing local realism with atomic-cascade experiments, due to the low-efficiency of optical photon detectors, Lo and Shimony[39] have proposed a molecular test. Here, a sodium molecule in a singlet state is dissociated, by laser light, into two atoms whose spin components along chosen axes can be measured with Stern–Gerlach analyzers. This kind of test solves, in principle, the two difficulties discussed above for the atomic-cascade tests. In fact, the molecular dissociation is a two-body problem, so that the resulting atoms travel in opposite directions with the result that if one enters the aperture of the first measuring system, the other atom of the pair is also likely to enter the corresponding aperture in the second system. In the second place, very efficient detectors are available and the Stern–Gerlach analyzers also have a high efficiency. A careful analysis, however, has shown that the experiment, as initially proposed, does not provide a reliable test of local hidden-variables models.[40]

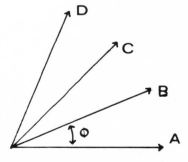

Figure 5. In atomic-cascade experimental tests of Bell's inequalities, proposition A (respectively B, C, D) is: "a light signal, coming from an atomic two-photon decay, is detected after crossing a polarizer with axis at angle 0 (respectively $\pi/8$, $\pi/4$, $3\pi/8$)." The quantum-mechanical prediction for the "distances" is (assuming ideal behavior for all devices except detectors) $d(X, Y) = \eta - \frac{1}{2}\eta^2(1 + \cos 2\phi)$, ϕ being the angle between X and Y and η the photon-detector efficiency. Then, the quadrilateral Bell's inequality (see Figure 4) could only be violated if $\eta \geq 2(\sqrt{2} - 1) \simeq 0.83$, while practical detector efficiencies are of order 0.15. The incorrect claim that local hidden variables have been refuted rests upon the empirical violation of other inequalities whose derivation involves additional, untestable, assumptions.

The present status of empirical tests of Bell's inequalities can therefore be summarized as follows. *No experiment performed until now has provided a valid test of non-contextual hidden-variables theories.* In fact, high-energy experiments should be interpreted using quantum theory if the results are to be considered as violations of Bell's inequalities, but this invalidates them as tests of (classical-like) models. Atomic-cascade experiments are useless due to the low efficiency of optical photon detectors. Also, no suitable test seems to be in preparation for the near future. The lack of a true empirical test of local realism versus quantum mechanics 20 years after Bell's discovery suggests that the contradiction of principle may not exist in practice. In conclusion, noncontextual (and, *a fortiori*, local) hidden-variables theories are still possible.

7. Conclusions

In the first section it is shown that the EPR[1] statement remains valid, namely, that the quantum-mechanical description of physical reality is not complete. Consequently, the search for hidden-variables theories should be a priority of theoretical physics. In Section 2, it is argued that hidden-variables theories should preferably be of noncontextual type. In Section 3 it is shown that, although noncontextual theories are incompatible with the full quantum formalism, they are compatible after some, possibly minor, modifications of it. Hence, the relevant question is whether such theories are compatible with experiments. After a necessary introduction to physical axiomatics, it is shown in Section 5 that empirical tests of noncontextual theories are possible using Bell's inequalities. Finally, after a short review of the performed experimental tests, it is concluded that no evidence exists that noncontextual theories of the microworld are impossible.

References

1. A. Einstein, B. Podolsky, and N. Rosen, *Phys. Rev.* **47**, 777 (1935). Reprinted in Ref. 2.
2. J. A. Wheeler and W. H. Zurek, *Quantum Theory and Measurement*, p. 20, Princeton University Press (1983).*
3. N. Bohr, *Phys. Rev.* **48**, 696 (1935). Reprinted in Ref. 2.
4. E. Schrödinger, *Naturwissenshaften* **23**, 807, 823, 844 (1935). English translation reprinted in Ref. 2.
5. J. von Neumann, *Mathematical Foundations of Quantum Mechanics*, Princeton University Press (1955) (German original, 1931).
6. E. A. Hylleraas, *Z. Phys.* **65**, 209 (1930).

---

* This book contains reprints of the most relevant papers dealing with the interpretation of quantum mechanics.

7. J. G. Cramer, *Rev. Mod. Phys.* **58**, 647 (1986).
8. L. E. Ballentine, *Rev. Mod. Phys.* **42**, 358 (1970).
9. A. Einstein, *Investigations on the Theory of the Brownian Movement* (translated by A. D. Cowper) Dover Publ., New York (1956).
10. M. Jammer, *The Conceptual Development of Quantum Mechanics*, McGraw-Hill Book Co., New York (1966); *The Philosophy of Quantum Mechanics*, Wiley, New York (1974)*
11. D. Bohm, *Phys. Rev.* **85**, 166, 180 (1952). Reprinted in Ref. 2.
12. F. J. Belinfante, *A Survey of Hidden Variables Theories*, Pergamon Press, Oxford (1973).
13. C. A. Hooker (ed.), *The Logico-algebraic Approach to Quantum Mechanics*, Vols. I and II, D. Reidel, Dordrecht (1975, 1979).
14. G. W. Mackey, *Mathematical Foundations of Quantum Mechanics*, W. A. Benjamin, New York (1963).
15. A. M. Gleason, *J. Math. Mech.* **6**, 885 (1957). Reprinted in Ref. 13.
16. J. M. Jauch, *Foundations of Quantum Mechanics*, Addison-Wesley, Reading (1968).
17. C. Piron, *Foundations of Quantum Physics*, W. A. Benjamin, Reading (1976).
18. J. S. Bell, *Rev. Mod. Phys.* **38**, 447 (1966). Reprinted in Ref. 2.
19. J. S. Bell, *Physics* **1**, 195 (1965). Reprinted in Ref. 2.
20. J. F. Clauser and A. Shimony, *Rep. Prog. Phys.* **41**, 1881 (1978).
21. F. Selleri and G. Tarozzi, *Riv. Nuovo Cim.* **4**, 1 (1981).
22. S. P. Gudder, *J. Math. Phys.* **11**, 431 (1970).
23. A. Fine, *Phys. Rev. Lett.* **48**, 291 (1982).
24. L. de la Peña, A. M. Cetto, and T. A. Brody, *Lett. Nuovo Cim.* **5**, 177 (1972).
25. W. M. de Muynck, *Phys. Lett. A* **144**, 65 (1986).
26. A. Aspect, P. Grangier, and G. Roger, *Phys. Rev. Lett.* **47**, 460 (1981); **49**, 91 (1982); A. Aspect, J. Dalibard, and G. Roger, *Phys. Rev. Lett.* **49**, 1804 (1982); A. Aspect and P. Grangier, *Lett. Nuovo Cim.* **43**, 345 (1985).
27. S. Kochen and E. P. Specker, *J. Math. Mech.* **17**, 59 (1967). Reprinted in Ref. 13.
28. J. S. Bell, *Rev. Mod. Phys.* **38**, 447 (1966). See also Ref. 27.
29. L. Cohen, *J. Math. Phys.* **7**, 781 (1966).
30. S. P. Gudder, in: *The Logico-algebraic Approach to Quantum Mechanics* (C. A. Hooker, ed.), Vol. II, p. 323, D. Reidel, Dordrecht (1979).
31. G. Birkhoff and J. von Neumann, *Ann. Math.* **37**, 823 (1936). Reprinted in Ref. 13.
32. E. Santos, Can quantum-mechanical destruction of physical reality be considered complete?, report to Conference on Microphysical Reality and Quantum Formalism, Urbino, Italy, 1985 (to appear).
33. E. Santos, *Phys. Lett. A* **155**, 363 (1986).
34. J. F. Clauser and M. A. Horne, *Phys. Rev. D* **10**, 526 (1974).
35. H. Reichenbach, *Philosophic Foundations of Quantum Mechanics*, University of California Press, Los Angeles (1944). Partially reprinted in Ref. 13.
36. A. J. Legget, *Phys. Rev. Lett.* **54**, 857 (1985).
37. W. Perrie, A. J. Duncan, H. J. Beyer, and H. Kleinpoppen, *Phys. Rev. Lett.* **54**, 1790 (1985).
38. T. W. Marshall, E. Santos, and F. Selleri, *Phys. Lett. A* **98**, 5 (1983).
39. T. K. Lo and A. Shimony, *Phys. Rev. A* **23**, 3003 (1981).
40. E. Santos, *Phys. Rev. A* **30**, 2128 (1984); A. Shimony, *Phys. Rev. A* **30**, 2130 (1984).

---

* These two books discuss at length the historical development of the interpretation of quantum mechanics.

16

Variable Detection Probability Models for Einstein-Podolsky-Rosen-Type Experiments

SAVERIO PASCAZIO

1. Introduction

Paradox is a word of Greek origin: it means against expectation, against the common opinion (parà + dòxa). The paper of Einstein, Podolsky, and Rosen (EPR)[1] in 1935 shocked the whole physics community because its far-reaching conclusions were against expectation. The EPR ingredients were only two: a reality criterion (what is called today a realistic attitude) and a firm belief in what is known as Einstein's locality. Starting from this simple recipe they showed that quantum mechanics is not a complete theory: there exist elements of physical reality which have no counterpart in the quantum-mechanical formalism.

This conclusion was put in a quantitative form only 30 years later, when Bell[2] showed that in a local-realistic theory (LRT) the correlation function is constrained to obey an inequality that can be violated by quantum mechanics (QM). Bell's inequality was unfortunately not liable to experimental investigation, essentially because it relied strongly upon a perfect-correlation assumption that cannot be achieved in actual experiments. The first ones who solved this problem and derived an experimentally testable inequality were Clauser, Horne, Shimony, and Holt (CHSH)[3] in 1969 and then Clauser and Horne (CH)[4] in 1974. Unfortunately, there is a

SAVERIO PASCAZIO • Department of Mathematics, Free University of Brussels, 1050 Brussels, Belgium.

shortcoming in the derivations of both CH and CHSH inequalities: the so-called no-enhancement hypothesis (NEH).

The aim of the present work is to point out that the NEH, despite its "natural" appearance, is strange to the realm of locality and realism. The "naturalness" of a hypothesis is not a scientific problem and is not liable to experimental tests. Non-Euclidean geometries were "unnatural" to Greeks and atoms were "unnatural" to most physicists of the last century, but neither the former nor the latter were correct. As we noted at the beginning of this section paradox means "not expected" which is, roughly speaking, the synonym of "not natural." Therefore, a way out of the EPR paradox cannot be found in a "natural" way, nor in an obvious one. Among possible logical solutions to the puzzle is the denial of the "natural" NEH: We will show in the following that by assuming that individual photons behave differently when interacting with the photon detector, one can reproduce the experimental data within errors, in all the experiments so far performed.

We commence our considerations by reanalyzing CH and CHSH no-enhancement assumptions.

2. The No-Enhancement Hypothesis

We shall deal in the following with the atomic-cascade tests of Bell's inequality[5] in which two correlated photons are emitted by an excited atom and subsequently spin-analyzed by two analyzers (polarizers) and detected by two detectors (photomultipliers). Our local-realistic attitude allows us to write the probability of detecting the first photon, given its analyzer's orientation a, as $p_1(\lambda, a)$. Similarly, the probability of detecting the second photon given the second analyzer's setting b is $p_2(\lambda, b)$. Parameter λ describes the state of the source (emitting atom, neighbor atoms, possible fields, and so on), where λ spans the space Λ of the possible physical states with distribution function ρ. It is assumed that $\int_\Lambda d\rho = 1$.

We note that p_1 (p_2) does not depend on b (a), in agreement with our local philosophy which demands a measurement on the first (second) photon *not* to depend on the second (first) photon's analyzer.

The probability of detecting both photons given a, b, and λ is

$$p_{12}(\lambda, a, b) = p_1(\lambda, a)p_2(\lambda, b) \tag{1}$$

and the average probabilities are

$$p_{12}(a, b) = \int_\Lambda p_1(\lambda, a)p_2(\lambda, b)\rho(\lambda)\, d\lambda \tag{2}$$

for a coincidence count, and

$$p_1(a) = \int_\Lambda p_1(\lambda, a)\rho(\lambda) \, d\lambda, \qquad p_2(b) = \int_\Lambda p_2(\lambda, b)\rho(\lambda) \, d\lambda \qquad (3)$$

for a single count. CH proved that if equations (1), (2), and (3) hold, then

$$-1 \leq p_{12}(a, b) - p_{12}(a, b') + p_{12}(a', b) + p_{12}(a', b') - p_1(a') - p_2(b) \leq 0 \quad (4)$$

The problem with inequality (4) is that it involves single-count as well as double-count probabilities. Let us consider the $4p^2 \, {}^1S_0 - 4p4s \, {}^1P_1 - 4s^2 \, {}^1S_0$ calcium cascade. The QM predictions in this case (θ being the angle between the polarizer settings) are

$$p_{12}(\theta) = \tfrac{1}{4}\eta_1\eta_2, f_1 g(\varepsilon_+^1 \varepsilon_+^2 + \varepsilon_-^1 \varepsilon_-^2 F \cos 2\theta)$$

$$p_1(a') = \tfrac{1}{2}\varepsilon_+^1 \eta_1 f_1 \qquad\qquad\qquad (5)$$

$$p_2(b) = \tfrac{1}{2}\varepsilon_+^2 \eta_2 f_2$$

where $\varepsilon_\pm^i$ are the well-known polarizer efficiency parameters, η_i the quantum efficiencies of the counters, f_i the probability that the ith photon enters the ith optical apparatus, g the conditional probability that the second photon enters the second apparatus if the first photon entered the first one (three-body decay: weak angular correlation between the emissions), and F ($F \leq 1$) is a depolarizing factor taking into account the finiteness of the collecting lenses' angles. The last inequality in relation (4) takes the form

$$S(\theta) \leq 1 \qquad\qquad (6)$$

with $S(\theta) = [3p_{12}(\theta) - p_{12}(3\theta)]/(p_1 + p_2)$ if some rotational symmetries are taken into account; on the other hand, the QM prediction for S is

$$S_{QM} = \tfrac{1}{2}\eta g \left[2\varepsilon_+ + (3\cos 2\theta - \cos 6\theta)F\frac{\varepsilon_-^2}{\varepsilon_+} \right] \qquad (6')$$

where it has been assumed that $\eta_1 = \eta_2 = \eta$, $f_1 = f_2$, and $\varepsilon_\pm^1 = \varepsilon_\pm^2 = \varepsilon_\pm$.

We now follow Clauser and Shimony's[6] interesting analysis: if the value $\theta = \pi/8$ is chosen in equation (6'), then the condition to violate inequality (6) is

$$\eta g \varepsilon_+ \left[\sqrt{2}\left(\frac{\varepsilon_-}{\varepsilon_+}\right)^2 F + 1 \right] > 2 \qquad (7)$$

However in practical cases one can have, at most, $\varepsilon_\pm \simeq F \simeq 1$, but never $g \simeq 1$ nor $\eta \simeq 1$. In fact in a three-body decay (say an atom which emits two photons) one always has $g \ll 1$, and for existing photomultipliers $\eta \leq 0.2$. Therefore, with an atomic-cascade experiment, one can never violate condition (6) or (4). This is the reason why one is forced to introduce some additional hypotheses in order to eliminate the single-count probabilities in equations (4) and (6). Even with $g = 1$ (two-body decay), it should be noted that inequality (7) is a stringent condition on the detectors' and analyzers' efficiencies, something already remarked on by CHSH in 1969.

A loophole to this inconvenient situation was found by CHSH and then CH by means of their no-enhancement hypothesis (NEH). It reads:

  <u>NEH1</u> (CHSH version). If a pair of photons emerges from two polarizers, the probability of their joint detection is independent of the polarizers' settings a and b.

  <u>NEH2</u> (CH version). For every state λ, the probability of a count with a polarizer in place is less than or equal to the probability with the polarizer removed.

By NEH one usually refers to the CH version, which is somehow more general than the original CHSH version. We note that NEH2 is easily written in the form

$$0 \leq p_1(\lambda, a) \leq p_1(\lambda, \infty) \leq 1$$
$$\text{for every } \lambda$$
$$0 \leq p_2(\lambda, b) \leq p_2(\lambda, \infty) \leq 1$$

where ∞ denotes "absence of the polarizer." Both NEHs deserve some comments. In the first place CHSH,[3] CH,[4] and Clauser and Shimony[6] very much stressed the physical "plausibility" of their assumption. For instance, CHSH write that in case the outcome of an experiment favors QM against LRT, their assumption "could be challenged by an *advocate of hidden variable theories* · · · However, *highly pathological detectors* are required to convert hidden variable emergence rates into QM counting rates." And also: "both assumptions, in our opinion, are physically plausible."[6] Second, it is worth remarking that NEH2 is somehow both stronger and weaker than NEH1: it is stronger because it is stated for every λ while NEH1 applies only to those photons which passed through the polarizer. It is weaker because it is only an inequality while NEH1 is an equality.

Finally, we stress that NEH1 does not hold in the semiclassical radiation theory, where the amplitude of a wave after its passage through the polarizer depends on the polarizer's orientation, and that the obviousness of NEH2 (an obstacle cannot increase the probability of detection) "vacillates" if we think that, for instance, the probability of detection *does* increase if we insert a diagonally oriented polarizer between two perpendicular ones. This has been observed[6] but in our opinion, has not been given the attention it deserves.

It is similarly very important to stress that every Bell-type inequality so far tested experimentally relies upon NEH-type assumptions. For instance, in the second Orsay experiment[5] the correlation coefficient

$$E(a, b) = \frac{R_{++}(a, b) + R_{--}(a, b) - R_{+-}(a, b) - R_{-+}(a, b)}{R_{++}(a, b) + R_{--}(a, b) + R_{+-}(a, b) + R_{-+}(a, b)}$$

is measured, where $R_{\pm\pm}(a, b)$ are coincidence rates. The theoretical scheme for this four-coincidence experiment was first proposed by Garuccio and Rapisarda,[7] and a careful study of their paper shows that an additional NEH-type hypothesis is made there because the quantity $p_{++}(\lambda, a, b) + p_{--}(\lambda, a, b) + p_{+-}(\lambda, a, b) + p_{-+}(\lambda, a, b)$ is supposed to be independent of λ [$p_{\pm\pm}(\lambda, a, b)$ are the four-coincidence experiment equivalents of $p_{12}(\lambda, a, b)$ in equation (1)]. The Orsay group's remark concerning the experimental observation that the sum of the four coincidence rates $R_{\pm\pm}(a, b)$ is constant when changing the analyzers' orientation is, in our opinion, misleading if one does not stress the fundamental difference between statistical hypotheses [$\sum_{i,j=\pm} R_{ij}(a, b) = $ const] and individual hypotheses [$\sum_{i,j=\pm} p_{ij}(\lambda, a, b) = $ const]. After all, NEH is nothing but the requirement that a statistical property (the detector quantum efficiency η) be also valid at an individual level (for every emission λ).

In 1974, CH devised a model involving an angular hidden variable which could reproduce the QM predictions. Of course their model violated the NEH and was physically sensitive only for low values of the photomultiplier efficiencies. Moreover, the model had the unpleasant feature of dealing asymmetrically with the two photons of a correlated couple, which means that the analytical expressions for p_1 and p_2 in equation (2) were different. Clauser and Shimony[6] even expressed "some hope for a theorem to the effect that any model consistent with the experimental data will have anomalous features as does the CH model." As we shall see in the next section, this hope would be blighted by Marshall, Santos, and Selleri, who explicitly constructed such a model in 1983. We defer to a subsequent section some further considerations about Clauser and Shimony's "hope" in connection with some interesting papers explicitly taking into account the symmetry between the two EPR-correlated photons [$p_1 = p_2$ in equation (2)].

3. The Model by Marshall, Santos, and Selleri

In 1983 Marshall, Santos, and Selleri (MSS)[8] obtained a model able to reproduce the experimental data for atomic-cascade experiments within errors. The model involved just one angular hidden variable in such a way that the elementary detection probabilities in equations (1) and (2) are given by the Fourier expansions

$$p_1(a, \lambda) = \sum_{n=0}^{\infty} a_n \cos 2n(a - \lambda) \quad \text{and} \quad p_2(b, \lambda) = \sum_{n=0}^{\infty} b_n \cos 2n(b - \lambda) \quad (8)$$

MSS remarked that it is impossible to obtain even partial agreement with the QM predictions if only two terms are taken in expansion (8). However, if one more addendum is retained, then one arrives at

$$p_{12}(\theta) = c(1 + 2\alpha_1\beta_1 \cos 2\theta + 2\alpha_2\beta_2 \cos 4\theta) \quad (9)$$

where $\theta = (a - b)$ and c, α_i, and β_i $(i = 1, 2)$ are related to coefficients a_i and b_i $(i = 0, 1, 2)$ in expansion (8). By setting

$$c = \tfrac{1}{4}\varepsilon_+^1\varepsilon_+^2\eta_1\eta_2 f_1 g, \qquad \alpha_1 = \frac{\varepsilon_-^1}{\varepsilon_+^1}\sqrt{\frac{F}{2}}, \qquad \beta_1 = \frac{\varepsilon_-^2}{\varepsilon_+^2}\sqrt{\frac{F}{2}}$$

and

$$\alpha_2 = \frac{1}{2}\left[1 - \sqrt{1 - F\left(\frac{\varepsilon_-^1}{\varepsilon_+^1}\right)^2}\right], \qquad \beta_2 = \frac{1}{2}\left[1 - \sqrt{1 - F\left(\frac{\varepsilon_-^2}{\varepsilon_+^2}\right)^2}\right]$$

where the last two parameters are lower-bounded owing to positivity requirements on the probabilities (8) (the possibility $\alpha_2 = \beta_2 = 0$, which would give exact agreement with QM, is therefore ruled out), one obtains the same expression as QM for the coincidence count probability, plus a small term proportional to $\cos 4\theta$.

A natural interpretation of the model proposed is the following: If for the sake of simplicity, $F = 1$ and $f_i(\varepsilon_+^i/2)$ is the probability that the ith photon enters the ith lens system (is transmitted by the ith polarizer) $(i = 1, 2)$, then

$$\eta_i(\lambda) = \eta_i\left[\frac{1}{\sqrt{2\varepsilon_+^i}}(\sqrt{\varepsilon_M^i} + \sqrt{\varepsilon_m^i}) + \frac{1}{\sqrt{\varepsilon_+^i}}(\sqrt{\varepsilon_M^i} - \sqrt{\varepsilon_m^i}) \cos 2(\lambda - \alpha^i)\right] \quad (10)$$

(where α^i is the ith polarizer's setting) can be interpreted as the probability that the ith photon is detected by the ith counter. It is noteworthy that:

1. $\langle \eta_i(\lambda) \rangle_\lambda = \eta_i$, as expected on the basis of the definition of quantum efficiency η_i;
2. the model deals symmetrically with the two photons of an EPR-correlated couple.

Equation (10) rests on the idea of a variable detection probability (VDP), namely, different photons behave differently when interacting with photon detectors, the interaction being specified by the variable λ. Of course, λ is unknown within the bounds of the QM description of the physical reality, but it is not contradictory to QM by virtue of property 1. It is important to stress that VDP means that there are some photons carrying a $\bar{\lambda}$ such that

$$\eta(\bar{\lambda}) > \eta$$

If we interpret η as the probability of detecting a photon with the polarizer removed, then NEH is violated. The consequences of NEH are clear to us only in the light of the above-mentioned interpretation of the MSS model: roughly speaking, NEH forbids the possibility that, of a statistical ensemble, some photons have a higher *intrinsic probability* of being detected than others.

The interpretation (10) is physically sensitive only for some values of η; the ideal case $\eta = 1$, for instance, is *a priori* excluded because it implies the possibility that $\eta(\lambda) > 1$ for some λ [remember, $\eta(\lambda)$ is a probability]. In the model proposed, for example, one must require that

$$\eta < \sqrt{\varepsilon_+}/[(\sqrt{\varepsilon_M} + \sqrt{\varepsilon_m})/\sqrt{2} + (\sqrt{\varepsilon_M} + \sqrt{\varepsilon_m})]$$

This is a common feature of every VDP model, as we shall see. One ought to expect that when and if experiments are performed with (almost) ideal photon counters, different photons should behave (almost) identically when interacting with a counter, so that the λ-dependence in equation (10) would fade away. In other words equation (10), or any other VDP, provides a good approximation when describing the behavior of a photon only in the case of low detector efficiencies (say $\eta \leq 0.8$) but it cannot be expected to hold in the limit of ideal experimental situations. We will not extend our considerations in this direction for two reasons: First, the technical problems of achieving high values of η (for instance in photomultipliers) are so difficult that our discussion would be purely academic. Second, it seems to us that the possibility of an *intrinsic physical upper bound* on the value of

η should be seriously investigated. For if, for some reason, the value $\eta \simeq 1$ were not attainable even theoretically, the interpretation of VDPs would undergo a dramatic change in that the quantum efficiency η of a photon detector could be interpreted as a physical parameter whose value would be specified by physical sensitivity requirements on the probabilities.

4. The Distance Separating Quantum Theory from Local Realism

We saw in the previous section that the MSS model removes the unpleasant asymmetrical features that forced Clauser and Shimony to terminate their research on enhancement in the hope of a theorem that, as we have seen, could not be proven. The MSS model gives predictions which differ very little from the QM ones, so that no discrimination is possible on the basis of currently realized experiments.

The question now arises whether the QM prediction can be obtained in a local realist fashion. Of course we know that any local realist explanation we envisage must bear on VDP and must lead to a violation of NEH. We have been taught this lesson by the experimental violations of CH inequality[5] which leave no other way out. Some works by Marshall,[9] Caser,[10] and Corchero[11] have tried to answer this question. As we shall see it is impossible, within a certain class of local realist theories, to reproduce *exactly* the QM prediction for the coincidence count probability.

Marshall was the first to realize that, even dropping the NEH, a VDP model *involving one angular hidden variable* λ cannot give predictions identical to those of QM. Indeed, by defining the distance

$$d(p_{12}, p_{12}^{\text{QM}}) = \frac{2}{\pi} \int_0^{\pi/2} [p_{12}(\theta) - p_{12}^{\text{QM}}(\theta)]^2 \, d\theta \tag{11}$$

where p_{12} is given by equation (2), with $\theta = a - b$, and p_{12}^{QM} by equation (5), Marshall finds that this distance has a finite constant d_0 as lower bound. The analysis of Marshall[9] will not be presented because it requires some computer numerical evaluations, but it is worth emphasizing that Marshall concludes that a discrimination between QM and the class of VDP models considered by him is still possible if one makes use of better statistics than those usually achieved. Another step forward in this direction was made, independently, by Caser in 1984.[10] We will sketch his analysis, which is also the first (as far as we know) to develop the consequences of the explicit symmetry assumption $p_1 = p_2$ in integral (2).

Caser considers the case of one angular hidden variable λ and factorizes the counting probability p_i^+ (p_i^-) for an ordinary (extraordinary) ray which has crossed a two-channel polarizer in the following manner:

$$p_i^+(\lambda, \alpha_i) = p_i(\lambda - \alpha_i)f(\lambda - \alpha_i)$$

$$p_i^-(\lambda, \alpha_i) = p_i(\lambda - \alpha_i)[1 - f(\lambda - \alpha_i)] \tag{12}$$

where α_i is the ith polarizer setting, p_i is a detection probability and f_i the transmission probability for, say, light polarized along an "active" direction (ordinary ray). (We are modifying Caser's original discussion in order to deal with photons and not electrons: two-channel polarizers are the photon equivalent of Stern–Gerlach apparatuses.) It is necessary that

$$p_{12}^{++}(\theta) = p_{12}^{--}(\theta) = \tfrac{1}{2}C\cos^2\theta \quad \text{and} \quad p_{12}^{+-}(\theta) = p_{12}^{-+}(\theta) = \tfrac{1}{2}C\sin^2\theta \tag{13}$$

where

$$p_{12}^{++}(\theta) = \frac{1}{2\pi}\int_0^{2\pi} d\lambda\, p_1(\lambda - a)p_2(\lambda - b)f_i(\lambda - a)f_2(\lambda - b)$$

Similar expressions hold for p^{--}, p^{+-}, and p^{-+} (C is a constant allowing for renormalization). Caser shows that if $p_1 = p_2$ and $f_1 = f_2$, then equations (12) and (13) are in contradiction. His theorem can be generalized to an arbitrary number of parameter λ if they all lie in the plane of the analyzers.

In our opinion this result is very important, because it shows that no symmetric VDP model can reproduce the QM prediction (13) if it makes use of parameters λ restricted to the plane of the analyzers. It is anyway less general than one understands at first sight, because it has been shown by Home and Marshall[12] that, by means of a symmetrization attributed by them to Mermin, any asymmetric model may be turned into a symmetric one by the addition of one discrete variable s, in the following way. We set

$$\rho'(\lambda, s) = \tfrac{1}{2}\rho(\lambda), \quad \lambda \in \Lambda, \quad s = \pm 1$$

where ρ is the distribution function of λ in Λ [equation (2)] and ρ' the distribution function of (λ, s) in $\Lambda \otimes \{-1, 1\}$, and then introduce

$$p_1'(\lambda, 1, a) = p_1(\lambda, a) \quad \text{and} \quad p_1'(\lambda, -1, a) = p_2(\lambda, a)$$

$$p_2'(\lambda, 1, b) = p_2(\lambda, b) \quad \text{and} \quad p_2'(\lambda, -1, b) = p_1(\lambda, b) \tag{14}$$

where p_1 and p_2 are the same as in equation (2). Then the coincidence rates obtained by ρ', p_1', p_2' are identical with those obtained by ρ, p_1, p_2. In a correlated photon couple, one photon (say the first emitted) is assumed to be of type 1 and the other of type 2 (state $s = +1$), or vice versa ($s = -1$).

The model is then symmetrized by assuming the states $s = \pm 1$ to be equally populated. It is worth stressing that this "symmetrization procedure" does not completely invalidate Caser's theorem. In fact, the set of equation (14) describes an ensemble of *asymmetric* photon couples, the symmetry being recovered in a subsequent step, but only at a *statistical level*. Therefore, if one believes that in every single couple of EPR correlated photons the two photons behave symmetrically when interacting with analyzers and detectors, then Caser's results[10] still hold.

We conclude this section by mentioning the work of Corchero,[11] who somehow blends the two preceding authors' conclusions. Corchero considers the family of local hidden-variables theories which involve just one single parameter and are symmetric in the sense of Caser. He shows that a suitably defined distance S between this class of models and QM is an increasing function of the parameter $f = F\varepsilon_-^1 \varepsilon_-^2 / \varepsilon_+^1 \varepsilon_+^2$ [see expressions (5) for the notation]. The function $S(f)$ has a maximum S_0 at $f = 1$ (ideal case) and reaches its minimum at $f = 0.5$, being zero for $f \leq 0.5$. In practical cases, a discrimination is possible if one combines: a $J = 0$-1-0 cascade, for which $F \simeq 0.99$; calcite polarizers, for which $\varepsilon_m/\varepsilon_M < 10^{-4}$; relatively high statistics.

Moreover, in no experiment so far performed have these three conditions been fulfilled simultaneously. Corchero's work is noteworthy because it gives a finite numerical bound for the discrepancy between symmetric local theories and QM (Caser's work did not provide any) and because it shows that a lower statistics than that suggested by Marshall suffices in order to discriminate this class of models from QM.

The three works summarized in this section are extremely interesting, even though they share a certain limit with regard to their generality: indeed they all deal with the very particular case of one angular hidden parameter. But a door seems to be open for possible generalizations: Caser extended the validity of his theorem to an arbitrary number of parameters if they are restricted to the plane of the polarizers, while Ferrero, Marshall, and Santos[13] have apparently proved that the equivalent of Caser's theorem holds also in the general case of n arbitrary hidden variables.

5. Garuccio and Selleri's Proposal

An important step forward in VDP was taken by Garuccio and Selleri[14] and later by Selleri,[15] who studied an interesting class of VDP models restricted by some *physical assumptions*.

We begin with an example. A photon is assumed to carry a polarization vector l and a detection vector λ such that the probability that the photon crosses a polarizer set at a is given by

$$C = \cos^2(l - a) \tag{15}$$

(Malus' law) and the probability of being detected by a photomultiplier with quantum efficiency η is

$$D = \tfrac{8}{3} \cos^4(\lambda - a)\eta \tag{16}$$

It should be noted that the transmission probability C in expression (15) does not depend on λ, which is transmitted unchanged through the polarizer and then specifies the interaction of the photon with the photomultiplier [see equation (16)]. The photon, by crossing the polarizer, acquires the polarization a which "matches" λ to give the detection probability. It is easy to check that

$$\langle D(\lambda - a)\rangle_\lambda = \eta$$

so that interpreting η as the quantum efficiency of the photomultiplier is consistent with our scheme. If we assume η to be the detection probability when no polarizer is present, then we see that the VDP (16) exhibits enhancement, because there exist values of λ such that

$$D(\lambda - a) > \eta$$

in violation of NEH.

A straightforward calculation gives for the double count probability

$$p_{12}(\theta) = \tfrac{1}{4}(1 + \tfrac{8}{9}\cos 2\theta + \tfrac{1}{18}\cos 4\theta)\eta_1\eta_2$$

which agrees well with the result of the first Orsay experiment[5]:

$$p_{12}(\theta) = (0.249 + 0.218 \cos 2\theta)\eta_1\eta_2$$

The presence of a small term proportional to $\cos 4\theta$ is a characteristic of models of this type and yields experimentally testable discrepancies with QM. The example we have just presented is extremely useful for understanding the physical meaning of the class of models investigated by Garuccio and Selleri.[14,15] The polarization l looks much like a wave property of the photon field, while λ possesses rather the features of a "particle" property, because it does not interact with the polarizer, nor with any other optical devices such as a half-wave plate (as is assumed by Selleri[15]). It will be shown that this class of VDP models leads to large discrepancies with QM if other optical devices are inserted in the photon paths.

In order to elucidate the results obtained by Garuccio and Selleri, we will follow the more realistic approach of Garuccio *et al.*[16] in view of the interest shown by the Stirling and Catania groups in the experiments proposed by Garuccio and Selleri.[14,15] The class of VDP models that we consider can be characterized by the following assumptions: a photon is endowed with a linear polarization l such that $C(l - a)$ is the transmission probability through a polarizer set at a, and a detection variable λ, uniformly distributed between 0 and π such that $D(\lambda - a)$ is the detection probability after the photon has acquired the polarization a. The interaction with the polarizer does not change λ and the transmission probability C does not depend on λ. In the 0-1-0 calcium atomic cascade, the two correlated photons leave the source with the same λ and polarization vectors l, l' with normalized distribution function $\rho(l, l')$. We will first examine the case of ideal polarizers. The coincidence count probability is then

$$p_{12}(a - b) = T(a - b)R(a - b)$$

where

$$T(a - b) = \int_0^\pi dl \int_0^\pi dl' \, \rho(l, l')C_1(l - a)C_2(l - b)$$

is the double transmission probability and

$$R(a - b) = \frac{1}{\pi} \int_0^\pi d\lambda \, D_1(\lambda - a)D_2(\lambda - b)$$

is the double detection probability. In order to be close to the QM predictions, we require that

$$T(a - b)R(a - b) \simeq \tfrac{1}{4}\eta_1\eta_2[1 + \cos 2(a - b)] \tag{17}$$

Garuccio and Selleri's idea is to insert a third polarizer, set at a', between the first one, set at a, and the photomultiplier. The probability of a double count then becomes

$$\omega_3^{\text{VDP}}(a', a, b) = T(a - b)\cos^2(a' - a)R(a - b)$$

which is to be compared with the QM prediction

$$\omega_3^{\text{QM}}(a', a, b) = \tfrac{1}{4}\eta_1\eta_2[1 + \cos 2(a - b)]\cos^2(a' - a)$$

The ratio

$$\gamma^{\text{ID}}(a', a, b) = \omega_3^{\text{QM}}(a', a, b)/\omega_3^{\text{VDP}}(a', a, b)$$

can be written as

$$\gamma^{\text{ID}} \simeq \frac{1 + \cos 2(a - b)}{3[1 + \cos 2(a' - b)]} \frac{3T(a' - b)}{T(a - b)} \tag{18}$$

due to relation (17). But the transmission probability T satisfies the Bell-type inequality

$$3T(\theta) > T(3\theta)$$

so that, by choosing $a - b = 3(a' - b)$, relation (18) becomes

$$\gamma^{\text{ID}} \geqslant \frac{1 + \cos 6(a' - b)}{3[1 + \cos 2(a' - b)]} \tag{19}$$

This curve is plotted in Figure 1. Equation (19) is a Bell-type inequality for the class of VDP models considered. A discrimination with QM seems easily possible in the range $80° < a' - b < 90°$. Unfortunately, when polarizer

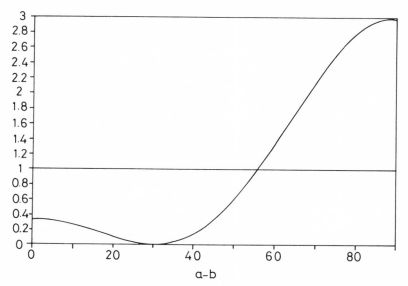

Figure 1. Lower limit of γ^{ID} for the VDP class considered (ideal polarizers).

imperfections are taken into account, it is not possible to repeat the simple analysis leading to relation (19). A rather different (and lengthy) approach[16] yields the result of Figure 2. The upper curve is relative to the Catania experiment, the lower curve to the Stirling one. It can be inferred from Figure 2 that a discrimination with QM is still possible, at $a - b \simeq 69°$ in the Catania case and $a' - b \simeq 68°$ in the Stirling case, but it is not that easy to carry out, experimentally, owing to the low values of the maxima.

Another kind of experiment has been proposed by Selleri,[15] who suggests inserting a half-wave plate between each polarizer and the respective detector. The idea in this case is that the detection variable λ does not interact with the half-wave plate, but the photon polarizations a and b do, so that the angles $\lambda - a$ and $\lambda - b$ can be varied in the detection probabilities $D_1(\lambda - a)$ and $D_2(\lambda - b)$. As a consequence, the average probability for a double detection

$$R(a - b) = \frac{1}{\pi} \int_0^\pi d\lambda \, D_1(\lambda - a) D_2(\lambda - b)$$

can be arbitrarily varied by changing the polarizations a and b into, say, a' and b' without modifying the polarizer settings, which are still a and b. On the other hand, no variation in the double count probability should be observed according to QM, and a discrimination is easily possible. It is noteworthy that Selleri also derives an inequality by making no additional

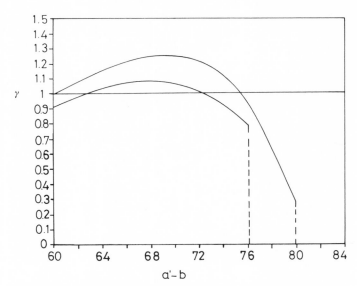

Figure 2. Upper bound to the lower limit of γ^{REAL} (real polarizers).

assumptions of NEH-type. This inequality is a constraint for the class of VDP models considered by him.

The painless derivation of theoretical discrepancies between QM and the class of VDP model considered led Garuccio and Selleri to conjecture that the insertion of a third polarizer in an EPR experimental setup should give rise to nonquantum-mechanical effects for *every* VDP model one can devise. The criticisms of this conjecture will be the subject of the next section.

6. The Counterexample by Ferrero, Marshall, and Santos

Garuccio and Selleri's claim, that the "anomalous" behavior of the ratio γ^{1D} [relations (18) and (19)] should be a feature of all VDP models, has been criticized by Marshall and Santos[17] and Ferrero and Santos.[18,19] We will not examine their papers in detail, because they are subject to a number of criticisms: for instance, Malus' law is sacrificed,[17,18] in order to approach closely the QM predictions for the coincidence count probabilities, while the probability of a coincidence count without polarizers is given by[19]

$$p_{12}(\infty, \infty) = \frac{8\pi}{(\pi + 2)^2} \eta_1 \eta_2 \simeq 0.951 \eta_1 \eta_2$$

so that some photons are lost somewhere in the experimental apparatus (the probability for a double detection without polarizers should be the simple product of the detector efficiencies). We will therefore sketch another recently proposed model[20,13] that overcomes all the above-mentioned difficulties. A photon is characterized by three parameters (l, φ, δ). The first two are similar to those used in classical optics to describe elliptic polarization:

$$-\pi/2 \leqslant l \leqslant \pi/2, \qquad -\pi/4 \leqslant \varphi \leqslant \pi/4$$

so that for a photon traveling along the z-axis, the electric vector is

$$E_x = A(t) \, \text{Re}[(\cos \lambda \cos \phi - i \sin \lambda \sin \phi) \, e^{i\omega t}]$$

$$E_y = A(t) \, \text{Re}[(\sin \lambda \cos \phi - i \cos \lambda \sin \phi) \, e^{i\omega t}]$$

The distribution function of l and φ is

$$\rho(l, \varphi) = \frac{1}{\pi} \cos 2\varphi \tag{20}$$

and these two parameters are assumed to obey the laws of classical optics when interacting with optical devices; for instance, $(l, \varphi) \rightarrow (a, 0)$ when the photon crosses a polarizer set at a. The variable δ has five possible values: δ_1 (δ_2) corresponds to the first (second) photon of a correlated pair, and when a photon crosses a polarizer

$$\delta_1 \text{ is changed in } \delta_3$$

$$\delta_2 \text{ is changed in } \begin{cases} \delta_4 & \text{if } |l - b| < \pi/4 \\ \delta_5 & \text{if } |l - b| > \pi/4 \end{cases}$$

$$\delta_3, \delta_4, \delta_5 \text{ do not change}$$

(The model can then be symmetrized by applying, for instance, the procedure suggested by Home and Marshall.[12]) The corresponding transmission and detection probabilities are given in Table 1. It is then a simple matter to show that

$$p_{12}(\infty, \infty) = \eta_1 \eta_2$$

$$p_{12}(a, \infty) = p_{12}(\infty, b) = \tfrac{1}{2}\eta_1 \eta_2$$

$$p_{12}(a, b) = \tfrac{1}{4}\eta_1 \eta_2[1 + \cos 2(a - b)]$$

in agreement with the QM predictions. Moreover, the prediction for γ^{ID} [relations (18) and (19)] is

$$\gamma^{ID} = 1$$

which shows that Garuccio and Selleri's conjecture was not correct. If it is assumed that the distribution function $\rho(l, \varphi)$ [equation (20)] does not change if quarter- or half-wave plates are inserted in the optical paths, the model gives predictions identical to the QM ones for any experiments of the type proposed by Garuccio and Selleri.[14,15]

The example discussed in this section is rather artificial but, in our opinion, extremely useful because it shows the versatility of VDP models. In spite of the theorems presented in Section 4, it has been possible to find a local realistic model which agrees with the QM predictions for all the correlation experiments so far contrived. Notwithstanding its unnatural features, we consider this model to be an important step forward in research on VDP. Nevertheless, we think that Garuccio and Selleri's proposals[14-16] are worthy of attention because the class of VDP models they studied, even though not exhaustive, has the nice feature of regarding the detection

Table 1. The Model by Ferrero, Marshall, and Santos

Incoming photon	Polarizer	Transmission probability		Average transmission probability	Transmitted photon	Detection probability
(l, φ, δ_1)	∞	1		1	(l, φ, δ_1)	η_1
	a	1 $\quad$ if $\lvert l - a \rvert < \pi/4$ 0 $\quad$ if $\lvert l - a \rvert > \pi/4$		$\frac{1}{2}$	$(a, 0, \delta_3)$	η_1
(l, φ, δ_2)	∞	1		1	(l, φ, δ_2)	η_2
	b	$\cos 2(l - b) \quad$ if $\lvert l - b \rvert < \pi/4$ $0 \quad$ if $\lvert l - b \rvert > \pi/4$	$1/\pi$	$\left. \begin{array}{c} \\ \\ \\ \\ \end{array} \right\} \frac{1}{2}$	$(b, 0, \delta_4)$	$\frac{1}{2}\pi\eta_2$
		$0 \quad$ if $\lvert l - b \rvert < \pi/4$ $1 + \cos 2(l - b) \quad$ if $\lvert l - b \rvert > \pi/4$	$\frac{1}{2} - 1/\pi$		$(b, 0, \delta_5)$	0
(l, φ, δ_3)	∞	1		1	(l, φ, δ_3)	η_1
	a	$\cos^2(l - a)$		$\frac{1}{2}$	$(a, 0, \delta_3)$	
(l, φ, δ_4)	∞	1		1	(l, φ, δ_4)	$\frac{1}{2}\pi\eta_2$
	b	$\cos^2(l - b)$		$\frac{1}{2}$	$(b, 0, \delta_4)$	
(l, φ, δ_5)	∞	1		1	(l, φ, δ_5)	0
	b	$\cos^2(l - b)$		$\frac{1}{2}$	$(b, 0, \delta_5)$	

parameter λ as a "particle attribute," in contrast to the wave parameter l, which is a polarization. It is indeed "fascinating that in trying to save Einstein locality one is pushed to give a more detailed *dualistic* description of the photon than provided by the usual quantum theory."[15]

7. Disguised VDPs

Other proposed solutions of the EPR puzzle can be related to the idea of VDP by a careful inspection of their characteristics. An example is the model proposed by Barut and Meystre,[21,22] who suggest measuring the spin correlation function on two separated subsystems of a classical spin-0 system. By defining the correlation function as

$$E(a, b) = (\langle AB \rangle - \langle A \rangle \langle B \rangle)/\sqrt{\langle A^2 \rangle \langle B^2 \rangle}$$

where $A = \mathbf{S}_1 \cdot \mathbf{a}$ and $B = \mathbf{S}_2 \cdot \mathbf{b}$ ($\mathbf{S}_1$ and $\mathbf{S}_2$ are classical spin vectors, with $\mathbf{S}_1 + \mathbf{S}_2 = 0$), Barut and Meystre succeeded in obtaining predictions identical to the QM ones. Several different interpretations have been suggested by Barut and Meystre in order to explain their result. In particular, they propose a discretization procedure of the classical spins that consists in recording each event with a weight factor of the type

$$\tfrac{1}{4}|\cos(\varphi - \varphi_a)| \qquad \text{or} \qquad \tfrac{1}{4}|\cos(\varphi - \varphi_b)| \tag{21}$$

according to the measure being performed on the first spin or the second spin (φ, φ_a, and φ_b being the angles between an arbitrary z-axis and the spin, a and b, respectively). Their conclusion is that, "Whether our system can be called local, in some generalized sense, is (...) probably a question of taste."[21] In our opinion Barut and Meystre's system is nonlocal, independently of questions of taste, if one does not identify VDP features in it. This has been done by Home and Marshall,[12] who put forward a model in which the weight factors (21) enter as a VDP. It is noteworthy that also Seipp[23] reaches the same conclusion, interpreting the factor (21) as a detection probability. It is remarkable that Seipp introduces in his discussion the probabilities $p_0(\lambda, a)$ and $p_0(\lambda, b)$ for the photons *not* to be observed, and estimates an upper bound of $(\pi - 1)^{-1} \simeq 0.47$ for the number of observed coincidences divided by the number of observed events.

This interesting result is closely related to a paper by Pearle[24] who first realized, in 1970 (before the CH paper had appeared!), that one can obtain the same results as QM if some of the data are "rejected," so that the recorded data are *not* a faithful sample of all pairs. Pearle also obtained a lower bound for the fraction of undetected particles; in our opinion, his results are "forerunners" of VDP. Moreover, we think that Pearle's and Seipp's bounds on the fraction of particles that *must* go undetected in order to reproduce the quantum-mechanical correlation have much to do with some estimates on how inefficient the detectors must be in order to not make the NEH. This problem has been tackled by Clauser, Horne, and Shimony,[4,6] Lo and Shimony,[25] and Garg and Mermin.[26] The best result in this direction was obtained by these last two authors who, improving a previous analysis,[25] have shown that the quantum efficiency η of a detector must satisfy the inequality

$$\eta > 2(\sqrt{2} - 1) \simeq 0.83$$

if one wishes to conduct a meaningful experiment without NEH.

This concludes our comments about the bounds on the fraction of "rejected data" and detector inefficiencies, but we conclude this section on "disguised" VDPs by presenting another example in which one can violate

NEH by starting from a different (and apparently completely unrelated) point of view.[27] We consider a hidden parameter λ controlling the interaction between a photon and a detector in such a way that if an a-polarized photon reaches the detector at time $t = 0$, it leaves (in the form of an electrical pulse) between t and $t + dt$ with probability

$$\rho(t)\, dt = e^{-t/\tau(\lambda - a)}\, dt/\tau(\lambda - a)$$

the "lifetime" τ of the process is therefore a function of the angle $\lambda - a$. It has been shown[27] that such a mechanism can lead to violations of Bell's inequalities. This model, which has apparently nothing to do with VDP, can nevertheless violate NEH. To see this, let us assume that a photon that has crossed no polarizer (say one belonging to an EPR-correlated couple) interacts with the detector with a "lifetime" τ_0:

$$\rho_0(t)\, dt = e^{-t/\tau_0}\, dt/\tau_0$$

If photons can be detected only within a coincidence window ω, then the probability of being detected is

$$p(\lambda, a) = \tfrac{1}{2} \int_0^\omega \rho(t)\, dt = \tfrac{1}{2}(1 - e^{-\omega/\tau(\lambda - a)})$$

for a polarized photon, and

$$p(\lambda, \infty) = \int_0^\omega \rho_0(t)\, dt = 1 - e^{-\omega/\tau_0}$$

for an unpolarized one. It is a simple matter to show that if $\tau_0 > \omega/\ln 2$, then

$$p(\lambda, a) > p(\lambda, \infty) \qquad \text{for some } \lambda$$

in violation of NEH.

This is the last example proposed here. The presence of a finite-coincidence window brings to light a disguised VDP and Bell's inequality can be violated.

8. Concluding Remarks

It has been claimed that the EPR problem is solved[28] and that the experimental results refute Einstein's locality and shatter our conception of physical reality.[29]

If the rate of scientific discoveries were as high as the rate of allegations, words like science and mystery would not be found in our dictionaries nowadays. But luckily (or unluckily, it is a matter of taste) science means facts and Nature is rather reluctant to give us unblurred facts. The situation with EPR-type experiments is not clear. The presence of *extra assumptions*, like NEH, in the derivation of experimentally testable Bell-type inequalities contaminates the local-realistic philosophy, which is the basis of Bell's original inequality derivation.[2] This leaves a reasonable (and peculiar) way out, via VDP models. We have seen in this chapter that a VDP can have many causes, can be the consequence of many different conceivable phenomena, and can have many important consequences. But above all, a VDP is a local-realistic explanation to the EPR puzzle. And this is sufficient reason to pursue the investigation in this direction.

ACKNOWLEDGMENTS. We acknowledge interesting remarks by F. Selleri: Section 6 is the result of several discussions with him. We are also grateful to E. Santos for having pointed out to us the papers of Corchero[11] and Ferrero *et al.*[13] This work was supported by the EEC twinning contract No. ST2J-0089-2-B.

Note Added in Proof

After reading the preprint of this paper, Marshall made the following comment: CHSH's hypothesis is not, strictly speaking, a no-enhancement hypothesis, and therefore its description as NEH1 in Section 2 is somehow misleading.

We emphasize that our purpose, when labeling CHSH's and CH's hypotheses as NEH1 and NEH2, was just to stress their common philosophy and that they both aim at a "painless" derivation of experimentally meaningful Bell-type inequalities. In addition models that violate CH's hypothesis usually (if not always) violate CHSH's hypothesis (even if the contrary is not true; see, for instance, semiclassical radiation theories). In conclusion, we think that it is worth defining CHSH's hypothesis as NEH1 for historical, more than strictly logical, reasons.

We are grateful to Trevor Marshall for having raised this "sore point."

References

1. A. Einstein, B. Podolsky, and N. Rosen, *Phys. Rev.* **47**, 777 (1935).
2. J. S. Bell, *Physics* **1**, 195 (1965).
3. J. F. Clauser, M. A. Horne, A. Shimony, and R. A. Holt, *Phys. Rev. Lett.* **23**, 880, (1969).
4. J. F. Clauser and M. A. Horne, *Phys. Rev. D* **10**, 526 (1974).

5. S. J. Freedman and J. F. Clauser, *Phys. Rev. Lett.* **28**, 938 (1972); R. A. Holt and F. M. Pipkin, Harvard University preprint (1974); J. F. Clauser, *Phys. Rev. Lett.* **36**, 1223 (1976); E. S. Fry and R. C. Thompson, *Phys. Rev. Lett.* **37**, 465 (1976); A. Aspect, P. Grangier, and G. Roger, *Phys. Rev. Lett.* **47**, 460 (1981); *Phys. Rev. Lett.* **49**, 91 (1982); A. Aspect, J. Dalibard, and G. Roger, *Phys. Rev. Lett.* **49**, 1804 (1982); W. Perrie, A. J. Duncan, H. J. Beyer, and H. Kleinpoppen; *Phys. Rev. Lett.* **54**, 1790 (1985).
6. J. F. Clauser and A. Shimony, *Rep. Prog. Phys.* **41**, 1881 (1978).
7. A. Garuccio and V. A. Rapisarda, *Nuovo Cim.* **65A**, 269 (1981).
8. T. W. Marshall, E. Santos, and F. Selleri, *Phys. Lett.* **98A**, 5 (1983).
9. T. W. Marshall, *Phys. Lett.* **99A**, 163 (1983); *Phys. Lett.* **100A**, 225 (1984).
10. S. Caser, *Phys. Lett.* **102A**, 152 (1984).
11. E. S. Corchero, Proposed Atomic Cascade Experimental Tests of Symmetric Local Hidden Variable Theories, Cantabria University preprint (1986).
12. D. Home and T. W. Marshall, *Phys. Lett.* **113A**, 183 (1985).
13. M. Ferrero, T. Marshall, and E. Santos, Chapter 19 in this volume.
14. A. Garuccio and F. Selleri, *Phys. Lett.* **103A**, 99 (1984).
15. F. Selleri, *Phys. Lett.* **108A**, 197 (1985).
16. A. Garuccio, S. Pascazio, and F. Selleri, in preparation.
17. T. W. Marshall and F. Santos, *Phys. Lett.* **107A**, 164 (1985).
18. M. Ferrero and E. Santos, *Phys. Lett.* **108A**, 373 (1985).
19. M. Ferrero and E. Santos, *Phys. Lett.* **116A**, 356 (1986).
20. M. Ferrero, Doctoral Thesis, Universidad de Oviedo (1986).
21. A. O. Barut and P. Meystre, *Phys. Lett.* **105A**, 458 (1984).
22. A. O. Barut, in: *Proceedings of the Joensuu Conference*, World Scientific, Singapore (1986); in: *Proceedings of the New York Conference* (1986), to appear.
23. H. P. Seipp, *Found. Phys.* **16**, 1143 (1986).
24. P. M. Pearle, *Phys. Rev. D* **2**, 1418 (1970).
25. T. K. Lo and A. Shimony, *Phys. Rev. A* **23**, 3003 (1981); see also E. Santos, *Phys. Rev. A* **30**, 2128 (1984) and A. Shimony, *Phys. Rev. A* **30**, 2130 (1984).
26. A. Garg and N. D. Mermin, *Phys. Rev. D* **35**, 3831 (1987).
27. S. Pascazio, *Phys. Lett.* **118A**, 47 (1986).
28. F. Röhrlich, *Science* **221**, 1251 (1983).
29. B. d'Espagnat, *In Search of Reality*, Springer-Verlag, Berlin (1983); *Sci. Am.* (November 1979); *Nature* **313**, 483 (1985); see also T. W. Marshall, *Nature* **308**, 669 (1984) and **313**, 483 (1985).

17

Stochastic Electrodynamics and the Einstein-Podolsky-Rosen Argument

Trevor Marshall

1. Introduction

The Einstein–Podolsky–Rosen (EPR) argument[1] was originally intended to demonstrate the incompleteness of quantum mechanics, but since the discovery of the Bell inequality[2] it has been used (see the introductory chapter of this book) to indicate a point at which we may expect the quantum theory to break down. The crisis comes about because, as foreseen by Einstein himself,[3] quantum theory predicts violations of the Principle of Local Action.[4,5]* This principle, which is a consequence of Special Relativity,[6] forbids the existence of any action propagating faster than light.

It is widely believed that every theory under threat has a protective set of auxiliary theories erected around it,[7] some of which may be abandoned in the face of unfavorable experimental evidence leaving the central theory intact. Quantum theory has a rather different kind of defensive system. It seeks to disarm the threat from the Principle of Local Action by attacking the scientificity of that principle. Since, so we are told,[8] quantum theory disproves the notion of objects existing independently of human consciousness, we cannot make statements about the actions such objects

---

* Born translates Prinzip der Nahwirkung as "Principle of Contiguity." I believe "Principle of Local Action" is more accurate.

Trevor Marshall • Department of Theoretical Physics, University of Cantabria, Santander 39005, Spain.

can exert on each other. In similar vein, the Sage of Copenhagen said[9] that no two objects which have once interacted can, at a later time, be observed separately.*

My view[10] is that it is the Bohr side of the argument which is unscientific. Such a theory of causation is characteristic of all prescientific cultures; it is magic. And magic, as Frazer[11] told us before quantum theory had been thought of, is "the bastard sister of science."

In this chapter, therefore, we simply assume Einstein locality as an inevitable feature of our world, and will use it both to interpret those experiments which have been done already and to suggest directions in which new experimental work should go.

The Principle of Local Action (PLA) is not a purely philosophical statement. Its scientific content was first pointed out, though somewhat obliquely, by Clauser and Horne.[12] As a result of the Clauser–Freedman[13] experiment and its subsequent refinements we have, by using the PLA, discovered a rather surprising property of atomic light signals ("photons"). This property is called *enhancement*, and is the subject of Section 2 of this chapter.

The new phenomenon of enhancement has given us what may be a very valuable clue to the oldest puzzle in quantum physics—the nature of light. I submit that, while we can now do formidable and accurate calculations in quantum electrodynamics, most of us are today no nearer to understanding wave-particle duality than Einstein was in 1905 (and, incidentally, he said so himself[19] toward the end of his life).†

The physics community, to its disgrace, has not until now taken the PLA seriously, which explains why so little effort has gone into examining enhancement. Recently two lines of investigation have been developed;[14,15] they may be considered as modern continuations of the ideas of Einstein and Planck, respectively. Einstein[16] considered that the primary objects of the electromagnetic field are strongly localized, the Maxwell description being valid only on the average. Planck,[17] on the other hand, considered that all electromagnetic signals should be treated as classical solutions of the Maxwell equations, and proposed that many of the apparently particle-like properties of light could be explained by the hypothesis of a real zero-point field.

I believe there is now some possibility of achieving a synthesis of these two approaches, but at the present stage it is better to see them developed

---

* The quote is: "any attempt at subdividing the phenomenon will demand a change in the experimental arrangements introducing new possibilities of interaction between objects and measuring instruments which in principle cannot be controlled."

† The quote is: "All these fifty years of conscious brooding have brought me no nearer to the answer to the question 'What are light quanta?' Nowadays every Tom, Dick and Harry thinks he knows it, but he is mistaken."

separately. The approach of *stochastic electrodynamics*, leading to *stochastic optics*, is the modern continuation of Planck's theory and will be described in the present chapter. Elsewhere in this book, Saverio Pascazio will describe other approaches to enhancement, including those based on Louis de Broglie's idea of particles carried by pilot waves, which is in the tradition of Einstein's approach.

2. Stochastic Realism: The Need for Enhancement

The experiments which have come closest to a realization of Bohm's version[18] of the EPR thought experiment are those using optical cascades,[13,20-24] and their close relative, the Stirling experiment,[25] which uses the two-photon decay of metastable atomic hydrogen. All except one[23] of these use one-channel analyzers, and we shall confine our discussion to such devices. This is for simplicity only; the analysis of two-channel devices, for all realist theories with enhancement, follows trivially from the one-channel analysis. An explicit example of such an extension is given in the model of Home and Marshall.[26]

We therefore consider the typical optical-cascade experiment (Figure 1). The source emits a succession of pairs of signals (one "red" and one "green"). Each pair is normally assumed to be independent of its predecessors and may be specified by a set of random variables λ. The polarizer-detector A is fitted with a polarizer of variable orientation θ_A, and B has a similar one of orientation θ_B. A registers a given red signal with probability $P_1(\theta_A, \lambda)$, while B registers independently a green signal with probability $P_2(\theta_B, \lambda)$. It is at this stage that we have imposed the requirement of the PLA; if action at a distance were possible, then P_1 could be a function of θ_B as well as θ_A. In the most recent version of the experiment,[24] very drastic steps were taken to ensure that the experimental conditions guaranteed such a decoupling of the two detection apparatuses.

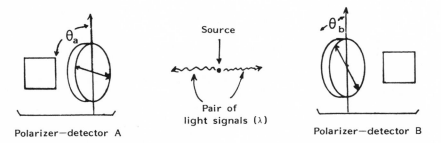

Figure 1. Diagram of a typical atomic-cascade experiment.

It now follows that the singles rate R_1 of registration at A is given by

$$R_1/R_0 = p_1 = \int \rho(\lambda) P_1(\theta_A, \lambda) \, d\lambda \qquad (2.1)$$

where R_0 is the rate at which pairs are emitted, and $\rho(\lambda)$ is the probability density of the signal variables. The singles rate R_2 at B is given by a similar expression, while the coincidence rate R_{12} is given by

$$R_{12}/R_0 = p_{12}(\theta_A - \theta_B) = \int \rho(\lambda) P_1(\theta_A, \lambda) P_2(\theta_B, \lambda) \, d\lambda \qquad (2.2)$$

the assumption that p_1 and p_2 are independent of θ_A and θ_B, while p_{12} is a function of the difference between these angles, is a natural consequence of rotational invariance, and is well verified experimentally. We shall assume in this chapter that P_1 and P_2 are given by the same function of their arguments, so that we may simply put $P_1 \equiv P_2 \equiv P$. This means we assume that the operations of the two detectors are at least approximately the same, with perhaps minor variation on account of the different colors of the two signals. It cannot be ruled out that the triggering of the counting mechanism by the first "photon" of the cascade produces a more substantial asymmetry.[27,28] however, it is my opinion that such an effect will eventually be shown to give a minor contribution. The explanation of enhancement, on the other hand, is greatly simplified by the above assumption. Hence

$$p_1 = p_2 = \int \rho(\lambda) P(\theta_A, \lambda) \, d\lambda \qquad (2.3)$$

and

$$p_{12} = \int \rho(\lambda) P(\theta_A, \lambda) P(\theta_B, \lambda) \, d\lambda \qquad (2.4)$$

Now according to quantum theory (29) the probability p_1 may be factorized into the probability of the red "photon" passing the polarizer times the probability of its activating the detector. If the polarizer is ideal and the detector efficiency is η, this gives

$$p_1 = p_2 = \tfrac{1}{2}\eta \qquad (2.5)$$

$$p_{12}(\theta_A - \theta_B) = \eta^2 q_{12}(\theta_A - \theta_B) \qquad (2.6)$$

where

$$q_{12}(\theta_A - \theta_B) = \tfrac{1}{2}\cos^2(\theta_A - \theta_B) \tag{2.7}$$

Equation (2.6) is interpreted as follows: p_{12} is the joint probability of a pair of "photons" both passing their respective polarizers and both being detected; *since it is considered that the processes of passing the polarizers and of being detected are independent*, we may regard q_{12} as the joint probability that a pair of "photons" both pass through their respective polarizers.

If we accepted the above italicized clause we would say that there exists a probability $Q(\theta, \lambda)$ such that

$$\int \rho(\lambda)Q(\theta, \lambda)\, d\lambda = \tfrac{1}{2} \tag{2.8}$$

and

$$\int \rho(\lambda)Q(\theta_A, \lambda)Q(\theta_B, \lambda)D\lambda = q_{12}(\theta_A - \theta_B) \tag{2.9}$$

Clauser and Horne[12] showed that no such probability can exist for any $\rho(\lambda)$. This is the modern form of the EPR "paradox." But within the local realist analysis there is no paradox: we simply infer that q_{12} cannot be a joint probability; *the above italicized clause is incorrect.*

Now for the experimental evidence. It is well verified that, after making some fairly small corrections for the imperfection of real polarizers, p_1, p_2, and p_{12} do indeed satisfy equations (2.5) and (2.6). However, this does not invalidate our conclusion; it is quite easy to find $\rho(\lambda)$ and $P(\theta, \lambda)$ satisfying equations (2.3) and (2.4). Here is a model, due to Watson,[5] which does so approximately:

$$\rho(\lambda) = \pi^{-1}(0 \leqslant \lambda \leqslant \pi) \tag{2.10}$$

$$P(\infty, \lambda) = \eta \tag{2.11}$$

$$P(\theta, \lambda) = \kappa\eta[\cos^2(\theta - \lambda) - \cos^2\mu]_+ \tag{2.12}$$

$$\kappa = \pi(\sin 2\mu - 2\mu\cos 2\mu)^{-1}, \qquad \mu = 0.3\pi \tag{2.13}$$

where the notation $(\ )_+$ means that the expression is zero if the bracket is negative and $P(\infty, \lambda)$ denotes the detection probability of λ when the polarizer is removed. In Table 1 we give the values of the experimental quantity

$$r(\theta_A - \theta_B) = \frac{p_{12}(\theta_A, \theta_B)}{p_{12}(\infty, \infty)} \tag{2.14}$$

Table 1. Values of the Angular
Correlation Function $r(\theta)$ for
the Watson Model in the
Perfect Polarizer Case[a]

θ	$r(\theta)$	$r_Q(\theta)$
0	0.522	0.500
$\pi/8$	0.427	0.427
$\pi/4$	0.233	0.250
$3\pi/8$	0.073	0.073
$\pi/2$	0.014	0.000

[a] $r_Q(\theta)$ is the quantum value.

for the most commonly observed angles, according to the quantum model and the Watson model. It has been shown[30] that none of the existing experiments provides us with data good enough to discriminate between the two models.

A significant feature of the Watson model is that, for $\lambda = \theta$, it gives

$$P(\theta, \theta) = \pi \sin^2 \mu (\sin 2\mu - 2\mu \cos 2\mu)^{-1} \simeq 1.341 \qquad (2.15)$$

and hence

$$P(\theta, \theta) > P(\infty, \theta) \qquad (2.16)$$

This means that, for a signal with $\lambda = \theta$, the probability of activating the photodetector is enhanced when a linear polarizer is interposed between the source and the detector. Clauser and Horne[12] showed that any local realist model capable of explaining the atomic-cascade data must have the general property of *enhancement*, in the sense that there must be values of λ for which

$$P(\theta, \lambda) > P(\infty, \lambda) \qquad (2.17)$$

It is possible,[26] with a little ingenuity, to devise a model which agrees *exactly* with the quantum prediction for $r(\theta_A - \theta_B)$. While such an exercise has some pedagogic value, it can cause us to lose sight of the main scientific issue. We are not seeking agreement with quantum theory, but rather explanation of experimental data. Indeed, since quantum theory gives us the quantity $q_{12}(\theta_A - \theta_B)$ which, as we saw above, is not a proper probability, we can claim that, notwithstanding all their ad hoc features, models like the Watson model provide the *only possible* explanation of the experimental data. (I should add that the very word "explanation" has become debased in Physics these last sixty years.)

We see, then, that enhancement is necessary to explain the data obtained in the atomic-cascade experiments. The earliest proof of the Bell inequality[2] was for local deterministic models only; all the probabilities like $P(\theta, \lambda)$ were zero or one. In such models enhancement was, literally, inconceivable. This emphasizes that we are dealing here with a realism more sophisticated than determinism, to which the name *stochastic* local realism* may be given. A signal λ has a probability $P(\infty, \lambda)$, lying between zero and one, of activating a detector. If a polarizer is interposed, so that the context is changed, then this probability is modified to $P(\theta, \lambda)$. All enhancement models, whether of the stochastic optics or of the wave-particle duality type, must assume that the modification of $P(\infty, \lambda)$ to $P(\theta, \lambda)$ goes beyond a simple multiplication by a polarizer-transmission factor.

I do not treat wave-particle duality models in this chapter, but it is a general feature of such models, which Selleri[14] calls *variable detection probability*, that $P(\infty, \lambda)$ does actually vary with λ. The Watson model may be regarded (see the next section) as the simplest example of a stochastic optical model, and from equation (2.11) we see that variable detection in this sense is not a necessary feature of all enhancement models. This means that some care will be needed in assessing the results of the Stirling three-polarizer experiment, described in A. J. Duncan and H. Kleinpoppen's chapter of this book.

3. Stochastic Optics: The Mechanism of Enhancement

Stochastic optics is a semiclassical radiation theory. In general such theories treat the pure electromagnetic field classically, while continuing to use nonrelativistic quantum mechanics to describe the motion of electrons in atoms. It has been known for a long time[31,32] that such theories are capable of explaining emission and absorption of light by atoms. The greatest difficulty for semiclassical theories[33] has been in explaining the interaction of light with macroscopic devices, such as linear polarizers and beam splitters.

In this and subsequent sections we shall show that the latter difficulties are overcome in a semiclassical theory which supposes the existence of a real zero-point field. The basic idea of stochastic optics is to treat the transmission of light (through lenses, mirrors, polarizers, and so on) exactly as in classical optics, but including the existence of a zero-point radiation present everywhere. The assumption is made that the zero-point radiation has the same nature as ordinary light but, for reasons to be discussed later,

---

* The world view represented by stochastic local realism seems very similar to that given by what Shimony[65] calls "contextual hidden variables theories of environmental type." The argument presented here shows that Shimony is incorrect in stating that the factorizability condition [his equation (2)] for such theories has been shown by experiment to be untenable.

it cannot be detected directly. In consequence, we assume for the transmission of light—including the zero-point—the same laws as in classical optics. On the other hand, we shall use ad hoc assumptions for the emission and absorption of radiation. We hope that these rules will one day be derived from more basic (classical, although stochastic) principles.

Thus stochastic optics is a less developed theory than either quantum optics or Jaynes[32] semiclassical radiation theory. It has its origin in a fairly well developed *classical* theory, namely, stochastic electrodynamics.[34-39] However, one certainly cannot claim that this latter theory gives an adequate treatment of atomic emission and absorption, though many of its results seem to point tantalizingly in such a direction. Its main successes to date lie precisely in the area where previous semiclassical radiation theories have failed: the interaction of radiation with macroscopic devices. An outstanding example is the explanation of the Casimir effect and other long-range van der Waals forces.[40,41]

In stochastic electrodynamics the radiation emitted "spontaneously" from an excited atom is considered to have been stimulated by an appropriate component of the zero-point field. Instead of being spread over a broad wave front, such radiation may be supposed to be concentrated in a pencil-shaped region, whose length is of the order of the natural coherence length (typically 1-10 m), and whose cross-section is of the order 10^{-4} to 10^{-3} m. The general description of the signal is similar to that originally proposed by Einstein[42] under the name of *needle radiation* (Nadelstrahlung), and the dimensions of the pencil are discussed elsewhere[43] within the general context of stochastic electrodynamics.

For present purposes it will suffice to describe the pencil, in its interior, as a plane wave, traveling in the z-direction, whose electric vector is

$$E_x = \mathrm{Re}[(E_1 + iE_2)\, e^{i\omega t}], \qquad E_1 + iE_2 = \beta e^{i\psi} \cos \varphi$$
$$E_y = \mathrm{Re}[(E_3 + iE_4)\, e^{i\omega t}], \qquad E_3 + iE_4 = \beta e^{i\chi} \sin \varphi \tag{3.1}$$

We shall suppose that β has a fixed value, which is the same for all such single-atom signals, and that the angles (φ, ψ, χ) are random variables whose distribution is dictated entirely by considerations of rotational invariance[44]:

$$\rho(\varphi, \psi, \chi) = (4\pi^2)^{-1} \sin 2\varphi \qquad (0 \leq \varphi \leq \pi/2, 0 \leq \psi, \chi < 2\pi) \tag{3.2}$$

From the standpoint of classical optics, (φ, ψ, χ) is simply the set of Stokes parameters[45] describing the signal, while from the standpoint of local realist theory it corresponds to the set of "hidden variables" traditionally labeled by λ, that is,

$$\lambda = (\varphi, \psi, \chi) \tag{3.3}$$

When such a signal, denoted $E = (E_1, E_2, E_3, E_4)$ interacts with a two-channel polarizer or beam splitter, fields E_t and E_r are transmitted. However, while in deterministic classical optics it would be natural to assume, on energy grounds, that $\overline{E^2} = \overline{E_t^2} + \overline{E_r^2}$, where the bar indicates time averaging, we must assume that, in stochastic optics, a contribution E_0 from the zero-point field is added to both channels, as depicted in Figure 2. The relevant part of the zero-point field is supposed to have the same plane-wave description as the signal, but with amplitude $\beta_0 = 1$ and Stokes parameters $\lambda_0 \equiv (\varphi_0, \psi_0, \chi_0)$ statistically independent of λ. For an ideal beam splitter the signals in the two channels will be

$$E_t = 2^{-1/2}(E + E_0), \qquad E_r = 2^{-1/2}(E - E_0) \qquad \text{(beam splitter)} \quad (3.4)$$

while for an ideal linear polarizer they will be

$$E_t = (E_1, E_2, E_{30}, E_{40}), \qquad E_r = (E_{10}, E_{20}, E_3, E_4) \quad \text{(linear polarizer)} \quad (3.5)$$

It is of interest that the description of a beam splitter in quantum optics is formally the same as equation (3.4),[46] though in that case the fields are interpreted as Hermitian operators rather than random variables. It would seem that both theories agree on the important role of the zero-point field; the parallel between operators and random variables in the two theories has been noted previously.[47,48] We shall return to the beam splitter in the next section and concentrate, for the moment, on the action of a linear polarizer, described by equation (3.5). The signals E_t and E_r cause photo-multiplier activations in their respective channels. The normal semiclassical

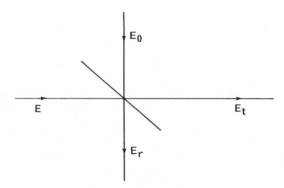

Figure 2. Action of a beam splitter or linear polarizer in stochastic optics. The incoming signal **E** is mixed with the zero-point component E_0 and split to produce the outgoing signals E_t and E_r.

assumption is[33] that the activation probability P for a time window of w is $2\xi w\overline{E^2}$, where ξ is the detector efficiency. We shall modify this assumption, taking account of the real zero-point field, to write

$$P = \xi w(2\overline{E^2} - \gamma)_+ \tag{3.6}$$

where the subscript $+$ means that P is zero if the bracket is negative and γ is a threshold greater than one (the zero-point intensity). Then the activation probabilities with and without a linear polarizer are

$$P(\infty, \lambda) = \xi w(\beta^2 - \gamma) \tag{3.7}$$

and

$$P(\theta, \lambda) = \xi w \int_0^{\pi/2} \sin 2\varphi_0 [f(\Theta) + \sin^2 \varphi_0]_+ \, d\varphi_0 \tag{3.8}$$

where

$$\cos 2\Theta = \cos 2\theta \cos 2\varphi + \sin 2\theta \sin 2\varphi \cos(\psi - \lambda) \tag{3.9}$$

and

$$f(\Theta) = \beta^2 \cos^2 \Theta - \gamma \tag{3.10}$$

The integral may be evaluated to give

$$P(\theta, \lambda) = \begin{cases} \xi w(f + \tfrac{1}{2}) & (f \geqslant 0) \\ \tfrac{1}{2}\xi w(f + 1)^2 & (0 \geqslant f \geqslant -1) \\ 0 & (f \leqslant -1) \end{cases} \tag{3.11}$$

and it is now clear that there are values of $\lambda = (\varphi, \psi, \chi)$ (such as $\varphi = \theta$, $\psi = \lambda$, giving $\Theta = 0$) for which equation (2.17) is satisfied, that is, enhancement occurs.

The Watson model of the previous section is now seen to be a primitive version of stochastic optics, in which the Stokes parameters were replaced by a single parameter λ representing linear polarization, so that Θ was simply equal to $(\lambda - \theta)$. The threshold parameter γ is represented, in the Watson model, by $\cos^2 \mu$. The main difference between the Watson model and that studied in this section is that we now have a mechanism for the enhancement process; the enhancement factor κ [see equation (2.13)] now no longer has to be put in "by hand." Enhancement occurs, for small Θ, because a single-atom signal is *superpolarized*. This means that, to extend a concept of Dirac, the signal has the same polarization both above and below the "sea" of zero-point radiation. On transmission through a linear polarizer the original intensity of β^2 is reduced to $\beta^2 \cos^2 \Theta$, in accordance with Malus' law, but, referring to Figure 2, this is enhanced by an intensity of $\frac{1}{2}$ coming from $\mathbf{E}_0$. We do not expect that the phenomenon of enhancement will easily be directly observed, since it always disappears after carrying out an ensemble average over λ. A direct observation of enhancement would require us to prepare an ensemble of signals all with the same λ, which means in the same polarization state. Referring again to Figure 2, it is evident that, according to our model of enhancement, the zero-point radiation field is precisely what prevents us from preparing such a pure polarization state.

It should be noted that, according to the interpretation now proposed for λ in the Watson model, the red and green signals may be considered to be perfectly correlated, that is, $\lambda_R = \lambda_G$. More generally, we should suppose that the set of pairs is described by a probability density $\rho(\lambda_R, \lambda_G)$ and, indeed, if the apertures of the light-collecting systems are finite, such a description would be necessary, leading to a classical calculation of the depolarization factor, denoted elsewhere[29] by F. The assumption of perfect correlation is appropriate in the limit of zero aperture. For a 0–1–0 cascade[13,22–24] we may then put $\lambda_R = \lambda_G$. For the 1–1–0 cascade[20,21] the corresponding assumption would be to put $\lambda_R = \pi/2 - \lambda_G$ which would simply result in $r(\theta_A - \theta_B)$ of Table 1 being replaced by $r(\pi/2 - \theta_A + \theta_B)$.

We may now make a similar assumption for the model of stochastic optics. Putting $\lambda_R = (\varphi, \psi, \chi)$, we assume that $\lambda_G = \lambda_R$ for the 0–1–0 cascade and $\lambda_G = (\pi/2 - \varphi, \psi, \chi)$ for the 1–1–0 cascade. The relation between the counting rates for the two types of cascade is then the same as above, and it is necessary to carry out the calculation only for the 0–1–0 case. The substitution of equations (3.7) and (3.11) in equations (2.3) and (2.4) is a tedious but straightforward matter.[43] The threshold parameter γ is chosen so that the singles rate is reduced by a factor of exactly one half with the polarizer in place, that is,

$$\int \rho(\lambda) P(\theta, \lambda) \, d\lambda = \tfrac{1}{2} P(\infty, \lambda) \tag{3.12}$$

Table 2 gives the results for the value 1.511 of parameter β. In the range $1.511 < \beta < 1.518$, the values of $r(\theta_A - \theta_B)$ come out just as close to the quantum values as they did in the Watson model. Since further refinements of this model are possible, I think it is best at this stage to draw attention to two qualitative features shared by Tables 1 and 2: (1) the values of $r(\pi/8) - r(3\pi/8)$ always exceed 0.25, thereby violating the Clauser-Freedman[29] inequality, and showing again that enhancement can result in a violation of any *homogeneous* Bell-type inequalities; (2) the value of

Table 2. Values of the Angular
Correlation Function in the
Perfect Polarizer Case[a]

θ	$r(\theta)$	$r_Q(\theta)$
0	0.499	0.500
$\pi/8$	0.417	0.427
$\pi/4$	0.241	0.250
$3\pi/8$	0.093	0.073
$\pi/2$	0.039	0.000

[a] The parameter β has the value 1.511 and the
threshold γ is chosen so that the linear
polarizer reduces the singles rate by exactly
one half; $r_Q(\theta)$ is the quantum value.

$r(\pi/2)$ is substantially greater than it is in quantum optics. In my opinion
both experimental and theoretical work should now be concentrated on the
latter of these two features. For too long now our efforts have been concentrated on discriminating between quantum theory and a rather narrow family
of realist models which satisfy homogeneous Bell inequalities. For a more
general discussion of the distinction between homogeneous and
inhomogeneous Bell inequalities, reference should be made to the article
of Emilio Santos in this book.

We shall also see in the following sections that the introduction of
specific new local realist models, such as those based on the ideas of
stochastic optics or of wave-particle duality, suggests further tests of quantum optics. These new tests may in some cases be easier to perform than
the optical-cascade tests. We can therefore justifiably claim that the program
of investigation opened up by the EPR argument is now becoming more
and more relevant to the rest of Physics.

4. The Beam Splitter in Stochastic Optics

I should explain why the present section is included in a book devoted
to the EPR argument. As indicated at the end of the previous section, optical
experiments other than those using linear polarizers and atomic cascades
can be used to discriminate, for example, between stochastic optics and
wave-particle duality. However, semitransparent mirrors and other types of
beam splitter play a much more central role than this in respect of the EPR
argument. It has become traditional to claim that all semiclassical radiation

theories must give counting statistics, in the two channels of a beam splitter, which satisfy certain inequalities.[33,49-51] More recently[52] Chubarov and Nikolayev have discovered that, according to quantum optics, a pair of independent "photons" from a monochromatic source, after passing through a beam splitter, have polarizations which are correlated (see Figure 3) in exactly the same way as a pair of "photons" from an atomic cascade. This means that, potentially at least, the beam splitter is a simpler EPR source than an atomic cascade. It also means, as Chubarov and Nikolayev point out, that the description of the beam splitter, according to quantum optics, is in contradiction with all local realist theories. Of course, both stochastic optics and realist theories based on wave-particle duality fall in this latter category. That is why we have to study the beam splitter.

I propose to confine myself in this chapter to stochastic optics, so the first task is to show why the traditional inequalities[33,49-51] of semiclassical radiation theory do not apply in this case. Indeed, in view of the experimental evidence,[49,51] we must show how a theory based on a purely wave description of light can account for a strong particle-like anticorrelation between the counts in the two channels of a beam splitter.

To see how this can occur, we refer again to Figure 2. The stochastic addition of E and E_0 gives, in both the transmitted and reflected channels, a signal whose intensity, according to equation (3.4), lies between $\frac{1}{2}(\beta - 1)^2$ and $\frac{1}{2}(\beta + 1)^2$, where β^2, as explained in the previous section, is the signal-to-noise ratio. The actual values of these intensities depend on the phases of the λ_0-variables relative to the λ-variables, but for all such phases the sum of these intensities is $\beta^2 + 1$. Now in both channels the detectors have a

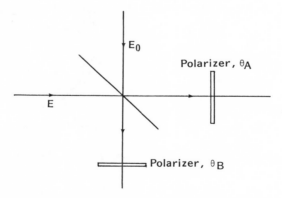

Figure 3. Thought experiment of Chubarov and Nikolayev showing the beam splitter as a source of EPR-correlated photons. As in Figure 2, the zero-point component E_0 produces the correlation.

threshold intensity of γ. This means that, if $\gamma > \frac{1}{2}(\beta^2 + 1)$, at most one detector can be activated by each single-atom signal. In order that activations do occur for some values of λ and λ_0, we must have $\gamma < \frac{1}{2}(\beta + 1)^2$. It follows that we will obtain the same anticorrelation statistics as in quantum optics, provided that

$$(\beta + 1)^2 > 2\gamma > \beta^2 + 1 \tag{4.1}$$

With a purely wave theory, such as we are considering, there is no difficulty in interpreting wave-recombination experiments of the Janossy-Naray type.[53,51] Here reference should be made to Figure 4, where it will be seen that the second beam splitter recombines the signals $\mathbf{E}_t$ and $\mathbf{E}_r$ from the first beam splitter, giving the transmitted signal

$$\mathbf{E}_t' = 2^{-1/2}[\mathbf{E}_t + \exp(i\omega l / c)\mathbf{E}_r] \tag{4.2}$$

l being the difference between the two path-lengths. Now on substituting for $\mathbf{E}_t$ and $\mathbf{E}_r$ from equation (3.4), we find that, for $l = 0$, $\mathbf{E}_t' = \mathbf{E}$, while for $l = \pi c / \omega$, $\mathbf{E}_t' = \mathbf{E}_0$. This means that, if the path lengths differ by an integral number of half-wavelengths, the signal $\mathbf{E}$ and the noise $\mathbf{E}_0$ are perfectly reproduced in the two outgoing channels. Since the threshold is above the zero-point level, the fringe visibility, as defined by Grangier *et al.*,[51] is 100%, which is in good agreement with the 98.5% reported by these authors. It should be especially noted that the outcome of this experiment is,

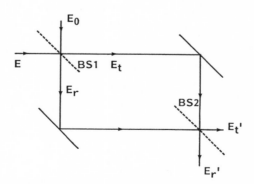

Figure 4. A typical recombination experiment of Janossy–Naray type to illustrate single-photon interference.

according to our analysis, the same even if it is carried out in a "delayed choice" manner, with the second beam splitter inserted after the signals E_t and E_r have left the first beam splitter. This is one prediction in agreement with orthodox quantum theory, but in disagreement with certain "alternative interpretations"[54] of that theory. The property that, for most values of λ_0, one of the two signals E_t and E_r is subthreshold, but nevertheless retains its phase coherence so that the two signals interfere on recombination, may be considered a form of *latent order.*[55]

Although stochastic optics explains qualitatively the phenomena generally considered to exhibit most clearly the wave-particle-duality nature of light, the actual coincidence rates predicted are not identical with quantum optics. We saw in the previous section that, to explain the atomic-cascade data, the parameter β may lie in the range $1.511 < \beta < 1.518$. Also, for a given β, the threshold parameter γ was fixed by requiring that the singles rate with a polarizer in place is exactly one half of its value without the polarizer. While β is certainly a property of the signal alone, and must therefore be given the same value for both analyzing devices, we could plausibly suppose that γ takes a different value for different devices. The counting rate in one channel of the beam splitter is given[43] by

$$N_t = \xi w \pi^{-1} \int_0^\pi d\psi \int_0^{\pi/2} \sin 2\varphi \, d\varphi (\tfrac{1}{2}\beta^2 + \tfrac{1}{2} + \beta \cos \varphi \cos \psi - \gamma)_+ \quad (4.3)$$

and if we put it equal to $\tfrac{1}{2}(\beta^2 - \gamma)\xi w$, this determines γ in terms of β. A calculation[43] then shows that the γ so obtained for the beam splitter differs by no more than five percent from the value obtained for the linear polarizer in the previous section. With the values of γ corresponding to the β at the two ends of the above range, one obtains

$$0 < \beta^2 + 1 - 2\gamma < 0.28 \quad (4.4)$$

This means that the second inequality (4.1) is not necessarily satisfied. There is a small range of values of λ_0 for which the beam splitter gives signals of around the same intensity in the two channels, resulting in a nonzero activation probability in both channels. *Stochastic optics predicts that the photon can be split.*

The above conclusion certainly seems shocking, in that it contradicts what is probably the most elementary assumption of quantum optics. This is therefore an appropriate point for me to say again that the locality argument of Einstein, Podolsky, and Rosen is a basis not for *reinterpreting*, but for *superseding*, quantum theory. Realist models incorporating the locality property will inevitably lead to predictions different from those of quantum theory, and that is the proper basis for the design of experimental tests. If we look at the best existing data[51] we find[56,57] that they are not good enough to discriminate between stochastic optics and quantum optics. This is because there is always a background of accidental coincidences between the two channels of a beam splitter, arising from the possibility that two atoms emit their signals within the same time window of *w*. Hence the counting statistics must be sufficient to separate the true "photon-splitting" events given by equation (4.4) from such an "accidental" background.

5. Future Experiments

In the previous two sections we have established that in respect both of polarization correlation and of beam-splitter counting rates, the conventional wisdom is at fault. Neither a violation of a homogeneous Bell inequality nor a violation of Clauser's[49] beam-splitter inequality is evidence of nonlocality in nature. Nevertheless, given a slightly improved method of statistical analysis, both these techniques can be used to obtain valuable evidence about the nature of the light. Data from polarization correlation counts should concentrate on the value of $r(\pi/2)$ rather than on that of $r(\pi/8) - r(3\pi/8)$, while data from beam splitters should be concentrated on source insensities for which $Nw \ll 1$, where N is the rate at which signals are emitted.

The improvement required in the quality of data has been estimated in both these areas[30,57] and it is almost certainly achievable. But the aim of such experiments should be clearly defined. Having abandoned the immature idea that all classical or local realist theories must be deterministic, it should be clear that the program of testing the whole family of local realist theories is impossibly overambitious. Instead we should look at stochastic realist ralternatives to quantum theory, and see where they differ from quantum theory in their predictions. If they agree for a wide range of phenomena and disagree in a region not yet investigated, then we know where we should look.

The proposal of Garuccio and Selleri[58] leading to the design of the new Stirling experiments[59] falls entirely within the above experimental philosophy. It considers a class of models, based on the de Broglie version of wave-particle duality, which agrees with quantum optics over a very wide range of phenomena—what the authors call "single-photon physics," and which also fits the hitherto existing atomic-cascade data. Such models predict rather large departures from quantum optics if, in Figure 1, a third polarizer is added between the polarizer and detector at A. The experimental results so far obtained do not yet definitely rule out Garuccio–Selleri-type models, but they do indicate this as a likely outcome.

The analysis of the Stirling experiments according to stochastic optics is rather different from the Garuccio–Selleri analysis. As we saw in the previous section, stochastic optics does not necessarily give the same predictions as quantum optics for single-photon processes, and it is not a trivial matter to devise experiments which discriminate between these predictions. According to quantum optics, a second polarizer reduces the counting rate by a factor of $\cos^2 \theta$, where θ is the angle between the polarizers. This is Malus' law. Stochastic optics gives, for "single-photon" signals, small but, in principle, measurable deviations[43] from Malus' law. But there seems to be no previous experimental evidence on this point. The Stirling experiment may well be the first to check the predictions of Malus' law for single-atom signals. Here then is another field of investigation suggested by stochastic optics. It could be that, with two polarizers at A, just as much will be learned with no polarizer at all at B (see Figure 1), using the detection at B only to trigger the opening of the window w at A. Such an experiment would be similar in scope to the beam-splitter experiment of Grangier, Roger, and Aspect[51] discussed in the previous section.

The other area which, I would suggest, has been badly neglected is the study of the effects of using real instead of ideal polarizers. It has been considered adequate[29] to characterize linear polarizers, such as piles of plates and calcite crystals, by the two quantities ε_M and ε_m, which give the transmission rates when light polarized parallel and perpendicular to the polarizer's axis is incident on it. For classical light, whose polarization is described by the Stokes parameters (see Section 3), we need[60] twice as many polarizer parameters as this. It is particularly important in this connection that the Holt–Pipkin[20] experiment be repeated. This, the only atomic-cascade experiment to use calcite polarizers ($\varepsilon_m \simeq 10^{-4}$), was also the only one to find significant deviations from the quantum predictions. The best pile-of-plates polarizers have $\varepsilon_m = 0.02$, which means that the *amplitude* of the perpendicular component is fifteen percent of the parallel. According to stochastic optics, this certainly makes an important difference to Malus' law investigations.[43] It may well be that the biggest deviations of stochastic optics from the quantum prediction for cascade coincidences are for near-ideal polarizers. This has been a feature of most ad hoc enhancement models; the original one of Clauser and Horne[12] breaks down altogether for calcite polarizers, while that of Marshall, Santos, and Selleri[61] gives a deviation about twice as big for calcite as for piles of plates.

6. Conclusions

This chapter, taken in conjunction with that of S. Pascazio, should convince anyone that concrete local realist models of the electromagnetic field can generate an interesting and growing research program. Although its motivation is largely philosophical, such a program shows us how to pose new problems and reexamine old ones in traditional areas of Physics such as optics.

If the stochastic optics approach leads us to a more profound understanding of the electromagnetic interaction, there will be a kind of historical justice done. As I remarked in the first section of this chapter, the most serious "paradox" (I prefer to say inconsistency) of the quantum theory has its origin in quantum optics—in the quantization of the electromagnetic field. Neither Einstein,[19] the originator of this quantization, nor Dirac,[62] who gave us its modern formalism, considered it fully satisfactory, especially as its nonlocal features became more apparent. Furthermore, it now begins to look as if some of the earliest opponents were rejecting field quantization precisely on the grounds of what we now recognize as its nonlocality. I referred in the introduction to the criticisms of Planck[17] which led him to introduce the zero-point field, an idea also developed by Nernst.[63] When Einstein first introduced Bose statistics of "photons" he was criticized by Ehrenfest, and Einstein's reply to this criticism was an admission that the interaction between "photons," causing them to satisfy Bose statistics, must be "very mysterious."[64] We now know how mysterious!

Viewed in this way, the Einstein–Podolsky–Rosen argument, and Einstein's precise statement of the Principle of Local Action, show Einstein's rather late conversion to the school of Planck, Nernst, and Ehrenfest. Because the argument of the EPR article was posed for particles, it was natural that its rediscovery by Bell should also be so posed. It took another ten years for the argument to be extended, by Clauser and Horne, to cover both particles and waves, but their treatment was flawed by an over-hasty dismissal of the enhancement possibility. In this chapter I have tried to show that Einstein was correct in his championship of locality. But also I have tried to show that a full development of the EPR argument, in the light of recent experimental evidence, leads us to conclude that those who criticized Einstein's quantum theory during its earliest period were also correct. Far from being an EPR-type particle, the "photon" is not a particle at all.

References

1. A. Einstein, B. Podolsky, and N. Rosen, *Phys. Rev.* **47**, 777 (1935).
2. J. S. Bell, *Physics* **1**, 195 (1964).
3. A. Einstein, Autobiographical Notes, in: *Albert Einstein: Philosopher-Scientist* (P. A. Schilpp, ed.), Tudor, New York (1949).
4. A. Einstein, *Dialectica* **2**, 320–324 (1948). English translation in *The Born–Einstein Letters,* Macmillan, London (1971).
5. T. W. Marshall, in: *Microphysical Reality and Quantum Formalism* (G. Tarozzi and A van der Merwe, eds.), D. Reidel, Dordrecht (1988).
6. A. Einstein, *Ann. Phys.* **23**, 371 (1907).
7. I. Lakatos, *Falsification and the Methodology of Scientific Research Programmes* (I. Lakatos and A. Musgrove, eds.), p. 100, Cambridge University Press (1970).
8. B. d'Espagnat, *Scientific American* (November 1979).
9. N. Bohr, Discussion with Einstein on epistemological problems in atomic physics, in: *Albert Einstein: Philosopher-Scientist* (P. A. Schilpp, ed.), Tudor, New York (1949).
10. T. W. Marshall, in: *Microphysical Reality and Quantum Formalism* (G. Tarozzi and A. van der Merwe, eds.), Introduction, Reidel, Dordrecht (1988).
11. J. G. Frazer, *The Golden Bough*, 2nd edition, p. 52, Macmillan, London (1900).
12. J. F. Clauser and M. A. Horne, *Phys. Rev. D* **10**, 326 (1974).
13. S. J. Freedman and J. A. Clauser, *Phys. Rev. Lett.* **28**, 938 (1972).
14. F. Selleri, *Phys. Lett. A* **108**, 197 (1985).
15. T. W. Marshall and E. Santos, Stochastic optics: a reaffirmation of the wave nature of light, *Found. Phys.* (in proof) (1988).
16. A. Einstein, *Phys. Z.* **10**, 817 (1909).
17. M. Planck, *Verh, Dtsch. Phys. Ges.* **13**, 138 (1911).
18. D. Bohm, *Quantum Theory*, Prentice Hall, London (1951).
19. A. Einstein (1951), quoted on title page of P. L. Knight and L. Allen, *Concepts of Quantum Optics*, Pergamon Press, Oxford (1983).
20. R. A. Holt and F. M. Pipkin, Quantum mechanics versus hidden variables: polarization correlation measurements on an atomic mercury cascade, preprint, Harvard University (1974).
21. E. S. Fry and R. C. Thompson, *Phys. Rev. Lett.* **37**, 465 (1976).
22. A. Aspect, P. Grangier, and G. Roger, *Phys. Rev. Lett.* **47**, 460 (1981).
23. A. Aspect, P. Grangier, and G. Roger, *Phys. Rev. Lett.* **49**, 91 (1982); A. Aspect and P. Grangier, *Nuovo Cim. Lett.* **43**, 345 (1985).
24. A. Aspect, J. Dalibard, and G. Roger, *Phys. Rev. Lett.* **49**, 1804 (1982).
25. W. Perrie, A. J. Duncan, H. J. Beyer, and H. Kleinpopper, *Phys. Rev. Lett.* **54**, 1790 (1985).
26. D. Home and T. W. Marshall, *Phys. Lett. A* **113**, 183 (1985).
27. S. Pascazio, *Phys. Lett.* **118A**, 47 (1986).
28. E. Santos, *Phys. Lett. A* **101**, 379 (1984).
29. J. F. Clauser and A. Shimony, *Rep. Prog. Phys.* **41**, 1881 (1978).
30. T. W. Marshall, *Phys. Lett. A* **100**, 225 (1984).
31. M. D. Crisp and E. T. Jaynes, *Phys. Rev.* **179**, 1253 (1964).
32. E. T. Jaynes, *Coherence and Quantum Optics IV* (L. Mandel and E. Wolf, eds.), p. 495, Plenum Press, New York (1978).
33. L. Mandel, *Prog. Opt.* **13**, 27 (1976).
34. T. H. Boyer, *Ann. Phys. (N.Y.)* **56**, 474 (1970).
35. P. Claverie and S. Diner, *Int. J. Quantum Chem.* **12**, Suppl. 1, 41 (1977).

36. L. de la Peña, in: *Stochastic Processes Applied to Physics and Other Related Fields* (B. Gomez, S. M. Moore, A. M. Rodriguez-Vargas, and A. Rueda, eds.), World Scientific, Singapore (1983).

37. T. H. Boyer, in: *Foundations of Radiation Theory and Quantum Electrodynamics* (A. O. Barut, ed.), Plenum Press, New York (1980).

38. T.H. Boyer, *Scientific American* (August 1985).

39. E. Santos, in: *Open Questions in Quantum Physics* (G. Tarozzi and A. van der Merwe, eds.), D. Reidel, Dordrecht (1985).

40. T. W. Marshall, *Nuovo Cim.* **38**, 206 (1965).

41. T. W. Marshall, *Nuovo Cim.* **41**, 188 (1966).

42. A. Einstein, *Phys. Z.* **18**, 121 (1917). English translation in: *Sources of Quantum Mechanics* (B. L. van der Waerden, ed.), Dover New York (1968).

43. T. W. Marshall, report to conference "Quantum Uncertainties," Bridgeport, Connecticut (1986), Plenum Press, New York (W. M. Honig, ed.), to appear.

44. T. W. Marshall and E. Santos, Stochastic optics: a reaffirmation of the wave nature of light, Section 4, *Found. Phys.* (in proof) (1988).

45. M. Born and E. Wolf, *Principles of Optics*, pp. 30–32, Pergamon Press, Oxford (1984).

46. R. Loudon, *Opt. Commun.* **45**, 361 (1983).

47. E. Santos, *Nuovo Cim. B* **22**, 201 (1974).

48. T. H. Boyer, *Phys. Rev. D* **11**, 809 (1975).

49. J. F. Clauser, *Phys. Rev. D* **9**, 853 (1974).

50. R. Loudon, *Rep. Prog. Phys.* **43**, 913 (1980).

51. P. Grangier, G. Roger, and A. Aspect, *Europhys. Lett.* **1**, 173 (1986).

52. M. S. Chubarov and E. P. Nikolayev, *Phys. Lett. A* **110**, 199 (1985).

53. L. Janossy and Z. Naray, *Acta Phys. Hungar.* **7**, 403 (1957).

54. J. A. Wheeler, in: *Quantum Theory and Measurement* (J. A. Wheeler and W. H. Zurek, eds.), pp. 182–213, University Press, Princeton (1983).

55. D. Greenberger, in: *Microphysical Reality and Quantum Formalism* (G. Tarozzi and A. van der Merwe, eds.), Reidel, Dordrecht (1988).

56. T. W. Marshall and E. Santos, *Ann. N. Y. Acad. Sci.* **480**, 400 (1986).

57. T. W. Marshall and E. Santos, *Europhys. Lett.* **3**, 293 (1987).

58. A. Garuccio and F. Selleri, *Phys. Lett. A* **103**, 99 (1984).

59. A. J. Duncan, in: *Microphysical Reality and Quantum Formalism* (G. Tarozzi and A. van der Merwe, eds.) Reidel, Dordrecht (1988).

60. R. C. Jones, *J. Opt. Soc. Am.* **46**, 528 (1956).

61. T. W. Marshall, E. Santos, and F. Selleri, *Phys. Lett. A* **98**, 5 (1983).

62. P. A. M. Dirac, The development of quantum mechanics, in: *Contributi del Centro Linceo Interdisciplinario Roma*, Anno CCCLXXI, No. 4 (1974).

63. W. Nernst, *Verh. Dtsch. Phys. Ges.* **18**, 83 (1916).

64. A. Einstein, Sitzungsber, *Preuss Akad. Wiss., Phys.-math.* K1, 3 (1925).

65. A. Shimony, *Brit. J. Phil. Soc.* **35**, 34, 35 (1984).

18

Explicit Calculations with a Hidden-Variable Spin Model

A. O. BARUT

1. Introduction

The basic problem we study here is whether the two-spin correlation experiments together with the theoretical Bell inequalities have already excluded the possibility of introducing hidden variables into quantum theory, as is often concluded. This question is answered negatively by explicitly reproducing the quantum-mechanical two-spin correlation function by a classical model, where the spin is associated with a classical dipole-moment vector. We then study the behavior of single events in classical and quantum models and conclude that the detector efficiency may be a fundamental limitation rather than just a technical problem to be overcome by better techniques. We further show that the assumptions underlying the derivation of Bell inequalities involve statements about single events which are consistent with neither the classical nor quantum models. It is important therefore to work with explicit physical situations rather than with abstract assumptions.

2. Classical Spin

In a nonrelativistic domain, by a *classical spin* we mean a dynamical variable **S** which satisfies in the presence of a magnetic field **B** the equation

A. O. BARUT • International Center for Theoretical Physics, Trieste, Italy. *Permanent address*: Department of Physics, University of Colorado, Boulder, Colorado 80309, United States.

$$\frac{d\mathbf{S}}{dt} = a\mathbf{S} \times \mathbf{B} \tag{1}$$

where a is a constant. The set of dynamical variables of the particle $(\mathbf{x}, \mathbf{p})$, namely the phase space, is augmented now by the vector $\mathbf{S}$.

The first point to emphasize, which is perhaps not generally realized, is that the conjugate variable to a spin component is another spin component even in classical theory. To see this, we note first that the magnitude of spin vectors $\mathbf{S}$ is constant even in an inhomogeneous field, because

$$2\mathbf{S} \cdot \frac{d\mathbf{S}}{dt} = \frac{d}{dt}(\mathbf{S}^2) = 0$$

using equations (1). This leaves us with two independent spin components only, and not three. Since equations (1) are of first order, we expect that the phase space is two-dimensional. Indeed, we have a two-dimensional symplectic or canonical system, and a pair of canonical dynamical variables can be chosen in at least three different ways. We require from a set of

Table 1. Three Choices of the Conjugate Pair of Variables and Corresponding Poisson Brackets

Conjugate pair	Hamiltonian and Poisson brackets
S_3, ϕ	$H = a\sqrt{S^2 - S_3^2}(B_1 \cos\phi + B_2 \sin\phi) + aS_3B_3$ $\{f, g\} = \left(\dfrac{\partial f}{\partial S_3}\dfrac{\partial g}{\partial \phi} - \dfrac{\partial f}{\partial \phi}\dfrac{\partial g}{\partial S_3}\right)$ $\{S_3, \phi\} = 1$
S_1, S_2	$H = a(S_1B_1 + S_2B_2 + \sqrt{S^2 - S_1^2 - S_2^2}\,B_3)$ $\{f, g\} = \sqrt{S^2 - S_1^2 - S_2^2}\left(\dfrac{\partial f}{\partial S_1}\dfrac{\partial g}{\partial S_2} - \dfrac{\partial f}{\partial S_2}\dfrac{\partial g}{\partial S_1}\right)$ $\{S_1, S_2\} = (S_2 - S_1^2 - S_2^2)^{1/2}$
θ, φ	$H = aS(\sin\theta \cos\varphi B_1 + \sin\theta \sin\varphi B_2 + \cos\theta B_3)$ $\{f, g\} = \dfrac{1}{S \sin\theta}\left(\dfrac{\partial f}{\partial \theta}\dfrac{\partial g}{\partial \varphi} - \dfrac{\partial f}{\partial \varphi}\dfrac{\partial g}{\partial \theta}\right)$ $\{\theta, \varphi\} = \dfrac{1}{S \sin\theta}$

canonical conjugate variables (p, q) that both the Poisson bracket and Hamilton's equations are satisfied:

$$\dot{p} = \partial H / \partial q = \{H, p\} \quad \text{and} \quad \dot{q} = -\partial H / \partial p = \{H, q\}$$

For the Hamiltonian $H = a\mathbf{S} \cdot \mathbf{B}$, Table 1 shows the three choices of the conjugate pair of variables and the corresponding Poisson brackets.

The whole phase space of spin is thus the homogeneous space S_2, the sphere. The essential point is that even for classical spin there is only one coordinate degree of freedom (one spin component or one angle) and one conjugate generalized momentum. Thus in the canonical formalism we can specify only one component of spin at a time for a given magnitude S of canonically conjugate pairs so that although, for example, S_1 and S_2 do not "commute" (nonzero Poisson bracket) they could be, classically, simultaneously measured at a given time.

With this spin model, first the single spin measurement will be discussed and then the two-spin correlation experiment.

3. The Stern–Gerlach Experiment with Classical Spins

This is the experiment par excellence, the prototype of a quantum measurement, in which the eigenstates of an observable are separated, the reduction of the wave packet takes place, and the system is in an eigenstate after the measurement.

The naive classical theory of the Stern–Gerlach experiment proceeds as follows. If a classical spin or magnetic dipole moment $\boldsymbol{\mu} = g\mathbf{s}$ enters the inhomogeneous magnetic field, the deflection of the particle in the z-direction of the magnetic field after a length l is[1]

$$s = \frac{1}{6} \frac{\mu_z}{m} \frac{\partial B_z}{\partial z} \frac{l^2}{v^2} \cos \theta = s_0 \cos \theta \tag{2}$$

where θ is the angle between $\boldsymbol{\mu}$ and $\mathbf{B}$ (in the z direction). The number n of particles, in a repeated experiment, that arrives at s can be found from

$$\frac{dn}{ds} = \frac{dn}{d\theta} \frac{d\theta}{ds}$$

Now from equation (2), $ds/d\theta = -s_0 \sin \theta$; $dn/d\theta$ depends on the initial distribution of spin directions. If we assume that they are uniformly distributed over the sphere, we have $n(\theta)\, d\theta = n_0 \sin \theta\, d\theta$ or $dn/d\theta = n_0 \cos \theta$, hence

$$\frac{dn}{ds} = -\frac{n_0}{s_0} \frac{\cos \theta}{\sin \theta} = -\frac{n_0}{s_0} \frac{s}{(s_0^2 - s^2)^{1/2}}$$

or, by integration,

$$n(s) = \frac{n_0}{s_0}(s_0^2 - s^2)^{1/2}$$

So we would expect a broad, almost uniform distribution (Figure 1). The deflections of spin in the inhomogeneous fringe fields have been neglected. If the distribution of θ at the entrance to the Stern–Gerlach magnet is not like $n(\theta) = n_0 \sin \theta$, but a function which favors the two poles ($\theta = 0$, $\theta = \pi$), due to the magnetic field in the z-direction, then we would get a distribution peaked at the two values $s = \pm s_0$. Quantum mechanics assumes that the spin components in equation (1) take only two values, hence the deflection has two values $s = \pm s_0$. (The experimentally observed broad distributions can be explained by the thermal distribution of velocities.)

The theory of the Stern–Gerlach experiment is actually much more complicated both classically and quantum mechanically. This is because, in an inhomogeneous field, the equations of orbit and of spin are nonlinearly coupled. The full phase space consists now of $(\mathbf{x}, \mathbf{p}, S_1, S_2)$, for example, and we have the equations (nonrelativistically and for a *neutral* magnetic dipole moment)

$$\dot{\mathbf{x}} = \frac{1}{m}\mathbf{p}, \qquad \dot{\mathbf{p}} = -\boldsymbol{\mu} \cdot \nabla\mathbf{B}, \qquad \dot{\boldsymbol{\mu}} = \boldsymbol{\mu} \times \mathbf{B}(x) \qquad (3)$$

The quantum-mechanical equations are the same except that $\boldsymbol{\mu}$ is replaced by $\mu_0\boldsymbol{\sigma}$, where $\boldsymbol{\sigma}$ are the noncommuting Pauli matrices.

Furthermore, because $\nabla \cdot \mathbf{B} = 0$, $\mathbf{B}$ cannot just be chosen in the z-direction but it must have at least another component, say B_x. Hence the Stern–Gerlach instrument is not a pure device to measure μ_z, for example. Sometimes one considers a large constant field B_0 in the z-direction, and an inhomogeneous field $B(x, z)$ satisfying the condition div $B = 0$, and one averages out over the x-direction, for example, in order to have approximately a single equation

$$\dot{p}_z = -\mu_z \frac{\partial B_z}{\partial z} \qquad (4)$$

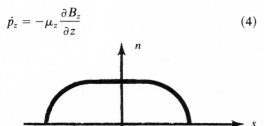

Figure 1. Number of particles $n(s)$ with a deflection s in the z-direction.

for the measurement of μ_z. All these imply that we do not have a simple and straightforward device such that if a polarized particle enters it, it will surely be deflected in the z-direction and its spin will be unchanged. It may also be deflected otherwise and its spin might get rotated along the way, both classically and quantum mechanically.

These considerations will be used to put into question one of the basic so-called natural assumptions underlying Bell inequalities, namely, that every single event that enters the Stern–Gerlach detector will be detected, and it has a definite outcome, either $+1$ or -1. In fact, it may not be detected at all, ending up at the walls of the magnet, for example, or it may be misdirected even with some small probability; but we can only say that, on the average, among all detected events, half have the value $+1$, the other half -1.

There is another interesting effect which can cause a change in the distribution of spin direction which we assumed to be isotropic, besides the fringe fields. If some *nonlinear*, radiative-reaction type of terms, are added to equation (1), then

$$\frac{d\mathbf{S}}{dt} = a\mathbf{S} \times \mathbf{B} + \lambda[\ddot{\mathbf{S}} - (\dot{\mathbf{S}} \cdot \mathbf{S})\mathbf{S}]$$

Then it turns out[2,3] that for $\mathbf{B}$ in the z-direction, the $\pm z$-directions are attractors so that all initial spins in the upper hemisphere end up after a long time in the direction of the north pole, and all spins in the lower hemisphere end up in the direction of the south pole. Although the classical value of λ is small, hence this process is very slow, this model shows nicely how an apparent quantization of spin or change of the isotropic angular distribution might come about.

4. On the Quantum Theory of the Stern–Gerlach Experiment

The ideal textbook Stern–Gerlach experiment is discussed in terms of the two equations

$$\dot{z} = \frac{1}{m} p_z \quad \text{and} \quad \dot{p}_z = -\mu \frac{\partial B}{\partial z} \sigma_z \tag{5}$$

only. And since σ_z has two eigenvalues and eigenvectors, a single polarized spin (in the z-direction) will be surely deflected into the $+z$ or $-z$ direction (with probability one) and an unpolarized spin beam, after repeated experiments, splits into two parts each with probability $\frac{1}{2}$. It is also tacitly assumed in many theoretical discussions that for each single event there is a definite outcome, $+1$ or -1. In other words, there are no undetected events, or false events (spin-up being deflected once in a while to the wrong side, and vice versa). Of course, we could have undetected and false events, yet, after repeated counts, still a splitting each with probability $\frac{1}{2}$, on the average. We note that quantum theory does not make statements about individual events, but only about probabilities in repeated events.

Now as in the classical case, the real equations for the Stern–Gerlach experiment are

$$\dot{\mathbf{x}} = \frac{1}{m}\mathbf{p}, \qquad \dot{\mathbf{p}} = -\mu_0 \boldsymbol{\sigma} \cdot \nabla \mathbf{B}, \qquad \dot{\boldsymbol{\sigma}} = \boldsymbol{\sigma} \times \mathbf{B} \qquad (6)$$

Again, because div $B = 0$, the magnetic field cannot be purely in the z-direction. There are three sets of coupled dynamical variables $\mathbf{x}$, $\mathbf{p}$, and $\boldsymbol{\sigma}$ which lead to a highly complicated nonlinear problem, even for the simplest possible B-field geometry. It can be shown that the propagator for this problem is not diagonal in spin variables, so that there is a probability of a spin-up to end in the spin-down direction.[4] If one puts a large constant magnetic field B_0 in the z-direction and averages over the other precessions, one could perhaps say approximately that we measure the z-component of the magnetic moment, but of course only on the average, and not for each individual event.

It is actually sufficient in quantum theory to make only probability statements and avoid making mental pictures about individual events. This is in fact what quantum theory tells us to do. In so doing, we also avoid the difficulties associated with *the reduction of the wave packet*. The difficulty comes if one says that an individual state which is a superposition of eigenstates will be found after the measurement in an eigenstate with probability one. Instead we should say that the measurement produces eigenstates only on the average; the individual state is as undetermined after the measurement as before the measurement. The measurement can only tell us how many eigenstates there are, and with what probabilities a state is associated.

Another way of putting this discussion is to question the existence of an ideal measuring device which monitors every incoming particle and produces a definite final result for each particle.

The detection efficiency η of a measuring device many not be just a temporary technical limitation eventually to be overcome by a perfect detector, but of a more fundamental nature. In addition, we have the quantum or probabilistic nature of the source of the particles which we assume always prepares the right initial states (e.g., always a singlet state).

If we remain within quantum theory and work with repeated events and probabilities, there is no problem; all the indeterminacies of individual events go into the probabilities. However, the difficulties arise when we come to the problem of hidden variables, or try to show the impossibility of introducing hidden variables, or reality of states. In so doing some assumptions are introduced about single events which seem natural, but actually contradict quantum theory, such as the assumption underlying Bell inequalities that each individual spin event has a definite outcome $+1$ or -1 with efficiency one. We now turn to a discussion of these problems.

5. Two-Spin Correlation Experiments

Again we shall compare calculations with both classical and quantum spins. In both cases we have two observables A and B:

$$A = \mathbf{S}_1 \cdot \mathbf{a} \qquad \text{and} \qquad B = \mathbf{S}_2 \cdot \mathbf{b}$$

where $\mathbf{a}$ and $\mathbf{b}$ are the two directions of the polarizers, $\mathbf{S}_1$ and $\mathbf{S}_2$ two (correlated) spins, and we are interested in the correlation function

$$E(a, b) = \frac{\langle AB \rangle - \langle A \rangle \langle B \rangle}{\langle A^2 \rangle \langle B^2 \rangle} \tag{7}$$

In the quantum case $\mathbf{S}_1 = \frac{1}{2}\boldsymbol{\sigma}_1$, $\mathbf{S}_2 = \frac{1}{2}\boldsymbol{\sigma}_2$, and $\langle \cdot \rangle$ means the quantum-mechanical expectation value in a state ψ. For the singlet state $|\psi_s\rangle$ of two spins, in particular, $\langle A \rangle = \langle B \rangle = 0$, and equation (7) simplifies to $E(a, b)_{\text{singlet}} = \langle \psi_s | \boldsymbol{\sigma}_1 \cdot \hat{a}\boldsymbol{\sigma}_2 \cdot \hat{b} | \psi_s \rangle$, which has the well-known value

$$E(a, b)_{\text{singlet}} = -\hat{a} \cdot \hat{b} \tag{8}$$

Classically, $\mathbf{S}_1$ and $\mathbf{S}_2$ are the spin vectors and $\langle \cdot \rangle$ means an expectation-value integral over all angles with measure $(1/4\pi) \int d\varphi \sin \theta \, d\theta$. We have $\mathbf{S}_1$, say, in the direction (θ, φ) and $\mathbf{S}_2$ in the direction $(\theta - v, \varphi + u)$ with v and u fixed. Thus the two spins are correlated. We take the component of $\mathbf{S}_1$ along $\mathbf{a}$, and that of $\mathbf{S}_2$ along $\mathbf{b}$. We repeat the experiment, assuming that $\mathbf{S}_1$ is uniformly distributed over the sphere, and evaluate the expectation values in equation (7). Clearly $\langle A \rangle = \langle B \rangle = 0$, and we find[5]

$$E(a, b) = \cos v \cos(\theta - u) \tag{9}$$

with $\cos \theta = \hat{a} \cdot \hat{b}$.

For antiparallel spins, $\mathbf{S}_1 = -\mathbf{S}_2$, $u = 0$, and $v = \pi$; hence we have[6]

$$E(\hat{a}, \hat{b})_{\text{antiparallel}} = -\hat{a} \cdot \hat{b} \tag{10}$$

the same result as equation (8). For parallel spins, $u = 0$, $v = 0$, and $E(a, b) = \hat{a} \cdot \hat{b}$, while for two perpendicular spins $E(a, b) = 0$. (In these calculations we have assumed vectors $\hat{a}$ and $\hat{b}$ to be in the equatorial plane.)

The final correlation function $E(a, b)$ measured in repeated experiments is thus exactly the same in quantum theory and for classical correlated spin vectors. There is no problem of reproducing this measured correlation,[7] which agrees with quantum mechanics, by a hidden-variable spin model. The hidden variables λ here are the angles (θ, φ) of the spin vector **S**. But although the finally evaluated correlation functions are the same, each individual event looks apparently quite different in the classical and in the quantum case. The behavior of individual events that we shall discuss in the next section is really the key to the entire controversial subject of hidden variables versus quantum mechanics. It goes beyond quantum mechanics, because quantum formalism cannot make statements about individual events.

6. Individual Events

In the analysis with hidden variables λ of the correlation function $E(\hat{a}, \hat{b})$, the following two main assumptions are made:[8]

1. $$E(a, b) = \int \rho(\lambda) \, d\lambda \, \tilde{A}(a, \lambda) \tilde{B}(b, \lambda) \qquad (11)$$

   where $\rho(\lambda)$ is a measure in the λ-space, while $\tilde{A}(a, \lambda)$ and $\tilde{B}(b, \lambda)$ are the two densities for the observables A and B for each setting of λ. Equation (11) is a form of "locality," and for our present purpose we do not need to investigate the meaning of this concept here, as it has been extensively analyzed in the literature.
2. The second assumption is that $\tilde{A}$ and $\tilde{B}$ have the values ± 1, or $\pm \frac{1}{2}$ for spins, or are bounded by 1 in absolute value.

With these assumptions (or with slight modifications of them), it is then possible to derive the so-called Bell inequalities, or that the correlation function (11) cannot be completely identical to the quantum correlation functions, hence the class of hidden-variable theories satisfying assumptions (1) and (2) cannot reproduce quantum theory.

Since we have already reproduced the quantum result by a hidden-variable model, we must see how it fares with the above assumptions. First, it is convenient to introduce the projection operators[5-6]

$$P_{\pm}^A = \frac{1}{2} \frac{\sqrt{\langle A^2 \rangle} \pm A}{\sqrt{\langle A^2 \rangle}} \qquad (12)$$

and their expectation values

$$P_{\pm}(a) = \langle P_{\pm}^A \rangle, \qquad P_{\pm}(b) = \langle P_{\pm}^B \rangle \qquad (13)$$

as well as the correlations

$$P_{ij}(ab) = \langle P_i^A P_j^B \rangle \qquad (i, j = \pm) \qquad (14)$$

in both the classical and quantum cases. We then have the identity

$$E(a, b) = P_{++}(a, b) + P_{--}(a, b) - P_{+-}(a, b) - P_{-+}(a, b) \qquad (15)$$

In the quantum case, $P_{\pm}^A = \frac{1}{2}(1 \pm \sigma_1 \cdot \hat{a})$ are the usual spin projection operators so that P_{++}, P_{--}, and so on, have the meaning of both spin components (relative to $\hat{a}$ and $\hat{b}$, respectively) being measured up, both down, and so on.

In the classical case, it can be seen that the projector operators $P_{\pm}^A$ project the spin vector **S** into the upper or lower hemisphere relative to the vector **a** pointing to the north pole. This gives us a natural discretization of the continuous classical spins.

In terms of these probabilities P_i, P_{ij} the two assumptions above are given by

$$P_i(a) = \int d\lambda \, \rho(\lambda) P_i(a, \lambda) = \frac{1}{2}, \qquad i = \pm$$

$$ \qquad (16)$$

$$P_{ij}(a, b) = \int d\lambda \, \rho(\lambda) P_i(a, \lambda) P_j(b, \lambda), \qquad i, j = \pm$$

For the singlet spin state we also have explicitly

$$P_{ij}(a, b) = \frac{1}{4}(1 - \mathbf{a} \cdot \mathbf{b}) \qquad \text{for } ij = ++, --$$

$$ \qquad (17)$$

$$= \frac{1}{4}(1 + \mathbf{a} \cdot \mathbf{b}) \qquad \text{for } ij = +-, -+$$

Now the general expression (7) with $\langle A \rangle = \langle B \rangle = 0$ always satisfies the "locality" assumption (11), as its form shows. For our classical spin vectors it can be written as

$$E(a, b) = \frac{1}{4} \int d\varphi \, d\theta \sin \theta \tilde{A}(a, \lambda) \tilde{B}(b, \lambda) \qquad (18)$$

with[6]

$$\tilde{A}(a, \lambda) = \sqrt{3} \cos \mu \qquad \text{and} \qquad \tilde{B}(b, \lambda) = -\sqrt{3} \cos \lambda$$

Here μ and λ are the angles between S_1 and a, and S_2 and b, respectively. Thus, clearly the second assumption that $\tilde{A}$ and $\tilde{B}$ are bounded in absolute value by 1 is not satisfied. Yet the integrated function $E(a, b)$ is the correct quantum-mechanical correlation function. For equations (16) we find

$$p_i(a) = \frac{1}{4\pi} \int d\varphi \, d\theta \sin \theta \, \tfrac{1}{2}(1 \pm \sqrt{3} \cos \mu) = \tfrac{1}{2}$$

(19)

$$p_i(b) = \frac{1}{4\pi} \int d\varphi \, d\theta \sin \theta \, \tfrac{1}{2}(1 \pm \sqrt{3} \cos \lambda) = \tfrac{1}{2}$$

and

$$p_{ij}(a, b) = \frac{1}{4\pi} \int d\varphi \, d\theta \sin \theta \, \tfrac{1}{4}(1 \pm \sqrt{3} \cos \mu)(1 \pm \sqrt{3} \cos \lambda) \qquad (20)$$

Again the integrated results are as in equations (16) and (17) but the densities $p_i(a, \lambda)$ and $p_i(b, \lambda)$ are not bounded by one, hence not probabilities.

The two-spin correlation experiment is often visualized as a black box with two outcomes on both sides, two light bulbs indicating the +, − outcomes on both sides (Figure 2). Thus one measures the coincidence ++, +−, −+, −− and evaluates equation (15). In this picture, each classical spin event would not light one and only one bulb on each side, but would contribute to all four bulbs, in general some bright some dim. Thus what we are testing here is not just the final correlations (they are the same), but whether an individual event lights up one lamp or two, three or four. We have assumed a discrete set of single events and try to model this discreteness by a hidden variable.

Let us continue to do this. The projection operators (12) introduced a natural discreteness to the classical spin. We denote by Ω_a^+ the upper

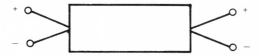

Figure 2. Two-spin coincidences with discrete ± outcomes indicated by light bulbs.

hemisphere with **a** being the north pole and Ω_a^- the lower hemisphere. Similarly for $\Omega_b^\pm$. If $\mathbf{S}_1 \in \Omega_a^+$ we call the outcome $+1$, and if $\mathbf{S}_1 \in \Omega_a^-$ we call the outcome -1. We have further to project $\mathbf{S}_1$ on **a** so that $\hat{S}_1 \cdot \hat{a} = \cos \mu$. We then obtain a discrete model with outcomes ± 1 subject to the following rules:

$$p_i(a) = N \int d\lambda \, \cos \mu \chi_a^i, \qquad i = \pm 1 \qquad (21)$$

where χ_a^i is the characteristic function of the hemisphere $\Omega_a^\pm$. After integration with respect to θ, this can be written as

$$p_i(a) = \tfrac{1}{4} \int d\varphi |\cos(\varphi - \varphi_a)| \chi_a^i \qquad (22)$$

and gives correctly $p_\pm(a) = \tfrac{1}{2}$.

For the joint probabilities we must count the events in the intersection of the hemispheres $\Omega_a^i \cap \Omega_b^i$, and with the projection factor $\cos \mu$. This gives the rule

$$p_{ij}(a, b) = \tfrac{1}{4} \int d\varphi |\cos(\varphi - \varphi_a)| \chi_i^a \chi_j^b \qquad (23)$$

For example,

$$p_{+-}(a, b) = \int_{\varphi_b - \pi/2}^{\varphi_a + \pi/2} \tfrac{1}{4} d\varphi |\cos(\varphi - \varphi_a)| = \tfrac{1}{4}(1 + \hat{a} \cdot \hat{b})$$

In this version of the model we have reproduced the discrete outcomes, but the joint probability (23) does not quite fit the assumption (16), which we shall critically reexamine at the end.

We remark that the factor $\tfrac{1}{4}|\cos(\varphi - \varphi_a)|$ arises because our observables A and B are the components of spins in the directions $\hat{a}$ and $\hat{b}$. There are two other observables A' and B', with discrete ± 1 outcomes,

$$A' = \text{sign}(\mathbf{S}_1 \cdot \hat{a}) \qquad \text{and} \qquad B' = \text{sign}(\mathbf{S}_2 \cdot \hat{b})$$

which measure just whether spins point into the upper or lower hemispheres but do not take into account the length of their projections. For these the correlation function is quite different.[6]

We give a third version of the hidden-variable spin model, which satisfies both the assumptions (16); however, this model assumes the detector efficiency in a fundamental way. The factor $\tfrac{1}{4}|\cos(\varphi - \varphi_a)|$ in equation (23) will be related to the detector efficiency.[9]

Let $p_\pm(a, \lambda)$ be the probability densities as before, but introduce another density $p_0(a, \lambda)$ measuring the probability that the event is not recorded. Similarly, we will have in addition to P_{++}, P_{+-}, P_{-+}, P_{--} the probabilities P_{+0}, P_{-0}, $P_{0,+}$, $P_{0,-}$ for single counts, and P_{00} for no count at all. We note that the single counts are undoubtedly present in the actual experiments. It is sufficient to consider the equatorial plane, hence one angle φ for the hidden variable λ. The model is then formulated as follows:

$$p_+(a, \varphi) = M\chi_a \cos(\varphi - \varphi_a)$$

$$p_-(a, \varphi) = M(\chi_a - 1) \cos(\varphi - \varphi_a)$$

$$p_0(a, \varphi) = 1 - p_+ - p_-$$

$$p_+(b, \varphi) = N(1 - \chi^b) \tag{24}$$

$$p_-(b, \varphi) = N\chi_b$$

$$p_0(b, \varphi) = 1 - N$$

Here, χ_a is the characteristic function on the circle such that

$$\chi_a = 1, \qquad \varphi_a - \pi/2 \le \varphi < \varphi_a + \pi/2$$

$$= 0, \qquad \text{otherwise}$$

The measure is $d\lambda = (1/2\pi) \, d\varphi$. The equation does not look symmetric, but will be symmetrized at the end.

We can now calculate all the integrated probabilities and obtain

$$p_\pm(a) = \frac{1}{\pi} M, \qquad p_0(a) = 1 - \frac{2}{\pi} M$$

$$p_\pm(b) = \tfrac{1}{2}N, \qquad p_0(b) = 1 - N \tag{25}$$

and also the joint probabilities

$$p_{++}(a, b) = p_{--}(a, b) = \frac{1}{2\pi} MN[1 - \cos(a - b)]$$

$$p_{+-}(a, b) = p_{-+}(a, b) = \frac{1}{2\pi} MN[1 + \cos(a - b)]$$

$$p_{+0}(a, b) = p_{-0}(a, b) = \frac{1}{\pi} M(1 - N) \tag{26}$$

$$p_{0+}(a, b) = p_{0-}(a, b) = \frac{1}{2} N - \frac{1}{\pi} MN$$

$$p_{00}(a, b) = 1 - N - \frac{2}{\pi} M(1 - N)$$

On the event space of measured coincidences, we define the reduced joint probabilities

$$\overline{p_{ij}}(a, b) = \frac{p_{ij}(a, b)}{p_{++} + p_{+-} + p_{-+} + p_{--}} = \frac{p_{ij}(a, b)}{C} \tag{27}$$

where C is the total number of all coincidences. We find $C = (2/\pi)MN$, and

$$\overline{p_i}(a) = \tfrac{1}{2}, \qquad \overline{p_i}(b) = \tfrac{1}{2}, \qquad i = \pm \tag{28}$$

$$\overline{p_{ij}}(a, b) = \tfrac{1}{4}(1 - \hat{a} \cdot \hat{b}) \qquad \text{for } ij = ++, --$$

$$= \tfrac{1}{4}(1 + \hat{a} \cdot \hat{b}) \qquad \text{for } ij = -+, +- \tag{29}$$

as in equation (17). The results (28) and (29) are independent of the constants M and N of the model (24). It becomes symmetric, if we choose $M = \tfrac{1}{2}\pi N$. Further, we have the restriction $0 < N < 2/\pi$.

The ratio of coincidences to the sum of total counts is

$$\frac{C}{1 - p_{00}} = \frac{(2/\pi)MN}{1 - N - (2/\pi)M + (2/\pi)MN} = \frac{N}{2 - N} \qquad \text{for } M = \frac{\pi}{2} N$$

which must be less than 0.47. Or, the ratio of single counts to coincidences is greater than $\pi - 2 = 1.13$. According to these numbers the maximum efficiency for a symmetric model is about 64%. Only if the experimental efficiency becomes larger than 64% can such a model perhaps be ruled out. The efficiency of the present experiments is about 2 or 3%.[7] Furthermore, it is important to measure the single counts as well as the coincidences; there seem to be more of them.

In all three versions of our hidden-variable two-spin correlation model, some of the assumptions underlying Bell inequalities are not satisfied: In version I, the continuous spin, the assumption $|A(a, \lambda)| \leq 1$; in version II, the "locality"; and finally in version III, the existence of $p_0(a, \lambda)$—the accounting for detection efficiency.

The detector efficiency problem has been discussed extensively by other authors[10,11] as well.

7. Conclusions

When discussing hidden variables in general terms, we have to assign them some properties. The assumptions underlying Bell inequalities are not as obvious as they appear when compared to our explicit model. They assign to the hidden variables partly quantum-mechanical and partly classical properties: The spin projections are supposed to have discrete values, either $+1$ or -1, as in quantum mechanics, but, unlike quantum mechanics, they have definite values in *every direction* simultaneously. Furthermore, *every single event* is assumed to be detected and have a definite outcome, while quantum mechanics makes only statements about probabilities in repeated experiments. One could require other properties from hidden variables. Our purely classical explicit hidden variables for spin reproduce exactly quantum spin correlations in three different forms.

References

1. O. Stern, *Z. Phys.* **7**, 249 (1921); W. Gerlach and O. Stern, *Z. Phys.* **9**, 353 (1922); *Ann. Phys.* **74**, 673 (1924).
2. A. F. Ranada and M. F. Ranada, *J. Phys. A* **12**, 1419 (1979).
3. A. O. Barut, in: *Fundamental Questions in Quantum Mechanics* (L. Roth and A. Inomata, eds.), p. 33, Gordon and Breach, New York (1986).
4. M. O. Scully, A. O. Barut, and W. E. Lamb, Jr., *Found. Phys.* **17**, 575 (1987).
5. A. O. Barut and P. Meystre, *Phys. Lett.* **105A**, 458 (1984).
6. A. O. Barut, in *Symposium on the Foundations of Modern Physics* (P. Lahti and P. Mittelstaedt, eds.), p. 321, World Scientific, Singapore (1985).
7. A. Aspect, J. Dalibard, and G. Roger, *Phys. Rev. Lett.* **49**, 1804 (1984).
8. J. S. Bell, *Physics* **1**, 195 (1964); J. F. Clauser and M. A. Horn, *Phys. Rev. D* **10**, 526 (1974); J. F. Clauser and A. Shimony, *Rep. Prog. Phys.* **41**, 1861 (1978).
9. H. P. Seipp, *Found. Phys.* **16**, 1143 (1986).
10. F. Selleri and G. Tarozzi, *Riv. Nuovo Cim.* **4**, 1 (1981).
11. D. Mermin, in: *New Techniques and Ideas in Quantum Measurement Theory* (D. M. Greenberger), Vol. 480, Annals of New York Academy (1986); A. O. Barut, *ibid.*, p. 393.

19

Symmetric and Asymmetric Models for Atomic Cascade Experiments

MIGUEL FERRERO, TREVOR MARSHALL,
AND EMILIO SANTOS

1. Introduction

It is pointed out elsewhere in this book (see, for example, the chapters of Santos and of Marshall) that the evidence from optical tests of the Bell inequalities cannot be used to support any claim[1,2] that quantum nonlocality is an experimentally established phenomenon. Such claims have been based on some confusion between the two principal types of Bell inequality: homogeneous and inhomogeneous. The first type of inequality is satisfied only in those local realist theories satisfying some kind of auxiliary hypotheses, while the second must be satisfied in all local realist theories. This was already clear in the article of Clauser and Horne,[3] but it is only relatively recently that any serious study has been made of those local realist theories which do not satisfy the auxiliary hypotheses.

2. Experiments with Two Polarizers

Homogeneous Bell inequalities are the only type which have been tested. They relate coincidence rates between events, separated by a space-like interval, in two photomultipliers. There are two different experimental procedures, depending on whether the linear polarizers are of the one-

MIGUEL FERRERO • Department of Physics, University of Oviedo, Oviedo, Spain. TREVOR MARSHALL AND EMILIO SANTOS • Department of Theoretical Physics, University of Cantabria, Santander 39005, Spain.

channel[4-7] or two-channel type.[8,9] It is widely believed that the second of these gives the most complete test of the Bell inequalities. Indeed the article reporting this experiment was entitled "A realization of the Einstein–Podolsky–Rosen–Bohm Gedanken experiment," and it was claimed that its result refuted all possible local realist theories. In this experiment the quantity

$$E(a, b) = \frac{R_{++}(a, b) + R_{--}(a, b) - R_{+-}(a, b) - R_{-+}(a, b)}{R_{++}(a, b) + R_{--}(a, b) + R_{+-}(a, b) + R_{-+}(a, b)} \qquad (1)$$

was measured. The quantum theoretic value is

$$E(a, b) = \cos(2a - 2b) \qquad \text{(quantum)} \qquad (2)$$

(We are quoting here the value for ideal polarizers.)

The authors reporting this experiment made what they call the "highly reasonable" auxiliary assumption that the set of photon pairs detected is a faithful sample of all the pairs emitted. Given this assumption, they proved that no local realist theory can satisfy equation (2). They also claimed that they had demonstrated the truth of the auxiliary assumption by checking that the quantity $R_{++}(a, b) + R_{+-}(a, b)$ is independent of a and b.

The "reasonableness" of an assumption is not accessible to scientific analysis, and we do not discuss it here. It is, however, a simple matter[10] to find a local realist model reproducing all the quantities in the numerator and denominator of equation (1) in exact correspondence with their quantum values. This means finding $\rho(\lambda)$, $P_+(a, \lambda)$, and $Q_+(b, \lambda)$, and so on such that

$$R_{++}(a, b) = k \int P_+(a, \lambda) Q_+(b, \lambda) \rho(\lambda) \, d\lambda$$

$$= \tfrac{1}{2} k \eta^2 \cos^2(a - b) \qquad (3)$$

$$R_{+-}(a, b) = k \int P_+(a, \lambda) Q_-(b, \lambda) \rho(\lambda) \, d\lambda$$

$$= \tfrac{1}{2} k \eta^2 \sin^2(a - b) \qquad (4)$$

and so on. The functions achieving this are

$$\rho(\lambda) = 1/\pi \qquad (0 \leq \lambda \leq \pi) \qquad (5)$$

$$P_+(a, \lambda) = \begin{cases} \eta & (|a - \lambda| < \pi/4) \\ 0 & (\pi/4 < |a - \lambda| < \pi/2) \end{cases} \tag{6}$$

$$Q_+(b, \lambda) = \begin{cases} (\pi\eta/2) \cos(2b - 2\lambda) & (|b - \lambda| < \pi/2) \\ 0 & (\pi/4 < |b - \lambda| < \pi/2) \end{cases} \tag{7}$$

$$P_-(a, \lambda) = \begin{cases} 0 & (|a - \lambda| < \pi/4) \\ \eta & (\pi/4 < |a - \lambda| < \pi/2) \end{cases} \tag{8}$$

$$Q_-(b, \lambda) = \begin{cases} 0 & (|b - \lambda| < \pi/4) \\ (-\pi\eta/2) \cos(2b - 2\lambda) & (\pi/4 < |b - \lambda| < \pi/2) \end{cases} \tag{9}$$

The range of integration in equations (3) and (4) is $(0, \pi)$ and all functions P and Q are periodic in λ with period π. In equations (6)–(9) the efficiency of both photodetectors is denoted by η, and the fact that, for example,

$$Q_+(b, b) = \pi\eta/2 > \eta \tag{10}$$

expresses the property of enhancement, which is characteristic of any local realist model violating a homogeneous Bell inequality. For a fuller discussion of this point see the chapter of Marshall in this book.

3. Experiments with More than Two Polarizers

It has been suggested by Garuccio and Selleri[11,12] that all local realist models of enhancement type, and agreeing with quantum theory in the two-polarizer experiment, will necessarily give predictions different from quantum theory if a second polarizer or a half-wave plate is put between polarizer a and its detector. Their model has been tested experimentally, and the results are reported in Kleinpoppen's chapter of this book.

The conjecture is false and the opposite can be now proved:

Theorem. If all photon detector efficiencies are smaller than $2/\pi$, no contradiction exists between quantum mechanics and local realistic theories in atomic cascade experiments in which any number of one-channel optical devices are inserted between the source and each detector.

Proof. The theorem is proved by constructing a local realist model which agrees with quantum mechanics for all such experiments. For the sake of simplicity we construct the model assuming ideal behavior for all optical devices except detectors. (It is easy to change the model to take account of nonideal behavior.) The present model improves slightly a previous one.[13]

We assume that the experiment consists of measuring, with two detectors, the coincidence rate of photons coming from an atomic source suitably excited (e.g., with a laser). Between the source and each detector is inserted a system of lenses, sometimes a filter, and any number of additional devices.

We characterize each photon by three parameters (l, φ, δ) which correspond, together, to λ in Section 2. The parameters l and φ, with $-\pi/2 \leq l \leq \pi/2$ and $-\pi/4 \leq \varphi \leq \pi/4$, are similar to those used in classical optics for the specifications of elliptical polarization, i.e., if an electromagnetic wave propagates along the z axis, the electric field has components along x and y given by

$$E_x = A(t) \, \mathrm{Re}\{(\cos l \cos \varphi + i \sin l \sin \varphi) \, e^{i\omega t}\}$$
$$E_y = A(t) \, \mathrm{Re}\{(\sin l \cos \varphi - i \cos l \sin \varphi) \, e^{i\omega t}\} \tag{11}$$

where $A(t)$ is a slowly varying function of time. We assume perfect correlation in the photon pair so that both photons have identical values of l and φ. The photons emitted by the source are assumed to have a distribution in (l, φ) proportional to $\cos 2\varphi$. Therefore, the function $\rho(\lambda)$ of Section 2 becomes now such that

$$\int \rho(\lambda) \, d\lambda \rightarrow \frac{1}{\pi} \int_{-\pi/4}^{\pi/4} \cos 2\varphi \, d\varphi \int_{-\pi/2}^{\pi/2} dl \tag{12}$$

The parameter δ is assumed to have five possible values: $\delta = 1$ and $\delta = 2$ correspond to the two photons of a correlated pair before crossing a polarizer. These values are associated at random in such a way that half the time $\delta = 1$ ($\delta = 2$) corresponds to the left (right) photon of the pair, and vice versa the other half. On the other hand, when a photon crosses a

Table 1. Change of the Three Parameters and Transmission Probability of Photons Arriving at a Polarizer at Angle a

Incoming	Outgoing	Transmission probability
$l, \varphi, \delta = 1$	$a, 0, \delta = 3$	$1 \quad$ if $|l - a| < \pi/4$ $0 \quad$ if $|l - a| > \pi/4$
$l, \varphi, \delta = 2$	$a, 0, \delta = 4$ $a, 0, \delta = 5$	$\cos 2(l - a) \quad$ if $|l - a| < \pi/4$ $1 + \cos 2(l - a) \quad$ if $|l - a| > \pi/4$
$l, \varphi, \delta \geq 3$	$a, 0,$ same δ	$\cos^2(l - a)$

polarizer we suppose that the value of δ is changed according to Table 1. An essential assumption of the model is that photons with $\delta = 1$ or $\delta = 2$ are less likely to be detected than those with $\delta = 4$. This is related to the fact that the no-enhancement hypothesis must be violated, for some λ, if we want to agree with quantum mechanics. We shall assume that a photon with $\delta \leq 2$ is detected with probability η (where η is the efficiency of the detector), a photon with $\delta = 4$ is detected with probability $\eta\pi/2$, and photons with $\delta = 5$ cannot be detected. The transmission probability of the photons in a polarizer is also given in Table 1 (we note that all photons with $\delta \geq 3$ are linearly polarized).

In all other instances (i.e., when a photon arrives at a half-wave plate, a lens, etc.) it is assumed that the parameter δ does not change, the parameters l and φ change according to the usual laws of optics, and the transmission probability is as predicted by quantum theory.

It can be realized that this model predicts the same single and coincidence counting rates as quantum mechanics in all cases. For instance, when a beam of photons arrives at a polarizer for the first time, it consists of a mixture of 50% $\delta = 1$ and 50% $\delta = 2$. From the incoming photons, a fraction $\frac{1}{4}$ emerges with $\delta = 3$, $(2\pi)^{-1}$ with $\delta = 4$, and $(\pi - 2)/4\pi$ with $\delta = 5$, all linearly polarized. This mixture, which contains on average one half of the incoming photons, has a mean detection probability η. It therefore behaves as an ensemble of identical photons because the parameter δ remains hidden thereafter. In particular, the final fraction detected will be $\frac{1}{2}\eta$ if no more absorption exists. On the other hand, the predicted coincidence detection probability is [compare with equation (3)] $\frac{1}{2}\eta^2 \cos^2 (a - b)$. It is noteworthy that any quarter-wave or half-wave plate inserted between the source and the polarizer does not produce any change in the mixture (12) of polarization parameters.[14] The model is not possible if the photon efficiency of any detector is greater than $2/\pi \approx 0.64$, because in this case photons with $\delta = 4$ would have a detection probability greater than 1. (Actual efficiencies are about 0.15.) As mentioned in Section 1, no models with enhancement are possible with ideal detectors.

The model can be easily adapted to include two-channel polarizers, as well as beam splitters, giving rise to any number of branching processes, provided that no recombination of the beams is performed. However the model, being a purely corpuscular one, cannot account for typically wave behavior (e.g., interference effects). Therefore, it does not prove the compatibility between quantum theory and local realism for all conceivable optical experiments. We have recently developed purely wave models (stochastic optics, see the chapter by Marshall in this book). Such models give, at least qualitatively, the same type of behavior at a polarizer or beam splitter as the model discussed above, and they describe also the interference effects obtained by recombination.

4. Symmetric Local Realist Models

The models discussed in the previous two sections are important as counterexamples to the incorrect claims referred to in our introductory section. In this capacity we consider as irrelevant any feelings that may arise as to their implausibility or artificiality.

Nevertheless, it must be said that the central reason for studying local realist models is to obtain some idea of what an eventual replacement for quantum theory would look like. From that point of view, we consider that the least satisfactory feature of the models we have been considering is the enormous difference between the polarizer-transmission probabilities of the two types of photon emitted. The principal concrete models of the electromagnetic field are those based on wave-particle duality and those based on stochastic optics (see the chapters of Pascazio and Marshall). Both of these are essentially symmetric in their treatment of the two light signals coming from the source. Some slight asymmetry may arise owing to the different frequencies of these signals and also possibly because one signal of a cascade, being emitted before the other, acts as the start of the coincidence circuit. However, in at least one of the experiments we are considering, that of the Stirling group,[7] both of these possible asymmetric features are absent. It is therefore reasonable to pose the question: Can we still reproduce the coincidence rates predicted by quantum theory with models having the same polarizer transmission probabilities for both signals of a pair?

This question has so far been studied for the two polarizer situation only. Caser[16] has proved that it is *not* possible to find a probability $P(Q)$ such that

$$\frac{1}{\pi} \int_0^\pi P(\lambda - a)P(\lambda - b)\, d\lambda = R_Q(d - b)$$

where, when nonideal behavior of detectors, polarizers, etc., are taken into account, the quantum-mechanical prediction can be written as

$$R_Q(a - b) = \text{const}[1 + f \cos 2(a - b)] \tag{13}$$

Caser actually proved the theorem for a more general set of auxiliary variables. We present below a variant of that theorem.

Theorem. There do not exist a probability density $\rho(\lambda)$ and a probability $P(a, \lambda)$ such that

$$\int \rho(\lambda)P(a, \lambda)P(b, \lambda)\, d\lambda = \text{const}[1 + f \cos(2a - 2b)] \tag{14}$$

if $f > \frac{1}{2}$.

Here it has been assumed that the two signals of the cascade are perfectly correlated and that each signal is described by the same set of auxiliary variables λ.

Proof. The dependence on (a, b) through $(a - b)$ implies the property that $\int \rho(\lambda) P(a + \theta, \lambda) P(b + \theta, \lambda) \, d\lambda$ is independent of θ. It follows then that equation (14) implies

$$\frac{1}{\pi} \int_0^\pi d\theta \int \rho(\lambda) P(\theta, \lambda) P(a - b + \theta, \lambda) \, d\lambda = \text{const}[1 + f \cos(2a - 2b)] \quad (15)$$

Because $P(\theta, \lambda)$ must be periodic in θ with period π, we may write

$$P(\theta, \lambda) = a_0(\lambda) + \sum_1^\infty a_n(\lambda) \cos[2n\theta + \delta_n(\lambda)] \quad (16)$$

Then

$$\frac{1}{\pi} \int_0^\pi P(\theta, \lambda) P(a - b + \theta, \lambda) \, d\theta = a_0^2 + \tfrac{1}{2} \sum_1^\infty a_n^2 \cos(2na - 2nb) \quad (17)$$

Comparison with (14) shows that

$$\int \rho(\lambda) a_n^2(\lambda) \, d\lambda = 0 \qquad (n \geqslant 2) \quad (18)$$

and hence, since $\rho(\lambda) \geqslant 0$,

$$\rho(\lambda) a_n(\lambda) = 0 \qquad \text{for all } \lambda \qquad (n \geqslant 2) \quad (19)$$

Therefore

$$\rho(\lambda) P(\theta, \lambda) = a_0(\lambda) \rho(\lambda) + a_1(\lambda) \rho(\lambda) \cos[2\theta + \delta_1(\lambda)] \quad (20)$$

and since this must be nonnegative for all θ, it follows that

$$\rho(\lambda) a_0(\lambda) \geqslant 0 \qquad \text{and} \qquad \rho(\lambda)[a_0(\lambda) - |a_1(\lambda)|] \geqslant 0 \quad (21)$$

But we now find from equation (17) that

$$\int \rho(\lambda) P(a, \lambda) P(b, \lambda) \, d\lambda = \int \rho(\lambda)[a_0^2(\lambda) + \tfrac{1}{2} a_1^2(\lambda) \cos(2a - 2b)] \, d\lambda \quad (22)$$

The property (21) ensures that, as $(a - b)$ varies,

$$\text{Min} \int \rho(\lambda) P(a, \lambda) P(b, \lambda) \, d\lambda \geq \tfrac{1}{2} \int \rho(\lambda) a_0^2(\lambda) \, d\lambda \qquad (23)$$

and

$$\text{Max} \int \rho(\lambda) P(a, \lambda) P(b, \lambda) \, d\lambda \leq \tfrac{3}{2} \int \rho(\lambda) a_0^2(\lambda) \, d\lambda \qquad (24)$$

Since the results are not compatible with equation (14) the theorem is proved.

It should be noted that the definition of symmetry used in equation (14) presupposes that the two light signals are perfectly correlated. A wider definition of symmetry would use a source distribution $\rho(\lambda_1, \lambda_2)$, where λ_1 and λ_2 are the sets of auxiliary variables for the two signals. We would then require

$$R(a, b) = \int \rho(\lambda_1, \lambda_2) P(a, \lambda_1) P(b, \lambda_2) \, d\lambda_1 \, d\lambda_2$$

It is not possible to prove a theorem of the above type for such models, because the model of the previous section belongs to this family.

The above theorem indicates that we should expect to find divergence from the quantum prediction for the coincidence rates, using any symmetric local realist theory. But before this knowledge can be put to practical use, we need to know the possible magnitude of such divergence. In other words we need inequalities; the homogeneous Bell inequalities are now obsolete, having served the very useful purpose of leading us from no-enhancement[3] to enhancement.

The search for such inequalities is in its initial stages. Marshall[17,18] has proposed various criteria for measuring the "distance" separating the set of functions

$$R(\theta, \{P\}) = \int_0^\pi P(\theta + \phi) P(\phi) \, d\phi$$

$$(25)$$

$$[P(\phi) \geq 0 \text{ for all } \phi, \text{ and } P(\phi + \pi) = P(\phi)]$$

from $R_Q(\phi)$ in relation (13). One distance which certainly has a lower bound is

$$d_n(P) = \sum_{r=0}^{n} \left[R\left(\frac{r\pi}{2n}, P\right) - R_Q\left(\frac{r\pi}{2n}\right) \right]^2 \qquad (26)$$

for $n \geq 3$. Owing to the interest in the homogeneous Bell inequality, all the data so far obtained have been for $n = 4$, usually with a concentration on the values $r = 1$ and 3. With hindsight it would have been better to design experiments to measure d_3, but since it was anticipated that the first serious exercise would be a reexamination of existing data, the function P was found[17] which minimized d_4. This gave values of $R(r\pi/n)$ differing from $R_Q(r\pi/n)$ by less than 0.02, and a discussion based on the usual theory of significance tests[18] showed that none of the existing data was good enough to detect such a difference.

Recently Corchero[19] has proposed a modified distance

$$d'_n(P) = \sum_{r=0}^{n} \left[R\left(\frac{r\pi}{n}, P\right) - R_Q\left(\frac{r\pi}{n}\right) \right]^2 \Big/ R_Q\left(\frac{r\pi}{n}\right) \qquad (27)$$

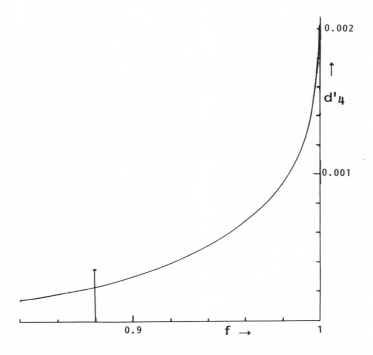

Figure 1. The lower bound of the distance d'_4 defined by equation (27) as a function of the parameter f appearing in the quantum prediction (13). With piles-of-plates polarizers f cannot be greater than about 0.95, while calcite polarizers would allow values as high as 0.99, so providing better tests. In the figure we display the 95% confidence interval for d'_4 obtained from the first Orsay experiment,[5] This is the only experiment for which sufficient published data to construct such an interval are available. The interval so obtained is compatible with both quantum mechanics and symmetric local hidden-variable theories.

This expression for the distance function has the effect of giving a greatly increased weight to the contribution from $R(\pi/2)$. The function P^+ which minimizes d'_4 is found to give substantially greater deviations from R_Q in the situation where the linear polarizers are calcite crystals than where they are piles of plates, because of the much smaller value taken by $R_Q(\pi/2)$ in the former case (see Figure 1). In our view this justifies a repetition of the Holt–Pipkin experiment[20] which is the only one so far to use calcites. These authors obtained the biggest divergence so far reported from $R_Q(\pi/8)$ and $R_Q(3\pi/8)$, but unfortunately did not report any measurements at all for $R(0)$, $R(\pi/4)$, and $R(\pi/2)$.

ACKNOWLEDGMENT. We acknowledge useful discussions with S. Caser.

References

1. B. d'Espagnat, *Scientific American* (November 1979).
2. A. Aspect, *The Ghost in the Atom* (P. C. W. Davies and T. R. Brown, eds.), p. 43, Cambridge University Press (1986).
3. J. F. Clauser and M. A. Horne, *Phys. Rev. D* **10**, 526 (1974).
4. J. F. Clauser and A. Shimony, *Rep. Prog. Phys.* **41**, 1881 (1978).
5. A. Aspect, P. Grangier, and G. Roger, *Phys. Rev. Lett.* **47**, 460 (1981).
6. A. Aspect, J. Dalibard, and G. Roger, *Phys. Rev. Lett.* **49**, 1804 (1982).
7. W. Perrie, A. J. Duncan, H. J. Beyer, and H. Kleinpoppen, *Phys. Rev. Lett.* **54**, 1790 (1985).
8. A. Aspect, P. Grangier, and G. Roger, *Phys. Rev. Lett.* **49**, 91 (1982).
9. A. Aspect and P. Grangier, *Lett. Nuovo Cim.* **43**, 345 (1985).
10. D. Home and T. W. Marshall, *Phys. Lett. A* **113**, 183 (1985).
11. A. Garuccio and F. Selleri, *Phys. Lett. A* **103**, 99 (1984).
12. F. Selleri, *Phys. Lett. A* **108**, 197 (1985).
13. M. Ferrero and E. Santos, *Phys. Lett. A* **166**, 356 (1986).
14. T. W. Marshall and E. Santos, *Phys. Lett. A* **107**, 164 (1985).
15. S. Pascazio, *Phys. Lett. A* **107**, 164 (1985).
16. S. Caser, *Phys. Lett. A* **102**, 152 (1984).
17. T. W. Marshall, *Phys. Lett. A* **98**, 5 (1983).
18. T. W. Marshall, *Phys. Lett. A* **100**, 225 (1984).
19. E. S. Corchero, *Phys. Rev. D* **36**, 636 (1987).
20. R. A. Holt and F. M. Pipkin, preprint, Harvard University (1974).

Index

DATE DUE

DEMCO 38-297